NANOMAGNETISM AND SPINTRONICS

Fabrication, Materials, Characterization and Applications

NANOMAGNETISM AND SPINTRONICS

Fabrication, Materials, Characterization and Applications

Editors

Farzad Nasirpouri
Sahand University of Technology, Iran

Alain Nogaret
University of Bath, UK

World Scientific

NEW JERSEY • LONDON • SINGAPORE • BEIJING • SHANGHAI • HONG KONG • TAIPEI • CHENNAI

Published by

World Scientific Publishing Co. Pte. Ltd.
5 Toh Tuck Link, Singapore 596224
USA office: 27 Warren Street, Suite 401-402, Hackensack, NJ 07601
UK office: 57 Shelton Street, Covent Garden, London WC2H 9HE

British Library Cataloguing-in-Publication Data
A catalogue record for this book is available from the British Library.

NANOMAGNETISM AND SPINTRONICS
Fabrication, Materials, Characterization and Applications

ISBN-13 978-981-4273-05-3
ISBN-10 981-4273-05-8

Printed in Singapore by Mainland Press Pte Ltd.

PREFACE

The goal of *spintronics* is to manipulate individual magnetic moments to integrate logic functions and non-volatile information storage on the same platform. As is often the case in *condensed matter science*, advances are made through the synthesis of novel materials, high quality materials bringing new physics. *Giant magnetoresistance* and *dilute magnetic semiconductors* are two examples at hand. The remarkable potential of spintronics for quantum computation faces major challenges when it comes to controlling simultaneously several qbits encoded in magnetic moments.

After giving a brief introduction to concepts in Nanomagnetism and Spintronics, the present book reviews recent techniques and their achievements in the synthesis and fabrication of magnetic nanostructures in part two. The methods presented here emphasize bottom up or top down approaches for nanodots, nanowires and thin films. They include focused ion beam irradiation, electron beam induced chemical vapour deposition, chemical, and electrochemical methods together. The third part of the book entitled Materials and Characterisation reviews magnetoelectric materials, the giant magnetoresistance in magnetic superlattices, dynamics effects in spin transfer torque oscillators, dilute magnetic oxides, rare earth nitrides together with nuclear resonance scattering and Mössbauer spectroscopy in spintronics. Finally, the last part of this book discusses applications to magnetic storage and bio-magnetism.

The book will be useful to graduate students for whom the introductory chapter is intended, to researchers in the field of nanoscience and engineers.

Spring 2009

Farzad Nasirpouri
(Sahand University of Technology, Iran)
Alain Nogaret
(University of Bath, UK)

CONTENTS

PART III. MATERIALS AND CHARACTERISATION

Chapter 1

CONCEPTS IN NANOMAGNETISM AND SPINTRONICS

Farzad Nasirpouri

Department of Materials Engineering
Sahand University of Technology
Tabriz, Iran
E-mail: Nasirpouri@sut.ac.ir

Alain Nogaret

Department of Physics, University of Bath, Bath, UK
E-mail: A.R.Nogaret@bath.ac.uk

Reduced size and dimensions reveal novel phenomena in magnetism and electronics. We aim to give a brief description of fundamentals of Nanomagnetism and Spintronics to introduce concepts used in the following chapters.

1.1. Nanoscale Science and Technology

Since the initial proposal of Feynman,[1] much progress has been made in the understanding of Physics on the nanoscale. Based on these advances, an increasing number of technological applications, from magnetic read heads to automotive sensors, are now reaching the commercial market. Nanoscience commonly refers to physical phenomena in ultra-small structures which have one or more dimensions below 100 nm.[2] "Nanomagnetism and Spintronics" form an interdisciplinary sub-field linking ferromagnetism and electronics which holds wide scientific and technological prospects for the future.[3] This chapter gives a brief introduction to fundamental ideas in nanomagnetism and spintronics. Review articles on more specific aspects have appeared in Refs. 4–18.

1.2. Nanomagnetism

Magnetic ordering in ferromagnetic materials[19] is a complex phenomenon involving competing energies over different length scales. On the scale of the interatomic distance, exchange interaction is responsible for the ordering of individual spins. In transition metals, ferromagnetic or antiferromagnetic spin alignment is determined by the symmetry of the orbital wavefunction and the position of the Fermi level relative to the centre of the d-band. On a larger scale, 10–100 nm, the finite dimensions of the sample cause the formation of magnetic poles which increase magnetic energy. This dipolar energy and the associated demagnetizing field are minimized by the formation of the magnetic domain structure. In rare earth based magnets, high atomic numbers cause preferential orientation of the magnetization due to spin orbit interaction. These effects explain the high magnetocrystalline anisotropy of rare earth magnets and their compounds. The interaction of the 4f orbitals with the lattice gives oblate (saucer-shaped) or prolate (cigar-shaped) orbitals which sets the magnetic easy axis in the direction that maximizes the angular momentum. This is the direction perpendicular to equatorial plane in oblate systems and in the plane of the equator in prolate ellipsoids. Magnetocrystalline anisotropy competes with exchange energy to set the width of magnetic domain walls. A domain wall consists of a spin chain that progressively tilts across the domain wall linking the magnetization in one domain to the magnetization in the next. Exchange interaction conspires to make the domain wall wider to minimize exchange energy. Conversely, the energy of magnetocrystalline anisotropy is minimized in vanishingly small domain walls. In the limit of a domain wall made of antiparallel spins, the energy of anisotropy is minimum because both spins align with the easy axis. The domain wall width corresponds to the optimum distance that minimizes the total energy of the domain wall. As a result, anisotropic magnets have thin domain walls, down to the interatomic distance, while magnets lacking anisotropy have domain walls as large as a few tens of nanometres. The length scales of magnetic ordering are shown in Fig. 1.1. The following section will explain the principles of magnetism which leads to explore nanomagnetic properties and devices.

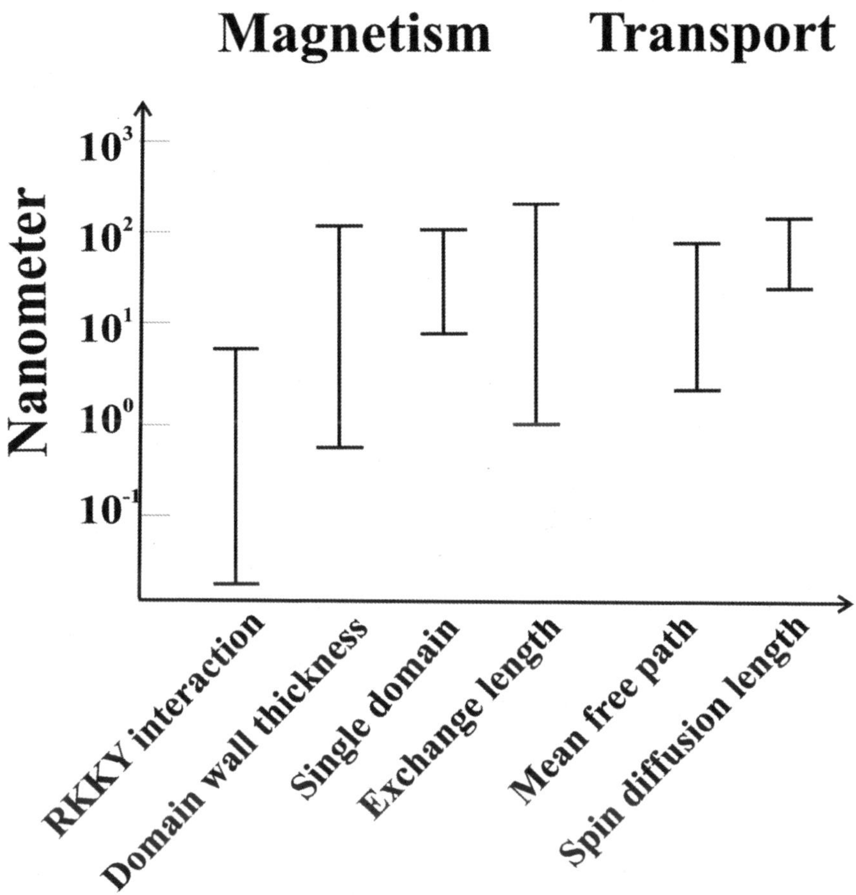

Fig. 1.1. Critical length scales in nanomagnetism and spintronics.

1.2.1. *Magnetic ordering on the nanoscale*

Ferromagnetic alignment is determined by Heisenberg exchange theory when J is positive in:

$$E = -2J_{exchange} S_1 . S_2 \tag{1}$$

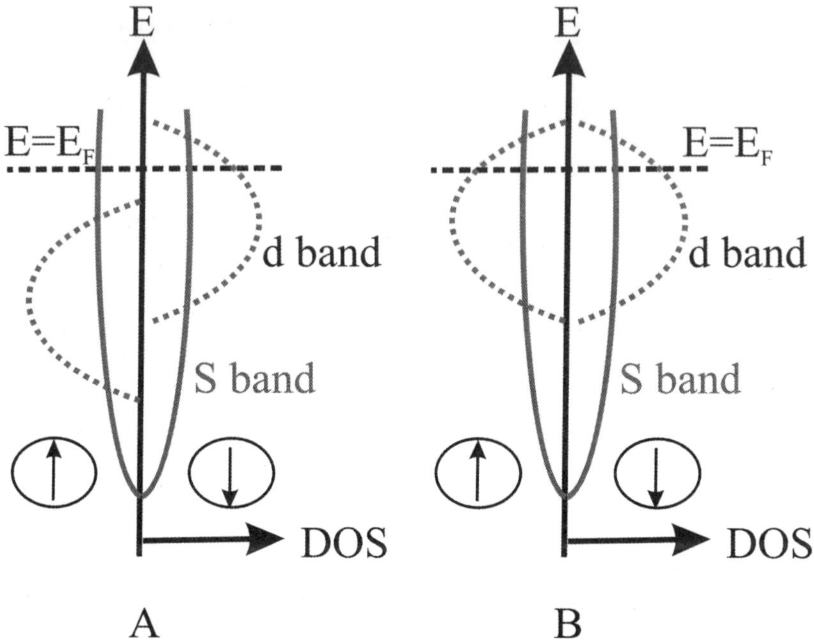

Fig. 1.2. Schematic of density of states as a function of energy in (A) ferromagnet and (B) a normal metal.

The consequence of the spin alignment due to the exchange interaction is that the material has a spontaneous magnetization giving a net magnetic moment per unit volume (magnetization). The magnetization of ferromagnets corresponds to the filling of incomplete energy shells: the 3d shell for transition metals (Fe, Ni, Co) and the 4f shell for lanthanides. These form the elemental ferromagnets. Magnetism in these metals arises from the filling of the 3d and 4f electron shells, respectively.[20]

In Fig. 1.2, the density of states of a typical 3d ferromagnet (a) is compared to a non-magnetic 3d metal (normal metal) (b), referred to as a Pauli paramagnet. At zero field the normal metal has an equal number of spin up and down electrons. Since the 4s orbitals are extended in space, the wavefunctions of neighbouring atoms strongly overlap giving a wide 4s-band (15–20 eV). In contrast, the 3d orbitals are more localized on the atomic sites giving a comparatively narrow energy band (4–7 eV). In the solid state, due to the strong hybridization of these two energy bands, one cannot make a clear distinction between the 3d and the 4s character. The 3d and 4s electrons are itinerant electrons, the former being less mobile[21] carry less current. Electrons fill the band up to the Fermi level according to the Pauli Exclusion Principle. Switching on exchange coupling transfers electrons from the spin down band to the spin up band. More parallel spins lowers the exchange energy of the system. The Pauli Exclusion Principle causes the transferred electrons to fill empty states above the Fermi level which has the effect of increasing the overall kinetic energy. Transition metals present the peculiarity of having a high density of states at the Fermi level due to the 3d electron band. As a result, the increase in kinetic energy is small comparatively to exchange interaction which has the net effect to reduce the total energy of the system and stabilize ferromagnetism. This is formulated by the Stoner criterion,

$$J_{exchange} \, . \, Z(E_F) > 1 \tag{2}$$

that requires either a large exchange integral, J, or a high Fermi density of states for ferromagnetism to be stable.

1.2.2. *Magnetization reversal*

The magnetization of ferromagnets is characterized by a magnetic hysteresis loop whose most important parameters are the coercive field (H_c) and the remanent magnetization (M_r) — see Fig. 1.3. The energy dissipated in one cycle corresponds to the M-H area enclosed by the hysteresis loop.

Magnetization reversal follows different processes depending on size. In small magnetic clusters, a few tens of nanometres in diameter, dipolar energy is smaller than other relevant energies. All magnetic microscopic moments bound by exchange interaction rotate coherently when the applied magnetic field is reversed. The energy barrier to magnetization reversal is due to anisotropy.

Clusters are very important to magnetic applications because the coercive field can be set precisely by engineering the magnetic anisotropy constant K. For this reason, a preferred method for engineering magnetic materials is to sinter magnetic powders instead of using bulk magnetic alloys.

On the mesoscopic scale, dipolar energy becomes comparable to the energy of formation of domain walls. Microscopic magnetic moments tilt parallel to interfaces lowering dipolar energy to the detriment of exchange energies. The magnetic structure exhibits complex patterns characterized by the formation of vortices, and is highly dependent on the geometry. Magnetization reversal occurs through the twisting or buckling of these vortices and lead to delicate steps in the magnetization curves.

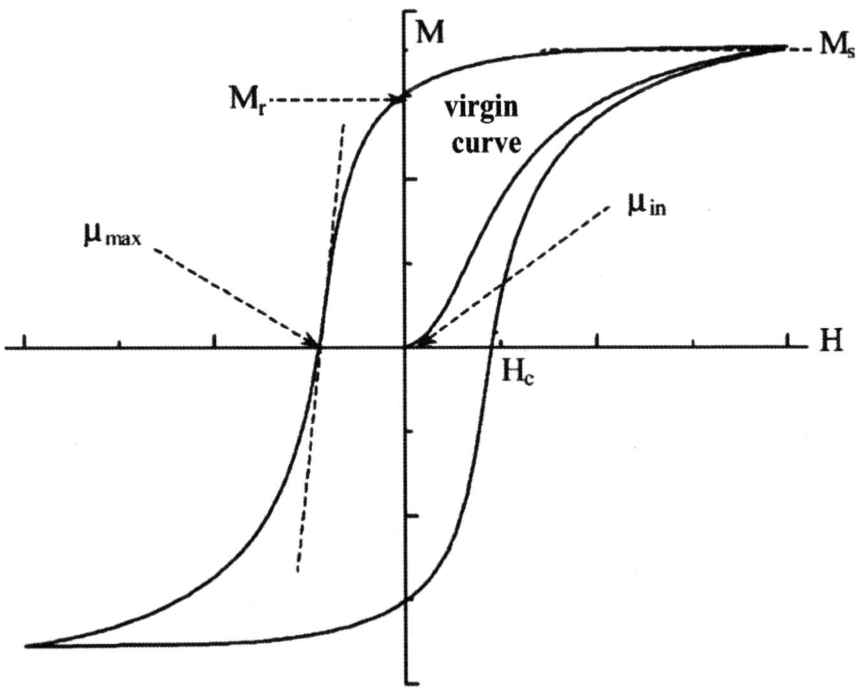

Fig. 1.3. Magnetic properties of materials as defined on the M-H plane of magnetization M versus magnetic field H. These include coercivity H_c, remanence M_r, initial permeability μ_{in}, maximum differential permeability μ_{max} and saturation magnetization M_s. (Virgin curve represents the response of the magnetic moments on the first sweep of magnetic field).

On the macroscopic scale, magnetization reversal is determined by the nucleation and propagation of magnetic domain walls in the bulk. One distinguishes the case of ideal crystals from dirty metals. In the former case magnetic domains nucleate at sites where exchange interaction is weakest. Subsequently, antiparallel magnetic domains rapidly expand allowing the magnetization to change sign. The coercive field in this case is controlled by the nucleation of magnetic domains. In dirty metals, defects pin magnetic domains as they expand. The coercive field is the field that enables magnetic domains to grow and overcome the pinning potential.

1.2.3. *Dimensionality in magnetism*

Dimensionality is a general term which is used to describe the effect of size confinement on the physical properties of matter. It becomes very important, when the properties of matter are discussed on the nanometre scale. In general, one divides nanostructures into three different groups including: thin films (2D structures), nanowires (1D structure), and nanodots or nanoparticles (0D structures). The magnetization of these different groups is significantly influenced by confinement.

1.2.3.1. *Thin magnetic films*

The electronic structure at the surface of a given material differs considerably from the bulk. Early studies of nickel monolayers on copper substrates show significant changes in the density of states as a function of surface energy. The different coordination number of atoms and the number of incomplete bonds at the surface modifies the local density of states. The d orbital is more localized and interacts more strongly with spin magnetic moments giving a higher net magnetization at the surface. This is an example of size effect in one dimension.

The magnetization reversal in thin ferromagnetic films is affected by such size confinement. The nucleation and propagation of magnetic domains which involve Bloch walls in the bulk is more likely to involve Néel walls when the thickness decreases. Comparison between the energy and thickness of Bloch wall and Néel wall as a function of film thickness confirms the tendency of the formation of Néel walls in the thinner film. In thin films with uniaxial in-plane anisotropy, magnetic domains form stripes and are separated by Néel walls for thicknesses below 50 nm. In thin films with perpendicular anisotropy, circular magnetic domains form which are highly mobile. These magnetic 'bubbles' hold potential for magnetic storage technology.[22]

1.2.3.2. *Nanowires or one dimensional magnets*

Fundamental interest in arrays of ferromagnetic nanowires lies in the emergence of novel magnetic and transport properties as the dimension approaches the length scale of a few nanometres to a few tens of nanometres. Current interest in research on ferromagnetic nanowires is stimulated by the potential application to future ultra-high-density magnetic recording media.[23]

The controlled production of magnetic nanowire arrays with outstanding characteristics is important to control the magnetization process. Free standing nanowires can be fabricated by different methods such as lithography and template electrodeposition.[24–28] The modes of magnetization reversal have been evaluated using a variety of 'nanowire' arrays fabricated by electrodeposition into lithographically made patterns with different aspect ratios of up to 3 and diameters below 180 nm. The smaller diameter wires have high out-of-plane remanence while larger diameter wires have low remanence. This is consistent with the predictions of a micromagnetic model, which indicates a change from a 'flower' (or single domain) to a 'vortex' remanent state with increasing diameter. The flower–vortex transition occurs at a diameter of 3.5 times of exchange length for cylinders without any magnetocrystalline anisotropy.[29] However, it is important to note that the remanence only decreases slowly as the vortex develops, and high aspect ratio vortex-state particles can still have significant remanence. Figure 1.4 depicts the change of magnetization modes for different geometries.

Previous studies[30–33] have indicated for particles with different aspect ratios that the reversal of magnetization is categorized into two different modes including coherent-rotation and curling. There exists a critical radius R_c: If $R < R_c$, reversal occurs by coherent-rotation, and, if

$R > R_c$, the reversal occurs by curling. The critical radius R_c is given by:

$$R_c = \left(\frac{kA}{\mu_o M_s^2} \right) \tag{3}$$

where k is a constant depending on the length/diameter ratio of the particles, ranging from 1.08 for an infinite cylinder to 1.38 for a sphere, A is the exchange constant in erg/cm and $\mu_o M_s$ is the magnetization at the saturation.[34]

The main factor which controls magnetization reversal in elongated ferromagnetic nanostructures is the domain wall. The nucleation and propagation of domain wall between opposing magnetic domains is of

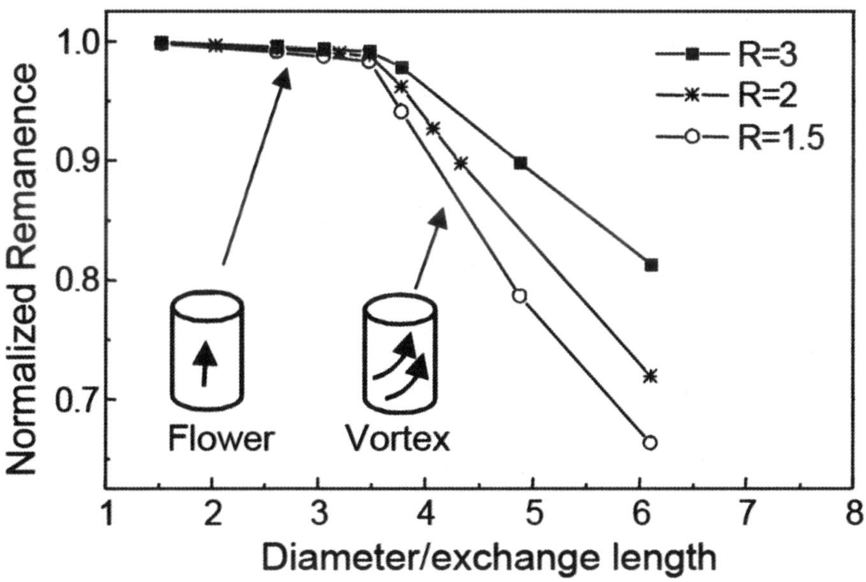

Fig. 1.4. The axial remanence of cylindrical particles as a function of diameter, for three different aspect ratios of R = 1.5, 2 and 3. Below 3.5 times of exchange length the vortex develops and the remanence gradually decreases, more slowly at higher aspect ratios. *Reprinted with kind permission from C.A. Ross.*[29]

basic interest and critical to the performance of spintronic and magnetic storage devices. A series of studies has been carried out on lithographically fabricated ferromagnetic nanowires[35,36] to investigate the magnetic domain wall dynamics for applications to logic spintronics devices which will be discussed in Chapter 14. In the other side, electrodeposited nanowire arrays stimulate great interest because they provide a relatively simple and inexpensive way to study the magnetic properties of this one dimensional structure. This type of fabrication method is also useful for making multilayered nanowires exhibiting giant magnetoresistance.[27,28] Figure 1.5 shows two well known templates in which magnetic nanowires are grown using electrochemical deposition.

To increase the density of magnetic storage, the spacing between magnetic domains (or single nanowires) must be minimized and the cross-talk between adjacent magnets must be avoided. Nanomagnets can be obtained by electrodepositing nanowires, whose typical dimensions may vary routinely from 500 nm down to 30 nm or even smaller. Generally, the factors determining the final magnetic response of nanowire arrays are, (i) the magnetic nature of individual nanowires, preferably having strong longitudinal magnetic anisotropy for data storage purposes, and (ii) the characteristics of the geometrical arrangement of the nanopore/nanowire array, which determines the strength of the magnetostatic interaction among neighbouring nanowires.[33]

Details of the magnetic properties of nickel and iron nanowires with diameters less than the domain wall have been explained by Kroll *et al.*[39] It is confirmed that magnetization reversal in nanowires indicates two stable orientations of the magnetic moments, one pointing parallel and one antiparallel to the wire axis. These two orientations are separated by an energy barrier giving a slight hysteresis in the in-plane magnetization curves. For Fe and Ni, shape anisotropy predominates giving rise to easy axes in the direction of antiparalel orientation, while the case for Co nanowires arrays is more complicated due to temperature-dependent and size-dependent magnetocrystalline anisotropy.[37–44]

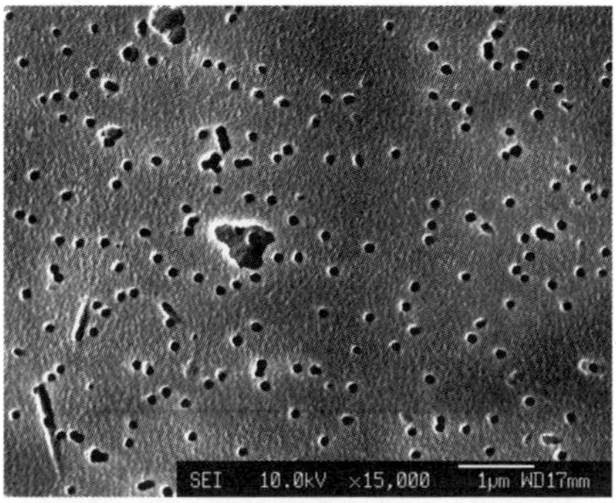

(a)

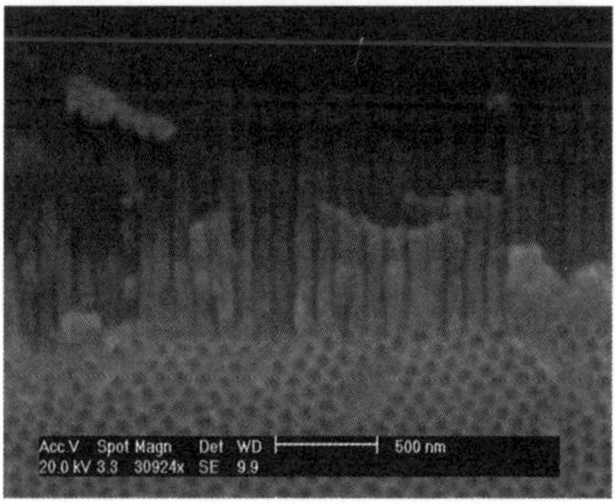

(b)

Fig. 1.5. (a) Top view of a polyester track-etched nanoporous membrane (Osmonics) and (b) naturally ordered anodic aluminium oxide membrane for template electrodeposition of magnetic nanowires in the pores.[28]

1.2.3.3. *Nanodots and superparamagnetism*

Competition between exchange and anisotropy constants[45] induces a critical size of nanodots. The state of lowest free energy of a ferromagnetic particle is one of uniform magnetization below a certain critical size. For larger particles, the magnetization is non-uniform. These two modes are known to correspond to single domain and vortex-states which have been studied extensively over last decade.[46,47]

Measurement of magnetization of single dot presents technical difficulties, though of high interest. Arrays of magnetic nanodots have been fabricated mainly by nanolithography techniques (i.e. electron beam, X-ray, etc.) and have been measured using small scale magneto-optical

Kerr effect. The magnetization curves obtained for the arrays of circular dots exhibiting the two reversal modes. The effect of diameter and thickness is quite important. As the size approaches the critical radius, the single domain mode is predominant.[48] Nanodots smaller than the critical radius, tend to give up ferromagnetic alignment, becoming superparamagnetic. This behaviour takes place when the fluctuations of magnetic moments caused by the thermal energy distort the intrinsic alignment of the ferromagnet. Considering a single domain particle with two distinct magnetic states, when the thermal energy becomes comparable with the energy barrier between the two states, superparamagnetism is established. This process obeys the Arrhenius-Néel equation:

$$f = f_o \cdot e^{\left(\frac{k_u V}{K_B T}\right)} \tag{4}$$

where k_u is the anisotropy constant, V is the volume of particle, K_B is the thermal energy, f is the frequency of switching between magnetization states.[49]

1.3. Spintronics

The idea of a resistance induced by spin scattering has first been put forward by Mott to explain the resistance drop observed in transition metals below their Curie temperature. Mott suggested that the electric current in the metal had a spin up and a spin down component. These formed independent conduction channels which only mixed via spin scattering on antiparallel layers. In the paramagnetic phase the resistance was therefore expected to be higher than in the ordered ferromagnetic phase, leading to the observed resistance drop. The development of metal epitaxy in the mid 1980s has allowed growing atomically thin layers of crystalline quality. This has opened the possibility of synthesizing sandwich structures such as Fe/Cr/Fe and Co/Cu/Co of width smaller than the electron mean free path. On the scale of the structure, a current flowing across it has two independent components corresponding to each spin polarization which mix depending on the relative orientation of the magnetizations. The ferro/normal/ferro multilayers thus formed the first man made device demonstrating the interplay between magnetic and electronic properties.

Giant magnetoresistance was first demonstrated in 1988[50,51] in sputtered Fe/Cr multilayers. A spin-electronic current was injected in the plane of a superlattice in the so-called current in-plane (CIP) configuration. The current perpendicular to plane (CPP) was investigated afterwards which showed the giant magnetoresistance to be isotropic.

GMR takes place because of spin-dependent electron scattering. The width of the trilayer structure is smaller than the mean free path so that momentum scattering has marginal effects compared to spin scattering. The band structure of the transition metal ferromagnet (Fig. 1.2) can explain the mechanism of GMR based on the properties of electrons with energies close to the Fermi energy (EF). The most mobile electrons are in the s-band while the spin polarized electrons are in the d-band. Current GMR models assume that s electrons mainly carry the current, whereas the d states are those responsible for spin scattering. From Fig. 1.2 it becomes obvious that electrons with spin down are scattered more than those with spin up.[52]

Giant magnetoresistance has been implemented in spin valves which are used in magnetic read heads. These are GMR layers in which one layer is pinned while the other is free to move. A large current driven through a spin valve induced high frequency oscillations of the magnetization through the spin torque effect. Namely the pinned layer polarizes a spin current which impinges on the free layer with opposite magnetization. The polarized spin current applies effective magnetic field

$$\vec{B}_{eff} = \frac{\hbar I}{(2eMAl)\vec{n}} \tag{5}$$

to the free layer. Here I is the drive current, M is the free layer magnetization, l is the thickness of the free layer and A is the junction area. This effective magnetic field tends to align the free layer magnetization with the magnetization of the pinned layer. This system becomes unstable as the resistance drops and the effective magnetic field increases further increasing the frequency of oscillations of the magnetization. Although small, the power emitted by these nano-oscillators presents an interest for on-chip wireless communications.

An attractive way to control the spin phase is to use an electric field via spin orbit coupling. The spin orbit coupling Hamiltonian is obtained in the classical limit of Dirac's equation. Special relavity stipulates that a charged particle moving at momentum $\vec{p}$ through electric field $\vec{E}$ will be subject to magnetic field:

$$\vec{B} = -\frac{\vec{p} \wedge \vec{E}}{mc^2} \tag{6}$$

This magnetic field induces spin precession proportional to the electric field. Dephasing induced by spin orbit coupling has been observed in the vacuum albeit under very large electric fields because the mc^2 term is large. In semiconductors, the energy difference between the conduction and valence bands replaces the Dirac gap $2mc^2$. The smaller semiconductor band gap allows spin orbit to induce sizable spin dephasing, an effect at the core of the proposal for a spin transistor by Datta and Das[53] and used in the Aharonov-Casher effect. A spin polarized current injected through the plane of a quantum well and subjected to a transverse electric field will change spin polarization at a rate controlled by the magnitude of the electric field. By choosing the electric field the electron spin polarization will either be transmitted of not, which is the basis of the spin transistor action.

References

1. R. P. Feynman, *Miniaturization* (Reinholds, New York, 1961).
2. G. M. Whitesides, *Nanotechnology: Art of the Possible, Technology Review* (Technology Review Inc., Cambridge, MA, Nov/Dec 1998).
3. Y.-H. Wu, *Encylopedia of Nanoscience and Nanotechnology*, Vol. 7 (American Scientific Pub., 2003), pp. 493–544.
4. J. Nogues, I. K. Schuller, *J. Magn. Magn. Mater.*, **192** (1999) 203.
5. R. P. Cowburn, *J. Magn. Magn. Mater.*, **242–245** (2002) 505.
6. S. D. Bader, *Rev. Mod. Phys.*, **78** (2006) 1, also see: S. Bader, K. S. Buchanan, S.-H. Chung, K. Y. Guslienko, A. Hoffmann, Y. Ji, V. Novosad, *Superlattices and Microstructures*, **41** (2007) 72.

7. C. L. Dennis, R. P. Borges, L. D. Buda, U. Ebel, J. F. Gregg, M. Hehn, E. Jouguelet, K. Ounadjel, I. Petej, I. L. Prejbeanu, M. J. Thornton, *J. Phys.: Condens. Matter*, **14** (2002) R1175.
8. G. Srajera, L. H. Lewisb, S. D. Bader, A. J. Epstein, C. S. Fadley, E. E. Fullerton, A. Hoffmann, J. B. Kortright, K. M. Krishnan, S. A. Majetich, T. S. Rahman, C. A. Ross, M. B. Salamon, I. K. Schuller, T. C. Schulthess, J. Z. Su, *J. Magn. Magn. Mater.*, **307** (2006) 1.
9. Y.-B. Xu, *Current Opinion in Solid State and Materials Science*, **10** (2006) 81.
10. C. Chappert, A. Fert, F. Nguyen Van Dau, *Nature Materials*, **6** (2007) 813.
11. M. R. J. Gibbs, *Current Opinion in Solid State and Materials Science*, **7** (2003) 83.
12. J. F. Gregg, I. Petej, E. Jouguelet, C. Dennis, *J. Phys. D: Appl. Phys.*, **35** (2002) R121.
13. S. A. Wolf, D. D. Awschalom, R. A. Buhrman, J. M. Daughton, S. Molnar, M. L. Roukes, A. Y. Chtchelkanova, D. M. Treger, *Science*, **294** (2001) 1488.
14. J. I. Martın, J. Nogues, K. Liu, J. L. Vicent, I. K. Schuller, *J. Magn. Magn. Mater.*, **256** (2003) 449.
15. M. R. Fitzsimmons, S. D. Bader, J. A. Borchers, G. P. Felcher, J. K. Furdynad, A. Hoffmann, J. B. Kortright, I. K. Schuller, T. C. Schulthess, S. K. Sinha, M. F. Toney, D. Weller, S. Wolf, *J. Magn. Magn. Mater.*, **271** (2004) 103.
16. S. D. Bader, *Surface Science*, **500** (2002) 172.
17. R. Skomski, *J. Phys.: Condens. Matter*, **15** (2003) R841.
18. C. A. Ross, *Annu. Rev. Mater. Res.*, **31** (2001) 203.
19. E. Wolf, *Nanophysics and Nanotechnology* (Wiley, 2004).
20. B. D. Cullity, *Introduction to Magnetic Materials* (Adison-Wesley, Reading, Mass., 1972).
21. Scientific Background on the Nobel Prize in Physics 2007. The Discovery of Giant Magnetoresistance Compiled by the Class for Physics of the Royal Swedish Academy of Sciences, October 2007.
22. R. O'Handley, *Modern Magnetic Materials* (Wiley, 2000).
23. H. Zeng, R. Skomski, L. Menon, Y. Liu, S. Bandyopadhyay, D. J. Sellmyer, *Phys. Rev. B*, **65** (2000) 134426.
24. M. Ghorbani, F. Nasirpouri, A. Iraji Zad, A. Saedi, *Materials and Design*, **27** (2006) 983.
25. F. Nasirpouri, M. Abdollahzadeh, M. J. Almasi, N. Parvini-Ahmadi, *Curr. Appl. Phys.*, **9** (2009) S91.
26. F. Nasirpouri, P. Southern, M. Ghorbani, A. I. Zad, W. Schwarzacher, *J. Magn. Magn. Mater.*, **308** (2007) 35.
27. F. Nasirpouri, Electrodeposition of magnetic nanowire arrays, in ed. A. El-Nemr, *New Developments in Electrodeposition and Pitting Research* (Research Signpost Pub., 2007), pp. 55–92.
28. W. Schwarzacher, D. Lashmore, *IEEE Trans. Magnetics*, **32** (1996) 3113.

29. C. A. Ross, M. Hwang, M. Shima, J.-Y. Cheng, M. Farhoud, T. A. Savas, H. I. Smith, W. Schwarzacher, F. M. Ross, M. Redjdal, F. B. Humphrey, *Phys. Rev. B*, **67** (2002) 144417, also see: C. A. Ross, M. Hwang, M. Shima, H. I. Smith, M. Farhoud, T. A. Savas, W. Schwarzacher, J. Parrochon, W. Escoffier, H. Neal Bertram, F. B. Humphrey, M. Redjdal, *J. Magn. Magn. Mater.*, **249** (2002) 200.
30. Y. Peng, T. H. Shen, B. Ashworth, *J. Appl. Phys.*, **93** (2003) 7050.
31. L. Cheng-Zhang, J. C. Lodder, *J. Magn. Magn. Mater.*, **88** (1990) 236.
32. G. T. A. Huysmans, J. C. Lodder, *J. Appl. Phys.*, **64** (1988) 2016.
33. R. Skomski, J. M. D. Coey, *Permanent Magnetism* (Institute of Physics Pub., Bristol, 1999).
34. M. Zheng, L. Menon, H. Zheng, Y. Liu, S. Bandyopadhyay, R. D. Kirby, D. J. Sellmyer, *Phys. Rev. B*, **62** (2000) 12282.
35. D. Atkinson, D. Allwood, G. Xiong, M. Cooke, C. C. Faulkner, R. P. Cowburn, *Nature Materials*, **2** (2003) 85.
36. D. Atkinson, D. A. Allwood, C. Faulkner, G. Xiong, M. D. Cooke, R. P. Cowburn, *IEEE Trans. Magnetics*, **39** (2003) 2663.
37. K. Ounadjela, R. Ferré, L. Louail, J. M. George, J. L. Maurice, L. Piraux, S. Dubois, *J. Appl. Phys.*, **81** (1997) 5455.
38. M. Hernandez-Velez, *Thin Solid Films*, **495** (2006) 51.
39. M. Kroll, W. J. Blau, D. Grandjean, E. R. Benfield, F. Luis, P. M. Paulus, L. J. de Jongh, *J. Magn. Magn. Mater.*, **249** (2002) 241.
40. P. M. Paulus, F. Luis, M. Kroll, G. Schmid, L. J. de Jongh, *J. Magn. Magn. Mater.*, **224** (2001) 180.
41. J. M. Garcıa, A. Asenjo, J. Velazquez, D. Garcıa, M. Vazquez, P. Aranda, E. Ruiz-Hitzky, *J. Appl. Phys.*, **85** (1999) 5480.
42. G. J. Strijkers, J. H. J. Dalderop, M. A. A. Broeksteeg, H. J. M. Swagten, W. J. M. Jonge, *J. Appl. Phys.*, **86** (1999) 5141.
43. H. Zeng, M. Zheng, R. Skomski, D. J. Sellmyer, Y. Liu, L. Menon, S. Bandyopadhyay, *J. Appl. Phys.*, **87** (2000) 4718.
44. H. R. Khan, K. Petrikowski, *J. Magn. Magn. Mater.*, **249** (2000) 458.
45. W. E. Brown Jr, *J. Appl. Phys.*, **39** (1968) 993.
46. R. P. Cowburn, D. K. Koltsov, A. O. Adeyeye, M. E. Welland, *Phys. Rev. Lett.*, **83** (1999) 1042.
47. K. Y. Guslienko, V. Novosad, Y. Otani, H. Shima, K. Fukamichi, *Phys. Rev. B*, **65** (2001) 24414.
48. F. Nasirpouri, A. Nogaret, D. Atkinson, M. Ghorbani, A. I. Zad, *J. Magn. Magn. Mater.*, **299** (2006) 356.
49. M. J. Bonder, Y. Huang, G. C. Hadjipanayis, in *Advanced Magnetic Nanostructures*, eds. D. Sellmyer, R. Skomski (Springer, 2006), p. 185.
50. M. N. Baibich, J. M. Broto, A. Fert, V. D. F. Nguyen, F. Petroff, *Phys. Rev. Lett.*, **61** (1988) 2472.

51. G. Binasch, P. Grunberg, F. Saurenbach, W. Zinn, *Phys. Rev. B*, **39** (1989) 4828.
52. P. Grutberg, in Magnetic Multilayers and Giant Magnetoresistance, ed. U. Hartmann (Springer, 2000).
53. S. Datta, B. Das, *Appl. Phys. Lett.*, **56** (1990) 665.

Chapter 2

ARTIFICIAL MAGNETIC DOMAIN STRUCTURES REALIZED BY FOCUSSED ION BEAM IRRADIATION

Simon Bending* and Simon Crampin

Department of Physics, University of Bath
Claverton Down, Bath BA2 7AY, UK
**E-mail: S.Bending@bath.ac.uk*

Atif Aziz

Department of Materials Science and Metallurgy
University of Cambridge, Cambridge CB2 3QZ, UK

We describe how artificial ferromagnetic domain patterns can be realized by focussed ion beam irradiation of Pt/Co/Pt trilayers with perpendicular magnetic anisotropy. This approach allows exquisite control of the coercive fields of 'patterned' regions with very high spatial resolution. Systematic extraordinary Hall effect studies of magnetic switching in artificial magnetic domains as a function of irradiation dose and size are described and interpreted. The potential of such structures is then illustrated in an investigation of the angle-dependent *intrinsic* domain wall resistivity in artificial superlattice structures. We find excellent agreement between these experimental results and a theory of domain wall resistivity based on spin-dependent potentials and scattering rates for majority and minority spin electrons due to Levy and Zhang.

2.1. Introduction

Intense efforts are currently being devoted worldwide to the goal of developing magnetic devices based on spin-polarised electronic currents. These so-called *spintronic* devices will combine the advantageous properties of magnetic and semiconductor materials, and are expected to be fast, non-volatile and versatile, capable of simultaneous data storage and processing, while at the same time consuming less energy. The first spintronic structures to be developed were magnetic sensors based on the giant magnetoresistance (GMR) effect. These devices rely on spin-dependent electronic transport in sandwiches composed of two ferromagnetic films separated by a very thin non-magnetic spacer. Spin scattering asymmetry of up and down spins in ferromagnetic metals leads to much lower scattering rates (lower resistance) when the magnetisation in the two layers is parallel as opposed to anti-parallel. GMR devices, and related structures based on tunnelling magnetoresistance, are now found in the read heads of nearly all magnetic hard drives and have been key in maintaining the trend for rapidly increasing data storage capacities. The discovery of GMR was also recognised by the award of the 2007 Physics Nobel prize. Magnetic random access memory (MRAM) is an 'active' second generation development in spintronics consisting of magnetic tunnel junctions where the magnetisation direction of one electrode is switched by the stray fields from current-carrying 'write' lines. These structures represent switchable non-volatile memory cells, which can be read via the resistance state of the junction. However, the *magnetic switching* used in MRAM technology is relatively inefficient and leads to unwanted interactions between neighbouring cells, and attention has now turned to third generation *current-switched* devices based on spin transfer torque due to spin-polarised current injection across a domain wall (DW) (see Refs. 1–6).

A key feature of all these sensor and memory cell technologies is the ability to prepare, pin and manipulate well-defined and reproducible domain wall structures. Traditionally shape anisotropy has been used in nanostructured thin films to control stable domain configurations, and domain walls have either been trapped at 'notches' (constrictions) or in curved wire segments. This reliance on *naturally occurring* domain structures imposes severe restrictions on the design of spintronic devices. Recently an entirely different approach to the generation and trapping of domain walls has began to be explored based on *artificial domain structures* created by low dose focussed ion beam (FIB) irradiation. This technique allows the local magnetic anisotropy and coercive fields of magnetic multilayers to be precisely tuned with nanoscale (≥10 nm) resolution. This allows far more flexibility in the generation and trapping of domain walls, and careful optimisation of the magnetic anisotropy in adjacent domains could potentially lead to much lower critical current densities in current-switched devices. The study of artificial domain structures in multilayer films with perpendicular magnetic anisotropy is currently most mature, and will be the focus of this article. There is, however, a growing body of literature on the influence of FIB irradiation on in-plane magnetised multilayer materials.

We will demonstrate how the exquisite control offered by artificial domain structures can be exploited to make an unambiguous *quantitative* investigation of *intrinsic* domain wall resistivity in magnetic "superlattices".

2.1.1. *Controlling magnetic anisotropy by irradiation*

Masked irradiation of ferromagnetic multilayers is becoming a widely used technique to realize planar structures with 'patterned' magnetic coercivity.[7] This approach raises the exciting prospect of directly imposing a desired 'artificial' domain structure onto ferromagnetic multilayers and removes the reliance on 'natural' magnetic domain structures that can be difficult to initialise or control.

Here we focus on the use of ion beam irradiation to control the magnetic behaviour of technologically relevant out-of-plane magnetic anisotropy systems. Co/Pt multilayers have perpendicular magnetic anisotropy (PMA) over a wide range of layer thicknesses.[8,9] The interface structure, namely the strain, roughness and degree of intermixing/alloy formation, plays a key role in the strength of the PMA, and this feature was exploited by Chappert *et al.*[9] to systematically reduce the coercive field of Pt/Co/Pt trilayers using 30 keV He^+ irradiation (the system reverts to in-plane magnetisation at very high doses for certain parameters). The induced changes arise from mixing/roughening at the Co/Pt interfaces combined with a relief of interface strain.[10] Remarkably the same magnetic control was demonstrated using a 28 keV Ga^+ FIB with extremely high spatial resolution (≥10 nm).[11] Moreover, most of the incident Ga^+ ions pass through the ultra-thin films used and implant into the substrate leading to minimal chemical contamination.

2.1.2. *Intrinsic domain wall resistivity*

Since the first measurements almost thirty years ago[12,13] of the resistance of single crystal Fe whiskers, which showed large changes as the multi-domain ferromagnetic state was eliminated in a small applied field, there has been considerable interest in the domain wall contribution to the resistivity of ferromagnetic materials. The topic enjoyed a dramatic revival with the discovery of giant magnetoresistance (GMR) in ferromagnetic/non-magnetic layered thin films[14,15] in which the relative orientation of the magnetisation in the thin ferromagnetic layers (effectively the domain structure) determines the spin-dependent electronic structure and scattering rates. Resistance changes as large as 35% have been observed[16] in Co/Cu multilayers at room temperature in an applied field of ~10 mT and, only ten years after the phenomenon was discovered, GMR sensors are now routinely used as read heads for data storage media. In GMR materials the non-magnetic layers break up the exchange interaction between the magnetic layers, while the chemical interfaces are almost atomically sharp. In contrast a conventional domain wall is chemically homogeneous, but the magnetisation varies gradually on the length scale of the domain wall width (typically ~1–100 nm). Despite the superficial differences between these two situations they share the common feature that the magnetisation changes on crossing a magnetic interface, and similar theoretical models have been applied to both.

Recent theories of domain wall scattering have been based on spin-dependent potentials and scattering rates for majority and minority spin electrons in ferromagnets.[17] Typically, in a uniformly magnetised material a large fraction of the electrical current is carried by one spin channel (i.e. a spin 'short circuit'). Where the current crosses a domain wall the spin channels are mixed leading to a partial suppression of the 'short circuit' effect and an *increase* in the resistance which depends quadratically on the ratio of the precession time for a Fermi surface electron in the exchange field to the time the electron takes to ballistically traverse the domain wall. It has also been shown theoretically that domain walls can *reduce* the resistance, either due to the suppression of weak localisation corrections to the conductivity[18] or due to the redistribution of charge among the majority and minority spin bands in the domain wall when its electronic structure is accounted for semiclassically.[19] *While some experimental observations can be quite well-described within one or other of these different theoretical frameworks, it is probably true to say that, until recently, the overall understanding of domain wall resistance was poor.*

Recent advances in the epitaxial growth of metals have allowed the preparation of a range of thin film samples for domain wall resistance investigations where the domain structure can be optimised by a judicious choice of magnetocrystalline anisotropy, film thickness and orientation and sample geometry (shape anisotropy). Many experiments have addressed the issue of domain wall resistivity in 4-probe measurements on micron or sub-micron wire width samples (e.g. Refs. 20–24). Typical values of the *total* magnetoresistance in ferromagnetic thin films are small, a few percent, and the domain wall contribution is generally much smaller still as the wall width is typically only a small fraction of the domain size. A major complication for such measurements arises from the fact that the domain wall magnetoresistance is readily masked by the extrinsic magnetoresistance associated with the domains, and an accurate picture of the magnetic microstructure is vital if the domain wall contribution is to be correctly extracted. There are two main contributions to ferromagnetic resistivity anisotropy in measurements well below the Curie temperature;[25] anisotropic magnetoresistance due to spin-orbit coupling and Lorentz magnetoresistance, which depends on the angle between the current density and the internal magnetic induction. The Hall effect can also lead to an *increase* in resistivity as the Hall field changes direction in alternating magnetic domains,[26] while diamagnetic effects influence the resistivity near a domain wall due to the modification of electron trajectories there.[27]

Full control of the properties of magnetic domains can, however, be achieved provided local anisotropies are changed in a systematic fashion. This can be realized in perpendicularly magnetised Pt/Co/Pt systems by local irradiation with a gallium FIB, which reduces the film coercivity due to relaxation and mixing of the interfaces[28] (see Sec. 2.2). Using Ga FIB irradiation it is now possible to *define* artificial static magnetic domains of arbitrary shapes and sizes in Pt/Co/Pt sandwich or multilayer structures with deep sub-micron precision. Unlike natural domains, artificial domains are fabricated by engineering the intrinsic spin-dependent behaviour of the multilayers and promise reproducible magnetic and magnetotransport properties. In this article we attempt to demonstrate the strong potential of this approach. In Sec. 2.3, the magnetic switching and area-scaling of the dosed regions are studied using the extraordinary Hall effect (EHE). Artificial domains can also be an important tool for studying spin transport properties *across* domain walls and in Sec. 2.4 we describe a quantitative investigation of DW resistance in a lateral magnetic 'superlattice'.

2.2. Fabrication of Artificial Domain Structures

A magnetic thin film sandwich structure (Pt(3.5 nm)/Co(0.5 nm)/Pt(1.6 nm)) was deposited using dc magnetron sputtering at room temperature and an Ar pressure of 2.7 mTorr. After deposition, the rms roughness of the Pt capping layer, as measured by AFM, was 0.4 nm. Exceptionally thin (1.6 nm) Pt capping layers have been used in these experiments to minimise the effect of current shunting out of the Co film in the magnetotransport structures. In addition, the very thin (0.5 nm) Co layer used in our experiment is probably not chemically continuous, and will contain some pin holes. However, since the Pt develops an induced magnetic moment in the proximity of Co, it will be magnetically continuous. The magnetic thin films were then patterned into conventional Hall transport structures based on the intersection of 2 μm wide 'wires' using optical lithography and reactive ion etching with a 1:1 mixture of SF_6 and Ar. The very thin Pt capping layers render the structures extremely sensitive to irradiation, and so thin SiO_2 layers with thicknesses in the range 0–24 nm have been deposited on the top to reduce the FIB Ga-ion energy and dose. Focussed ion beam irradiation was performed on the completed structure with a commercial FIB (FEI Strata 201). The properties of the ions after transmission through the

SiO_2 layer have been estimated using SRIM (The Stopping and Range of Ions in Matter) software.[29] The average energies (% of Ga-ions transmitted) of the Gallium ions after passing through 0, 8 nm, 16 nm and 24 nm SiO_2 are 30 keV (100%), 19.8 keV (99.6%), 11.7 keV (88.7%) and 7.1 keV (57%) respectively.

The incident energy of the Ga-ions was 30 keV and a 1 pA current was used to lightly dose the required area of the sample. The beam diameter was about 10 nm, the distance between neighbouring pixels 7.4 nm, and irradiation was performed at a magnification of 10,000×, yielding a write field of 30.4 × 28.5 μm^2. Ion doses of 0.004 pC/μm^2, 0.007 pC/μm^2, 0.011 pC/μm^2 and 0.015 pC/μm^2 were achieved using pixel dwell times of 0.2 μs, 0.4 μs, 0.6 μs and 0.8 μs respectively. Typical AFM and MFM images of a Ga-ion dosed region are shown in Fig. 2.1, where the test Pt (3.5 nm)/Co (0.5 nm)/Pt (1.6 nm)/SiO_2 (8 nm) sandwich structure has been irradiated with a 0.007 pC/μm^2 Ga-ion dose in a rectangular strip on the left hand side of the field of view. Under the measurement conditions (see caption) the dosed region switches, but not the undosed one, and the MFM image exhibits strong black contrast in

Fig. 2.1. (Left) AFM image of a test structure. (Right) MFM image of the same region with 50 nm tip lift-off. The perpendicular applied field was reversed to −50 Oe after saturation in a positive field. The dark area is the dosed region and the light regions undosed.[30]

the irradiated region while no detectable change in the surface topography is observed in the AFM image.

The magnetic properties of the dosed regions have been investigated using the Extraordinary Hall Effect (EHE), with the external magnetic field applied perpendicular to the plane of the film and the in-plane voltage measured perpendicular to the current.[30] In a perpendicular external magnetic field H, the Hall resistivity ρ_H can be written as[31]:

$$\rho_H = R_o H + R_e 4\pi M , \tag{1}$$

where the first term describes the contribution of the ordinary Hall effect, R_o is the ordinary Hall coefficient and H is the external magnetic field. The second term describes the Extraordinary Hall term where M is the perpendicular component of the magnetisation and R_e is the EHE coefficient, which can be expressed as a function of resistivity ρ: $R_e = a\rho + b\rho^2$. Here the first term represents the skew scattering process, and the second the side jump.

2.3. Magnetic Properties of Artificial Domain Structures

Figure 2.2 shows a family of hysteresis loops of the extraordinary Hall Effect (EHE) measured at 300 K on four separate probes fabricated with different thickness SiO_2 overlayers and various Ga-ion doses. For all samples an area of 12 μm × 2 μm has been irradiated with the Ga-ion beam. The EHE yields information about the perpendicular component of the magnetisation of the Pt/Co/Pt sandwich structures and, in each case, the EHE of the dosed element (solid line) is compared with a neighbouring undosed element (dotted line), which is 4 μm away and fabricated on the same 2 μm wide wire. Since the perpendicular anisotropy is very sensitive to the thickness of the Co film, this comparison helps to minimise errors introduced by small thickness inhomogeneities. All the undosed samples have very square hysteresis loops which indicate that they are all magnetically homogeneous.

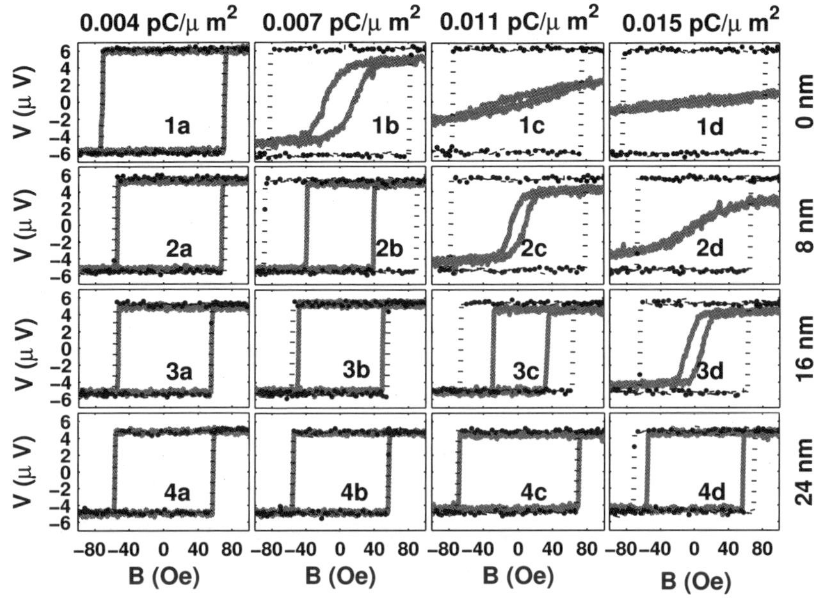

Fig. 2.2. EHE measurements of undosed (dotted lines) and dosed (solid lines) elements with different thickness SiO_2 overlayers at 300 K. The Ga-ion beam dose increases from left to right (see values at top edge of figure) and the SiO_2 thickness increases from top to bottom (see values at right edge of figure). From Ref. 30.

The Ga-ion beam dose increases from left to right (a to d) and the SiO_2 overlayer thickness increases (i.e. the effective energy of the Ga-ion beam decreases) from top to bottom (1 to 4). These results clearly show that the perpendicular anisotropy of the Pt/Co/Pt layer can be controlled not only by varying the Ga-ion dose, but also by varying the thickness of SiO_2 overlayer. The perpendicular magnetic anisotropy arises due to surface or interface anisotropy, which is a consequence of the reduced symmetry of the interface atoms, and is very sensitive to the interface roughness.[32] Ga-ion irradiation can induce three main changes in the Pt/Co/Pt sandwich structure: (1) It increases the roughness at the Pt/Co interface; (2) Due to the 10% lattice mismatch between Co and Pt, the thin Co layer is under tensile strain[33] which can relax after Ga-ion irradiation; (3) Ga-ion irradiation can mix the Co/Pt layers at the interface and result in the formation of a Co-Pt alloy.[34] Alloy formation requires a relatively high Ga-ion dose as compared to the first two processes. Since the Co film used in our experiments is very thin (0.5 nm) Co-Pt alloy formation would result in a significant reduction in the interfacial anisotropy. Therefore surface roughness and strain relaxation are believed to be the dominant phenomena that reduce the coercive field of our samples, whereas Co-Pt alloy formation destroys the perpendicular anisotropy entirely (see samples 1c, 1d and 2d). When no SiO_2 overlayer is present, the thin (1.6 nm) Pt capping layer allows the Ga-ion energy to be directly transferred to the Pt and Co atoms at the Pt/Co interface. Therefore the process of Pt/Co mixing can start at very low doses as observed in sample 1. When a SiO_2 overlayer is present,

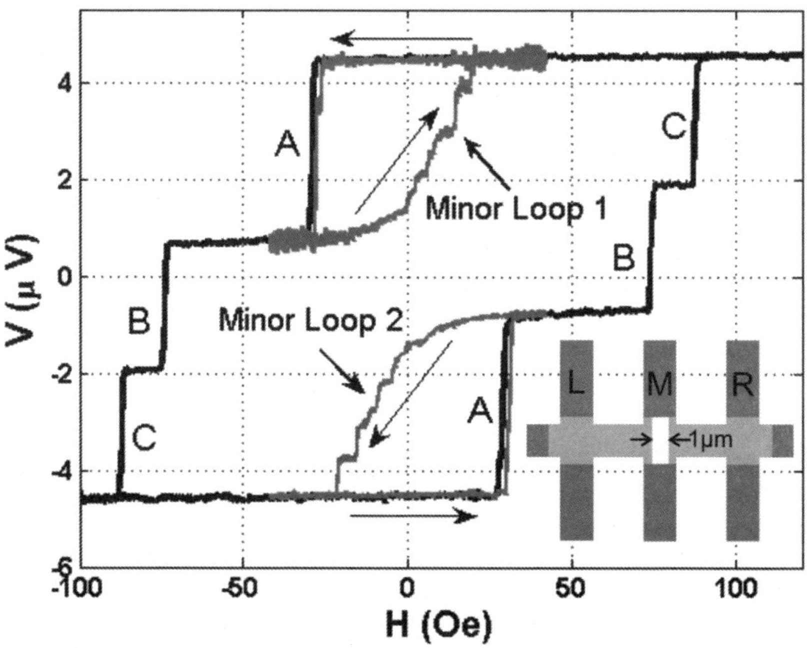

Fig. 2.3. EHE measurements of the central Hall cross (M) of the structure sketched in the inset which has a 3 μm × 0.5 μm dosed region at its centre. Step A corresponds to the switching field of the dosed region whereas steps B and C correspond to switching of the left and the right undosed regions respectively. Minor hysteresis loops 1 and 2 (gray) were obtained after saturating the sample in positive and negative magnetic fields respectively.[35]

however, it reduces both the energy absorbed by Pt and Co atoms at the Pt/Co interface and distributes the recoil energy more uniformly to the Pt/Co interface through collision cascades, and the mixing process occurs at a relatively high external dose allowing much better control of the Pt/Co interface roughness. Samples which have been irradiated with low effective ion energies have relatively square hysteresis loops indicating that they are homogeneously magnetised and have high irradiation homogeneity and low pinning site density.

The switching properties of perpendicularly magnetised dosed regions have been investigated in more detail using the extraordinary Hall effect.[35] Rectangular regions with dimensions $x \times 3\ \mu m^2$ (where $x = 250$ nm, 500 nm, 1 μm, 2 μm and 4 μm) were lightly dosed (0.007 pC/μm^2) using FIB in the middle of a Hall cross. A sketch of a typical device is shown in the inset of Fig. 2.3, where the ac current (I = 10 μA, f = 29 Hz) flows between contacts on the left and right, and L, M and R are labels for three adjacent Hall crosses. Figure 2.3 shows the EHE voltage measured at the middle Hall cross (M) (containing a 500 nm wide dosed stripe) which exhibits three steps. Changes in EHE voltage are directly related to the out-of-plane component of the magnetisation of the magnetic film in the Hall cross region. Step A corresponds to the switching of the dosed region whereas steps B and C are due to the undosed regions on the left and right sides. The minor loops 1 and 2 (gray lines) are obtained after saturating the sample in positive and negative magnetic fields respectively. These are shown in more detail in Fig. 2.4 which illustrates how the form of minor loops of identically dosed regions scales with their area after saturation at positive fields (the sweep rate for all minor loops was 2.6 Oe/s). The width of the dosed region was varied from 500 nm to 4 μm while the length was kept constant at 3 μm. (A 250 nm wide sample was also measured but its switching field lay very close to that of the undosed region making it difficult to measure a minor loop.) The following four features are clearly identifiable in Fig. 2.4. (a) Minor loops of dosed regions of width smaller than 2 μm are strongly asymmetric and (b) the asymmetry increases with the decreasing width of the dosed regions. (c) For data in the lower half of the figure, when H is swept in a positive sense, switching occurs at around 20 Oe, i.e. it is nearly independent of the width of the dosed region. (d) Discrete Barkhausen steps are also observable in the lower half of the minor loops.

The asymmetry of the minor loops shown in Fig. 2.4 indicates that the switching mechanism is different when the magnetic field is varied in a positive or negative sense. When all regions (dosed and undosed) are magnetised in the same direction (upper half), dosed regions switch sharply and the switching field increases strongly as the irradiated area decreases (c.f. Fig. 2.5). It shows the logical trend that when the area approaches zero, $H_c^d / H_c^{ud} \rightarrow 1$. Since this switching field is greater than the switching fields observed for the lower halves of the minor loops, it seems probable that reversal is governed by domain nucleation followed by domain wall motion. The increased switching field for small domains

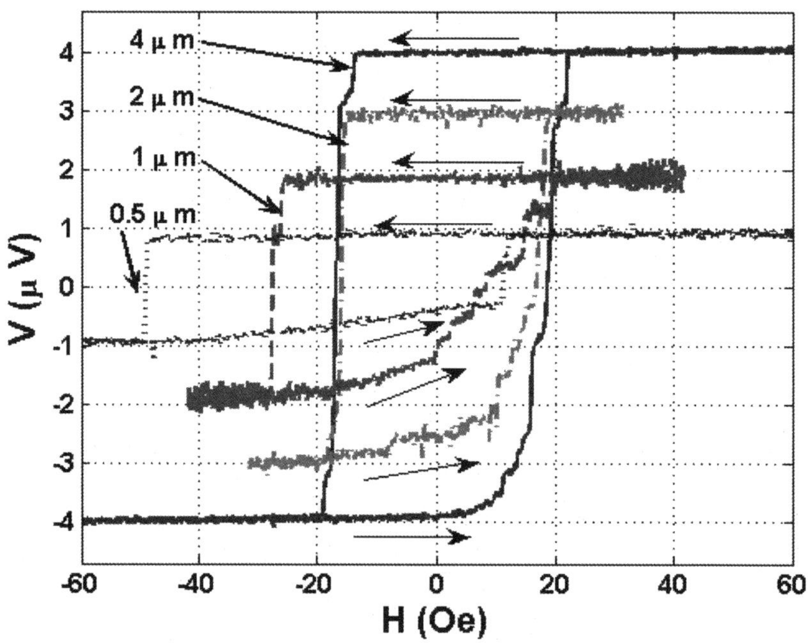

Fig. 2.4. EHE measurements at 300 K of minor loops of irradiated regions with widths varying from 500 nm up to 4 μm. Samples were first saturated in positive fields, and the external field then swept down until the dosed region reversed, at which point it was swept back up again. Adapted from Ref. 35.

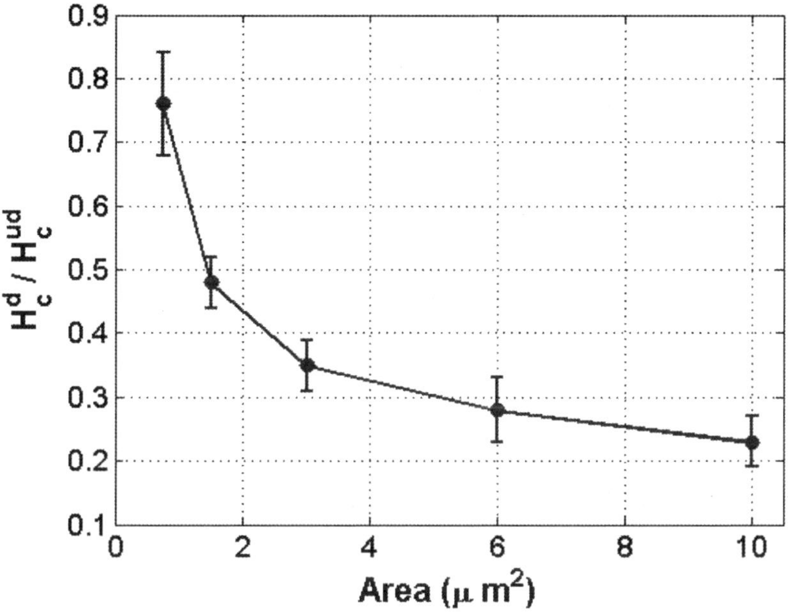

Fig. 2.5. Ratio of the coercive field of the dosed region (H_c^d) and the average coercive field of the undosed regions either side (H_c^{ud}) as a function of the area of the dosed region (0.007 pC/μm^2 FIB dose). Adapted from Ref. 35.

(<2 μm) may be due to the decrease in number of available nucleation sites. When an external magnetic field is applied in this way, such that it switches the dosed but not the undosed region, a magnetic domain will form whose size is determined by the irradiated region, with a domain wall at its border. We observe that switching back to uniform magnetisation from this state (lower half) does not depend strongly on the width of the dosed region, suggesting that reversal is now controlled by DW motion alone.

In addition to the switching mechanism, minor loops also shed light on the local pinning potentials in the dosed region. Switching in the lower half of the minor loops is not as sharp as in the upper half and has many plateaux-like features, which can be seen particularly clearly in the 1 μm dosed sample. When the external field is increased from negative values up to zero, a gradual change in EHE voltage is observed in the narrower structures indicating that magnetisation of the irradiated region is already beginning to relax. Using the Hall effect to quantify this we estimate that areas of ~0.1 μm^2, ~0.5 μm^2, ~0.7 μm^2 and ~0.7 μm^2 are relaxed for 4 μm, 2 μm, 1 μm and 0.5 μm wide regions respectively at H = 0 (i.e. domains with width <4 μm have significantly less than 100% remanence after these minor loops).

For positively increasing fields the reversal of the 1 μm wide irradiated stripe (area 3 μm^2) occurs predominantly in six approximately equal sized steps, which can be associated with local pinning sites present in the Ga-ion dosed region. We estimate that an area $A \approx 0.4\ \mu m^2$ reverses in each step, which must be correlated to large scale interface irregularities. For a Co film of thickness 0.5 nm, the magnetisation volume reversed during each step is 2×10^{-16} cm^3 and, assuming that interface irregularities are distributed uniformly, the characteristic separation of strong pinning sites is estimated as $\sqrt{A} \approx 600\,\text{nm}$ for the Ga-ion irradiated region. This is fairly consistent with the observation that artificial domains of width ≤500 nm generally reverse in a single step after the relaxation phase (H < 0).

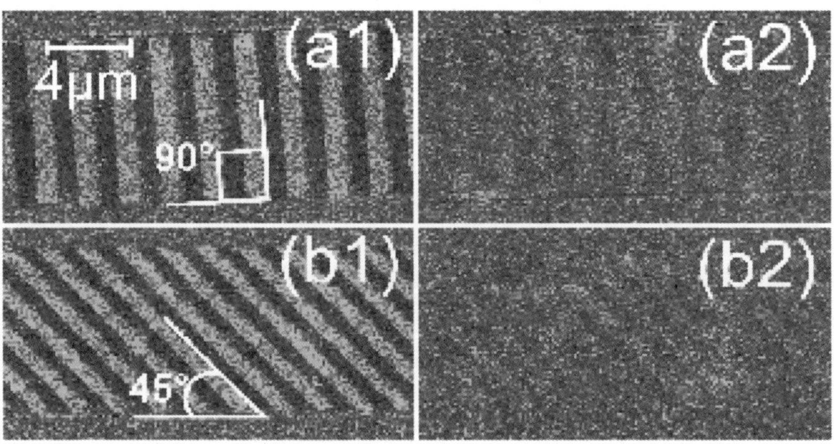

Fig. 2.6. (a1) and (b1) show MFM images of 90° and 45° domain 'superlattices' created by Ga-FIB irradiation at H = 0; (a2) and (b2) show MFM images of the same regions after saturation of the magnetisation at 300 Oe. Adapted from Ref. 36.

2.4. Angle-Dependent Domain Wall Resistivity Measurements

We now describe how unique insights into intrinsic domain wall resistivity can be obtained by studying the magnetotransport properties of *artificial* magnetic domain superlattice structures, formed in perpendicularly magnetised Pt/Co/Pt trilayers irradiated with a Ga focussed ion beam to locally modify the anisotropy.[34] As described above the deposition of thin SiO_2 capping layers prior to Ga-FIB irradiation can been used to precisely control the coercive fields of the artificial domains.[30,35,36] *In this way we have been able to define magnetic domain structures of arbitrary shape with <10 nm spatial resolution, in which the domain walls lie at precisely controlled angles*, as exemplified in Figs. 2.6(a) and (b). Other attractive features of this system which encourage its use for studying domain wall magnetotransport properties include the fact that closure domains, which can be a major source of anisotropic magnetoresistance,[37] are not energetically favoured in ultra-thin (0.6 nm) Co films with strong perpendicular anisotropy, whilst the relatively high film resistivity (~21 $\mu\Omega$ cm at 300 K) leads to short scattering times, τ, and $\omega_c\tau \sim 10^{-4} \ll 1$ (where ω_c is the cyclotron frequency), and allow Lorentz magnetoresistance and other effects associated with the wiggling of current lines at domain walls to be completely neglected.

Devices used in our transport studies were fabricated from a Pt(3.5 nm)/Co(0.6 nm)/Pt(1.6 nm) sandwich structure, which was deposited on a Si/SiO_2 substrate at 300 K using dc magnetron sputtering. Optical lithography and reactive ion etching were used to pattern the magnetic films in a Wheatstone bridge geometry based on 1 μm wide 'active' wires (Fig. 2.7(a)). After deposition of an additional 8 nm SiO_2 capping layer, the desired domain pattern was irradiated with a 30 keV Ga-ion dose of 0.009 pC/μm^2 in two asymmetric arms of the bridge (Fig. 2.7(b)) using an FEI Strata 201 Ga-FIB. To eliminate any extraordinary Hall effect that may originate from the Pt/Co/Pt underneath the voltage contacts, the Pt capping layer and Co layer were etched away in an Ar plasma after photolithography, and before evaporation of Ti/Au electrodes. It is possible that a very thin film of Co remained unetched after this step. The magnetic force microscopy (MFM) images shown in Figs. 2.6(a) and (b),

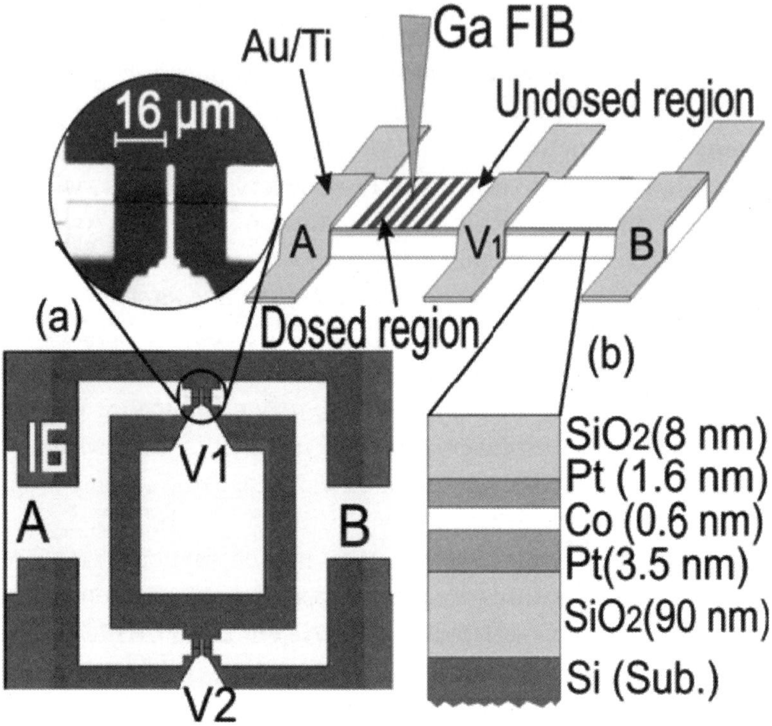

Fig. 2.7. (a) Micrograph of the Wheatstone bridge device consisting of four Pt/Co/Pt wires each 16 μm long and 1 μm wide. (b) Schematic of the top pair of Pt/Co/Pt wires of the Wheatstone bridge and sketch of the layer structure used. From Ref. 35.

demonstrate the precision with which 'superlattice' artificial domains can-be defined. Images a1 and b1 were captured at $H_\perp$ = 50 Oe when the magnetisation of the irradiated regions was reversed with respect to unirradiated regions, while a2 and b2 show the same regions after the magnetisation has been saturated everywhere at $H_\perp$ = 300 Oe. The very weak contrast in these latter images indicates that Ga-FIB irradiation causes a very small change in the saturation magnetisation due to Co/Pt mixing, and the resistivity of these regions also increases very slightly by ~2.5%. Measurements of our bridge devices were performed with a 10 μA ac current (f = 29 Hz) between current contacts A and B, and the output voltage measured with a lock-in amplifier between V1 and V2. An external magnetic field was applied perpendicular to the plane of the device with a solenoid.

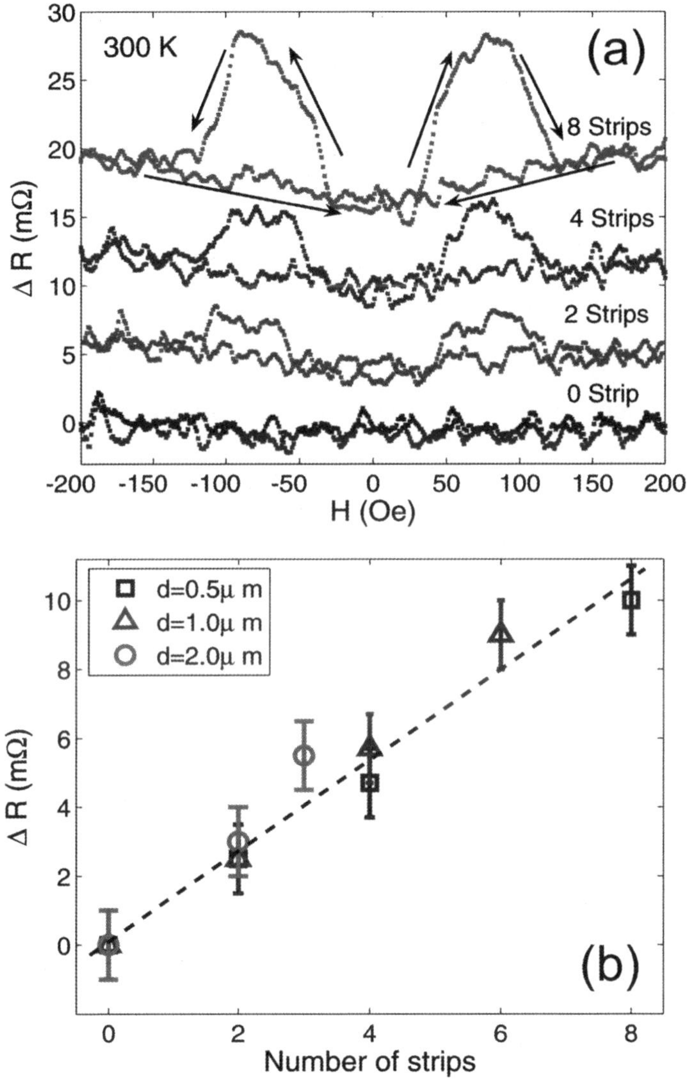

Fig. 2.8. (a) Magnetoresistance (offset vertically for clarity) of bridge structures containing 0, 2, 4 and 8 irradiated strips in CPW geometry at 300 K. (b) Maximum resistance change as a function of the number of irradiated strips of various widths, d.[36]

Figure 2.8(a) illustrates a set of magnetoresistance measurements recorded at 300 K on bridges with stripe domain 'superlattice' structures containing 0, 2, 4 or 8 artificial domains (all 500 nm wide in CPW geometry). Each measurement has been performed on a different, but otherwise identical, bridge structure fabricated on the same chip. For $H > +150$ Oe, both the irradiated and unirradiated domains are magnetised in the same direction and the resistance is at a low value. If the field is now reversed to $H \approx -50$ Oe, the magnetisation of the irradiated stripe domains, which have a smaller coercive field,[30,35] begin to reverse and the resistance increases to a maximum when domain walls are fully developed between irradiated and unirradiated regions. When the field is made more negative still it eventually reaches the coercive field of the unirradiated stripes (~100 Oe), which in turn start to reverse and the resistance drops back down to a minimum. The same behaviour is observed if the field is now swept back up in a positive sense. The weakly varying background signal is probably due to magnetoresistance arising from residual Co left under the voltage contacts after etching as described above. A separate Hall bar device was patterned in the same Pt/Co/Pt trilayer structure with Hall 'crosses' containing unirradiated and irradiated (with the same Ga-FIB dose) strips of the same width for EHE investigations.[30,35] The coercive fields inferred from these EHE measurements agree well with the values estimated from the magnetoresistance data. Figure 2.8(b) shows that the magnetoresistance of such domain structures increases linearly with the number of domain walls. Also included are data for structures with different irradiated domain widths, which confirm that the resistance is independent of the width of artificial domains, exactly as expected for intrinsic domain wall magnetoresistance at the border between the irradiated and unirradiated regions.

Figure 2.8(b) establishes the *intrinsic* domain wall origin of the magnetoresistance. Magnetic domains were then patterned at various angles with respect to the current flow in order to probe the perpendicular and parallel domain wall resistivities. Multiple artificial magnetic domains were defined, as described above, with the normal to the wall subtending an angle θ in the range $0°–60°$ with respect to the current direction. The geometry is shown in the inset of Fig. 2.10.

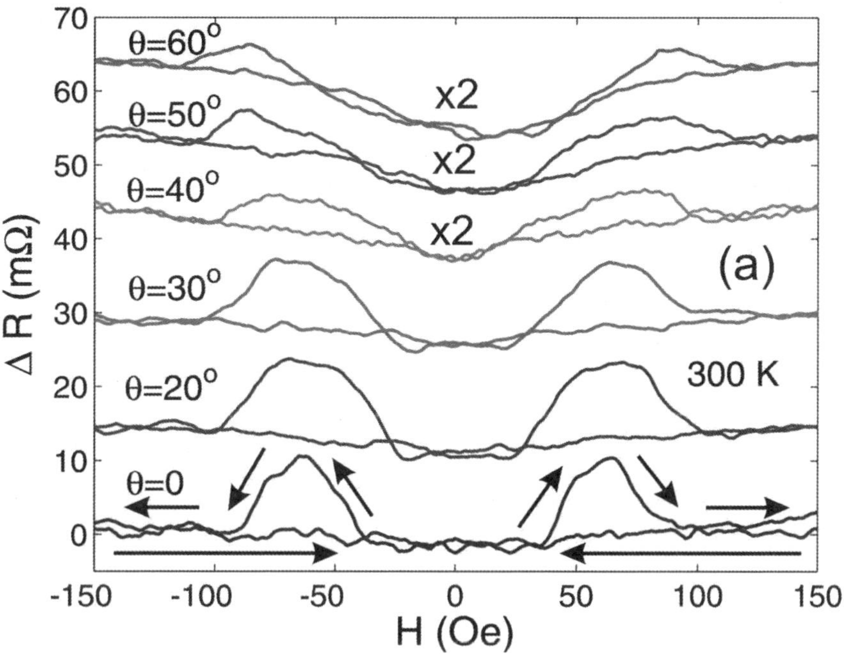

Fig. 2.9. 300 K magnetoresistance (offset vertically for clarity) of bridge structures containing irradiated strips patterned at different angles relative to the current flow. Adapted from Ref. 36.

For all samples the channel width was 1 μm, and the width of the irradiated artificial domains were kept invariant at 1 μm. The 16 μm long Pt/Co/Pt wires imposed geometrical constraints on the number of superlattice repeats that could be used. For $\theta = 0°$, 20°, and 30° six irradiated stripes were patterned, whereas for $\theta = 40°$, 50° and 60° only three irradiated stripes were possible. Therefore the magnetoresistance is due to twelve domain walls for the former set of angles, and six for the latter. Figure 2.9 illustrates the 300 K magnetoresistance of our bridge devices with different tilt angles. A steady decrease in the maximum resistance change is observed as the angle, θ, is increased. Also noticeable is a systematic broadening of the resistance peaks at larger angles. Since the domain width is the same in each case this can only arise from changes in the reversal dynamics due to angle-dependent

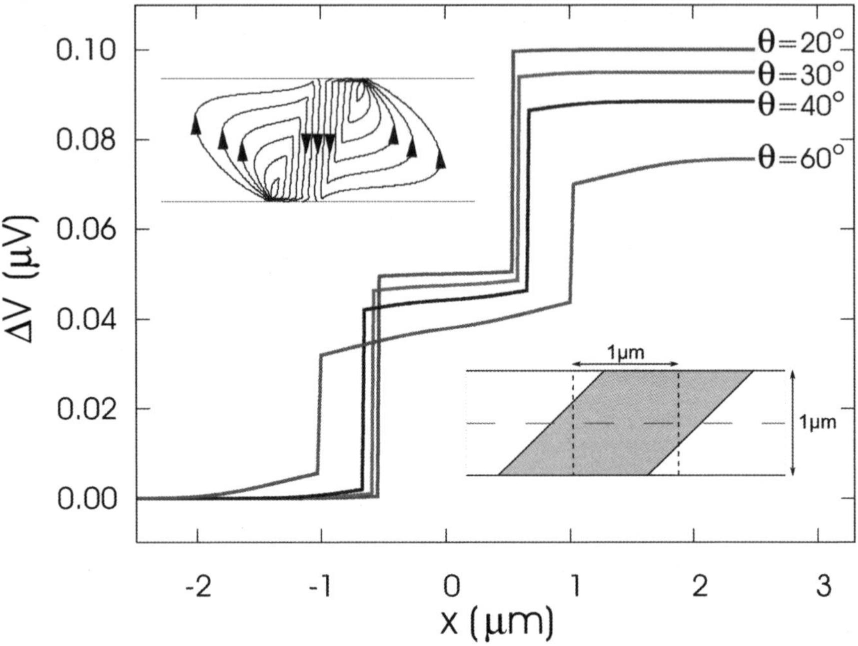

Fig. 2.10. Calculated potential profile along the middle of the device channel containing a single patterned strip 1 μm wide inclined at different angles (lower inset). Also shown are typical static eddy currents induced by the domain wall – a 30 nm region containing the domain wall has been expanded horizontally to enable the current lines there to be resolved. Adapted from Ref. 36.

changes in the magnetostatic interaction at the sharp and blunt corners of tilted domains. This will, for example, lead to a distribution of depinning fields for a domain wall trapped at the interface between an irradiated and an unirradiated domain.

To understand the angular dependence of the magnetoresistance, we model the current flow within the device channel, treating the problem as being two-dimensional. In steady state, the continuity equation requires that div $\boldsymbol{J} = 0$, and from Maxwell's equations curl $\boldsymbol{E} = 0$, where the current density $\boldsymbol{J}$ and electric field $\boldsymbol{E}$ are related via the resistivity tensor $\boldsymbol{E} = \boldsymbol{\rho J}$. On account of the high resistivity of our transport devices we neglect off-diagonal terms of $\boldsymbol{\rho}$ within each region, which is either (i) unirradiated, with $\rho_{xx} = \rho_{yy} = \rho_0$ (we take x to be along the device channel), (ii) irradiated, with $\rho_{xx} = \rho_{yy} = \rho_{\rm I}$, or (iii) a domain wall, for which

$$\rho_{DW} = \begin{pmatrix} \cos\theta & \sin\theta \\ -\sin\theta & \cos\theta \end{pmatrix} \begin{pmatrix} \rho_{CPW} & 0 \\ 0 & \rho_{CIW} \end{pmatrix} \begin{pmatrix} \cos\theta & -\sin\theta \\ \sin\theta & \cos\theta \end{pmatrix}, \quad (2)$$

with $\rho_{\rm CPW}$ and $\rho_{\rm CIW}$ the resistivities for current flow perpendicular and parallel to the domain wall respectively.[17] We solve the equations numerically in terms of a stream function, ψ, with $\boldsymbol{J} = (\partial\psi/\partial y, -\partial\psi/\partial x)$, assuming uniform current $\boldsymbol{J} = \boldsymbol{J}_0 = (J_0,0)$ at each end of the device channel and no current flow through the sides. The equations are discretised on a piecewise uniform finite difference grid, and solved iteratively using multigrid relaxation with continuity of the normal current and tangential electric field boundary conditions at each interface. Iterations are terminated when the residual is less that 10^{-10}, and the potential V along the device channel then calculated from $\boldsymbol{E} = -{\rm grad}V$.

Figure 2.10 shows typical results for the change in the potential along the device channel created by the presence of domain walls, for different angles of the irradiated region. As well as a sharp increase across the domain wall itself, the potential also displays an increase on both sides, which is due to the formation of static eddy currents, shown in the inset. The size and range of this contribution increases with the angle θ, but our simulations confirm that for the structures investigated here the potential change is fully contained within the device channel and not responsible for the changes in Fig. 2.9.

To understand the longitudinal channel resistance, we note that the *additional* resistivity due to the domain wall, $\delta\rho_{\rm CPW} = \rho_{\rm CPW} - \rho_0$, $\delta\rho_{\rm CIW} = \rho_{\rm CIW} - \rho_0$, is small. Neglecting the difference in the resistivity of the irradiated and unirradiated regions (which changes the final result by at most a few percent) the electric field can be written as $\boldsymbol{E} = (\boldsymbol{\rho}_0 + \delta\boldsymbol{\rho})(\boldsymbol{J}_0 + \delta\boldsymbol{J})$ so that to first order $\delta E_x = \delta\rho_{xx}J_0 + \rho_0\delta J_x$. Since δJ_x integrates to zero along the channel, the additional *resistance* may be attributed to $\delta\rho_{xx}$ and hence reflects the angular variation of the resistivity tensor in Eq. (2), which in general contains both CPW and CIW contributions:

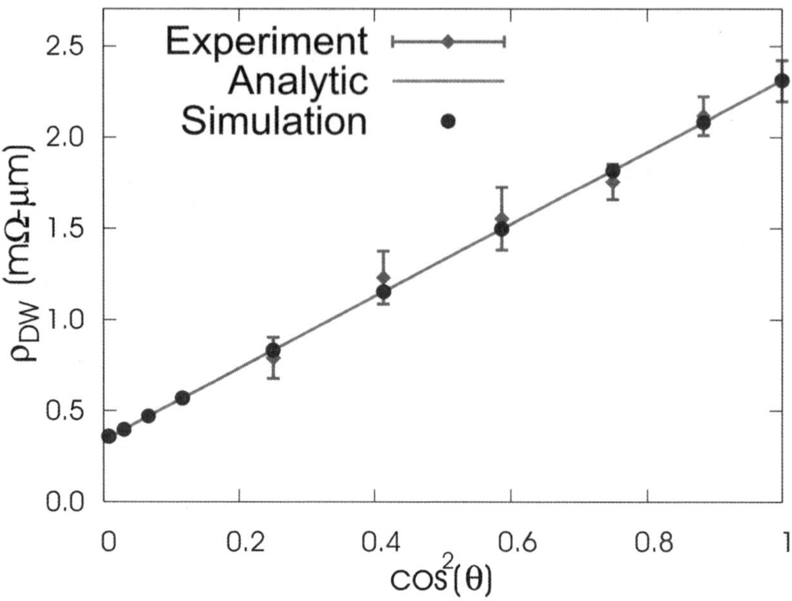

Fig. 2.11. Angular dependence of the total domain wall resistivity; diamonds: measured values from (a); circles: calculations such as in (b), using $\delta\rho_{\mathrm{CPW}} = 23 \times 10^{-3}$ $\mu\Omega$-cm, $\delta\rho_{\mathrm{CIW}} = 3.5 \times 10^{-3}$ $\mu\Omega$-cm, DW width 15 nm; line: Eq. (3) using the same values. Adapted from Ref. 36.

$$\delta\rho_{DW} = \delta\rho_{xx} = (\delta\rho_{CPW} - \delta\rho_{CIW})\cos^2\theta + \delta\rho_{CIW}\,. \tag{3}$$

In Fig. 2.11 we plot the domain wall resistivity, derived from curves of the type shown in Fig. 2.9, against $\cos^2\theta$. Also plotted are the results of our numerical simulations. Both measured and simulated results are accounted for very well by Eq. (3). In order to derive $\delta\rho_{\mathrm{DW}}$ from our measurements we have taken the domain wall width in our 0.6 nm thick Co layer to be 15 nm,[37] and assume a parallel resistor model for our trilayer films with resistivities uniformly scaled from the bulk values to give the correct total resistance.

Unlike previous attempts to study the anisotropy of domain wall resistivity,[38,39] which used naturally occurring reasonably well-ordered striped domain structures, these are the first measurements to exploit precise artificial control of domain wall orientation. This eliminates much of the uncertainty involved in analysis of the observed angular dependence. For example, it is not necessary to estimate the relative proportion of domain walls aligned perpendicular to the current.[39] As expected from Eq. (3) and the theory of Levy and Zhang, which predicts $\delta\rho_{CPW} > \delta\rho_{CIW}$, we find that the domain wall resistivity remains positive for all angles, with values for the perpendicular and parallel domain wall resistivities that may be inferred from our measurements of $\delta\rho_{CPW} = 23.1 \pm 1.1 \times 10^{-3}$ $\mu\Omega$-cm and $\delta\rho_{CIW} = 3.5 \pm 0.7 \times 10^{-3}$ $\mu\Omega$-cm respectively. We note that the results in Fig. 2.11 also limit the magnitude of any non-Hall-like contribution to the domain wall resistivity tensor – which has been assumed to be zero in Eq. (2) but which, if present, would make a contribution varying like $\sin 2\theta$ – to less than 0.5×10^{-3} $\mu\Omega$-cm. The magnetoresistivity ratio $MR_{CPW} = \delta\rho_{CPW}/\rho_0$ is 0.1%, which is comparable to values deduced for domain walls in other metallic systems.[38] The ratio of CPW to CIW resistivities is $\delta\rho_{CPW}/\delta\rho_{CIW} \approx 6.6 \pm 1.2$; according to LZ theory this is $3+10\sqrt{\rho_\downarrow / \rho_\uparrow} / (\rho_\downarrow / \rho_\uparrow + 1)$ where $\rho_\downarrow/\rho_\uparrow$ is the ratio of minority to majority spin carrier resistivities. The value $\rho_\downarrow/\rho_\uparrow \sim 5.5$ which we deduce is greater than in bulk Co ($\rho_\downarrow/\rho_\uparrow \sim 4$),[41] but consistent with the Fermi level density of states ratio that we find in band structure calculations of a Pt/3ML Co/Pt sandwich structure, where an enhancement of the minority-spin density of states arises due to band-narrowing.[42]

2.5. Conclusions and Outlook

We have described the magnetic properties of a new generation of magnetic devices in which artificial magnetic domains are introduced into perpendicularly magnetised Pt (3.5 nm)/Co (0.5 nm)/Pt (1.6 nm) trilayer structures by local Ga focussed ion beam irradiation. The switching properties and size-scaling of the dosed regions has been studied using the extraordinary Hall effect, allowing probable reversal mechanisms to be established. Discrete steps are observed in the minor loops of dosed regions which allow a direct estimate of the Barkhausen volume. Artificial domain walls are created at the border of dosed and undosed regions when an external magnetic field switches the former but not the latter, and this property has been used to create stripe-like domains. This approached has been exploited in a quantitative study of angle-dependent *intrinsic* domain wall resistivity in artificial magnetic superlattice structures. For the first time we are able to show that our experimental results are in truly quantitative agreement with a theory of domain wall resistivity based on spin-dependent potentials and scattering rates for majority and minority spin electrons due to Levy and Zhang.[17]

This model system also provides exciting opportunities for studying and exploiting domain wall physics, such as current-induced domain wall motion. The ability to control the exact domain structure that forms in nanowires and suppress extrinsic magnetoresistance effects eliminates difficulties that have plagued the domain wall resistance field for many years. These measurements have demonstrated that an unambiguous smooth mapping of the domain wall resistivity, from that for current flow perpendicular to the wall to that for current flow parallel, can be observed in carefully designed samples containing tilted domains. A natural development will be to study whether the magnitude of magnetoresistive effects can be enhanced by fabricating nanoconstrictions within the nanowire, or whether there are important effects arising from reducing the wire width/domain width ratio for tilted domains. Multiple-terminal "mesoscopic" measurements to probe transverse resistance *within* tilted domains will provide further insight into the validity of microscopic theories of domain wall scattering. Finally, our room temperature measurements point to applications in practical devices – one can envisage, for example, using the domain wall angle as a control parameter in multiple-state logic devices based upon spin-torque transfer.

Acknowledgments

This work was supported by the Leverhulme Trust, through grant no. F/00 351/F. We are grateful to Dr Peter Heard for giving us access to the Ga-FIB in the Interface Analysis Centre, University of Bristol, and for assistance with sample irradiation. We would also like to thank Dr Chris Marrows at the School of Physics and Astronomy, University of Leeds, for supplying the sputtered Pt/Co/Pt trilayers for this work.

References

1. L. Berger, *J. Appl. Phys.* **55**, 1954 (1984); L. Berger, *J. Appl. Phys.* **71**, 2721 (1992); L. Berger, *J. Phys. Chem. Solids* **35**, 947 (1974).
2. J. Slonczewski, *J. Magn. Magn. Mater.* **159**, L1 (1996).
3. J. Grollier *et al.*, *Appl. Phys. Lett.* **78** (2001); J. Grollier *et al.*, *Appl. Phys. Lett.* **83**, 509 (2003).
4. M. Klaui *et al.*, *Appl. Phys. Lett.* **83**, 105 (2003).
5. A. Yamaguchi *et al.*, *Phys. Rev. Lett.* **92**, 077205 (2004).
6. M. Hayashi *et al.*, *Phys. Rev. Lett.* **96**, 197207 (2006); M. Hayashi *et al.*, *Phys. Rev. Lett.* **97**, 207205 (2006); L. Thomas *et al.*, *Nature* **443**, 197 (2006); M. Hayashi *et al.*, *Nature Physics* **3**, 21 (2007); M. Hayashi *et al.*, *Phys. Rev. Lett.* **98**, 037204 (2007).
7. J. Fassbender *et al.*, *J. Phys D: Appl. Phys.* **37**, R179 (2004).
8. S. Hashimoto *et al.*, *J. Appl. Phys.* **67**, 4429 (1990).
9. C. Chappert *et al.*, *Science* **280**, 1919 (1998); T. Devolder *et al.*, *Appl. Phys. Lett.* **74**, 3383 (1999).
10. T. Devolder, *Phys. Rev. B* **62**, 5794 (2000).
11. T. Devolder *et al.*, *C. R. Acad. Sci. Paris, t.* **327** (Ser. II b), 915 (1999).
12. G. G. Cabrera and L. M. Falicov, *Phys. Status Solidi* (*b*) **61**, 59 (1974); *Phys. Status Solidi* (*b*) **62**, 217 (1974).
13. L. Berger, *J. Appl. Phys.* **49**, 2146 (1978).
14. S. S. P. Parkin, *Phys. Rev. Lett.* **71**, 1641 (1993).
15. T. Lalet and A. Fert, *Phys. Rev. B* **48**, 7099 (1993).
16. S. S. P. Parkin in *Magnetic Surfaces, Thin Films and Multilayers*, p. 211 (1992) (ed. S. S. P. Parkin, J.-P. Renard, T. Shinjo and W. Zinn).
17. P. M. Levy and S. Zhang, *Phys. Rev. Lett.* **79**, 5110 (1997).
18. G. Tatara and H. Fukuyama, *Phys. Rev. Lett.* **78**, 3773 (1997).
19. R. P. van Gorkom, A. Brataas, and G. E. W. Bauer, *Phys. Rev. Lett.* **83**, 4401 (1999).
20. K. Hong and N. Giordano, *Phys. Rev. B* **51**, 9855 (1995).
21. J. F. Gregg *et al.*, *Phys. Rev. Lett.* **77**, 1580 (1996).
22. Y. B. Xu *et al.*, *Phys. Rev. B* **61**, R14901 (2000).
23. A. D. Kent, J. Yu, U. Rüdiger, and S. S. P. Parkin, *J. Phys.: Condens. Matter* **13**, R461 (2001).
24. R. Danneau *et al.*, *Phys. Rev. Lett.* **88**, 157201 (2002).
25. I. A. Campbell and A. Fert in *Ferromagnetic Materials*, Vol 3 (1992) (ed. E. P. Wohlfarth, North-Holland, Amsterdam).
26. L. Berger, *Phys. Rev. B* **2**, 4559 (1970).
27. Y. I. Mankov, *Sov. Phys.-Solid State* **14**, 62 (1972).
28. C. Vieu, J. Gierak, H. Launois, T. Aign, P. Meyer, J. P. Jamet, J. Ferré, C. Chappert, T. Devolder, V. Mathet, and H. Bernas, *J. Appl. Phys.* **91**, 3103 (2002).

29. J. F. Zielgler, J. Biersack, and U. Littmark, computer software SRIM (2003). Available from www.srim.org.
30. Aziz, S. J. Bending, H. G. Roberts, S. Crampin, P. J. Heard, and C. H. Marrows, *J. Appl. Phys.* **98**, 124102 (2005).
31. L. Canedy, X. W. Li, and G. Xiao, *J. Appl. Phys.* **81**, 5367 (1997).
32. H. J. G. Draaisma, W. J. M. de Jonge, and F. J. A. Broeder, *J. Magn. Magn. Mater.* **66**, 351 (1987).
33. V. Mathet, T. Devolder, C. Chappert, J. Ferre, S. Lemerle, L. Belliard, and G. Guentherodt, *J. Magn. Magn. Mater.* **260**, 295 (2003).
34. T. Devolder, *Phys. Rev. B* **62**, 5794 (2000).
35. A. Aziz, S. J. Bending, H. G. Roberts, S. Crampin, P. J. Heard, and C. H. Marrows, *J. Appl. Phys.* **99**, 08C504 (2006).
36. A. Aziz, S. J. Bending, H. G. Roberts, S. Crampin, P. J. Heard, and C. H. Marrows, *Phys. Rev. Lett.* **97**, 206602 (2006).
37. I. Knittel and U. Hartmann, *J. Magn. Magn. Mater.* **294**, 16 (2005); U. Rudiger, J. Yu, L. Thomas, S. S. P. Parkin, and A. D. Kent, *Phys. Rev. B* **59**, 11914 (1999).
38. L. Klein *et al.*, *Phys. Rev. Lett.* **84**, 6090 (2000); M. Feigenson, L. Klein, J. W. Reiner, and M. R. Beasley, *Phys. Rev. B* **67**, 134436 (2003).
39. M. Viret *et al.*, *Phys. Rev. Lett.* **85**, 3962 (2000).
40. C. H. Marrows, *Adv. Phys.* **54**, 585 (2005); A. D. Kent, J. Yu, U. Rudiger, and S. S. P. Parkin, *J. Phys.: Condens. Matter* **13**, R461 (2001).
41. D. A. Papaconstantopoulos, *Handbook of The Band Structure of Elemental Solids* (Plenum Press, New York, 1986).
42. H. G. Roberts, PhD Thesis, University of Bath (2008).

Chapter 3

FABRICATION OF MAGNETIC NANOSTRUCTURES BY ELECTRON BEAM INDUCED DEPOSITION

Masaki Takeguchi

Nano-Characterization Center, National Institute for Materials Science
3-13 Sakura, Tsukuba 305-0003, Japan
E-mail: takeguchi.masaki@nims.go.jp

Masayuki Shimojo

Advanced Science Laboratory, Saitama Institute of Technology
1690 Fusaiji, Fukaya 369-0293, Japan
E-mail: mshimojo@sit.ac.jp

Electron beam induced deposition (EBID) is one of the promising techniques for fabrication of nanometer-sized structures with desired shapes at desired positions. Constitution of the nanostructures formed by EBID can be controlled by choosing the type of precursor element. This chapter describes the recent progress of fabrication of magnetic nanostructures using EBID.

3.1. Introduction

Magnetic nanostructures have attracted much attention because of interesting properties of such nanometer-sized magnetic materials and potential of application to not only ultrahigh density magnetic recording but also new nanometer-sized magnetic devices such as read heads and memory cells. The magnetic properties of the nanostructures strongly depend on their shape and size, and therefore fundamental study of magnetic properties of magnetic nanostructures with the desired shape is a significant issue.[1] For their application to the advanced magnetic devices, an understanding of the switching behavior of the magnetic nanostructures is important from the viewpoint of development of high density magnetic recording and magnetic random access memory. Their spatial distribution is also crucial because magnetic interaction between densely packed nanometer-sized magnetic elements such as rings and dots via magneto-static stray fields influences the switching behavior.[2] Thus, for both fundamental and application researches of magnetic nanostructures the fabrication of various desired shaped magnetic nanostructures at the desired positions has been required.[3–6]

Electron beam induced deposition (EBID) has become one of the most promising methods for such size- and position-controllable nanofabrication tools in point of view of high resolution, abundant types of deposit elements, and capability of three dimensional fabrications.[7–14] Constitution of the nanostructures formed by EBID can be controlled by selecting the type of precursor element. When magnetic elements such as iron and cobalt are employed as precursor in EBID, magnetic nanostructures are fabricated. Now, applications of EBID to fabrication of not only interconnections in semiconductor IC circuits and quantum dot- and wire-arrays in advanced electronic but also magnetic devices have been increasingly expected.

In EBID, metal-organic precursor gases are decomposed mainly by secondary electrons from a substrate surface and resultantly metal deposits are formed on the substrate surface. The formed deposits are composed of an amorphous phase or a mixture of nanocrystals and an amorphous phase. The amorphous phase contains a considerable amount of carbon in addition to metal elements.[15–17] This unfavorable carbon comes from ligands of the dissociated organic precursors. Decreasing the carbon concentration is a significant issue because the carbon rich nanostructures often degrade their electric and magnetic properties. For instance, the resistance of gold (or platinum)-containing nanowires formed by EBID is quite larger than that of pure gold (or platinum) wires.[18,19] In the case of magnetic materials, although iron- and cobalt-containing nanostructures were successfully produced by EBID with iron and cobalt carbonyl gases ($Fe(CO)_5$ and $Co_2(CO)_8$)[16,20–25] and by scanning tunneling microscope-assisted CVD with ferrocene gas ($Fe(C_5H_5)_2$),[26] quantitative magnetic characterization of individual nanostructures has rarely been carried out so far.

A postdeposition heat treatment is effective in the reduction of the carbon content of the EBID-fabricated nanostructures. For instance, the postdeposition heat treatment for the nanostructures formed with $Fe(CO)_5$ has demonstrated the transformation from a mixture of amorphous carbon and iron phases to crystalline alpha iron without significant shape change.[22,25] Che *et al.* have reported that ferromagnetic FePt nanostructures were obtained by EBID with a mixture gas of $Fe(CO)_5$ and $(CH_3)_3(C_5H_5)Pt$ followed by the heat treatment.[27] Mixing water vapor with metal-organic precursors is also effective to decrease the carbon content,[28,29] and recently carbon free magnetite (Fe_3O_4) nanostructures were successfully obtained by using a mixture gas of $Fe(CO)_5$ and water vapor.[30]

3.2. EBID Fabrication

EBID has been carried out using scanning electron microscopes (SEMs) and transmission electron microscopes (TEMs). The deposition is produced mainly by low-energy electrons, such as secondary electrons, and their spatial distribution becomes wider as the primary electron energy increases. It was thought that the resulting deposit would be larger for higher energy electrons, although the probe size of the electron beams decreases. Tanaka *et al.*[14] and Van Drop *et al.*[9] succeeded in fabrication of nanodots with a few nanometers in diameter using a

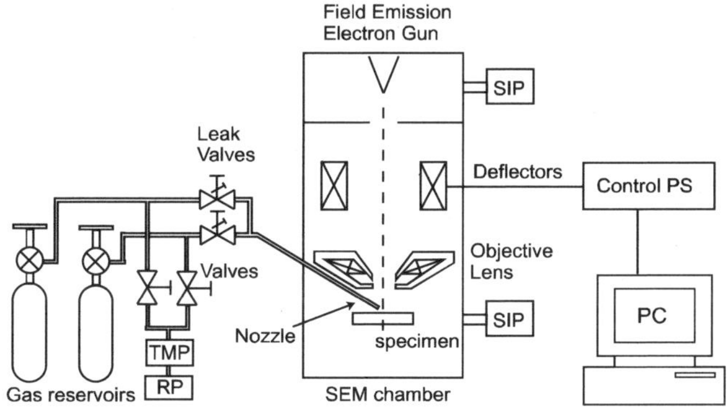

Fig. 3.1. Schematic drawing of a 30 kV FEG-SEM equipped with an introduction system for fabrication of the nanostructures by EBID.

200 keV electron beam in transmission electron microscope equipped with a field emission gun.

A 30 kV ultrahigh vacuum scanning electron microscope equipped with a field emission gun (JEOL JSM-7800FV), of which base pressure is about 2×10^{-6} Pa, was employed to carry out the EBID fabrication in the present work. A schematic drawing of the deposition experiment is shown in Fig. 3.1. A gas introduction system installed into the SEM is combined with two sets of gas source reservoirs and pipelines through leak valves to control the partial pressure of precursor gases. The precursor gases are introduced into a specimen chamber of the SEM through a nozzle with an inner diameter about 0.2 mm. Two kinds of gases are introduced simultaneously and their partial pressures can be independently controlled. The nozzle tip is about 1 mm from the electron beam position. In the present work, the nanostructures were fabricated on edges of carbon microgrid films, silicon films and molybdenum films at room temperature. The silicon and molybdenum thin areas were prepared by using a focused ion beam thinning system (JEOL JEM-9310FIB). The nominal electron beam current was about 0.8 nA with a beam diameter of 4 nm. The beam position was controlled by an external deflector voltage input using a computer with digital-analog converters. In order to fabricate free-standing nanostructures, the electron beam position was moved from the substrate into an open space.

3.3. Fabrication of Iron-Containing Nanostructures

Figure 3.2(a) shows a transmission electron microscopy (TEM) image of nanodots and free-standing nanorods formed on a carbon microgrid film at room temperature. The nanorods were fabricated by moving the beam from the carbon film edge to an open space at a speed of 2 nm/s. Arbitrary-shaped nanostructures were also fabricated by the subsequent electron beam patterning on the preliminarily formed nanorods. Figure 3.2(b) shows a TEM image of a square frame structure. The linewidths of the rods and frames are 30–50 nm. Higher-magnification TEM observation of the free-standing rods and frames showed that the surface was covered with nanocrystals and that their inside was amorphous. The diffraction pattern and electron energy loss spectroscopy

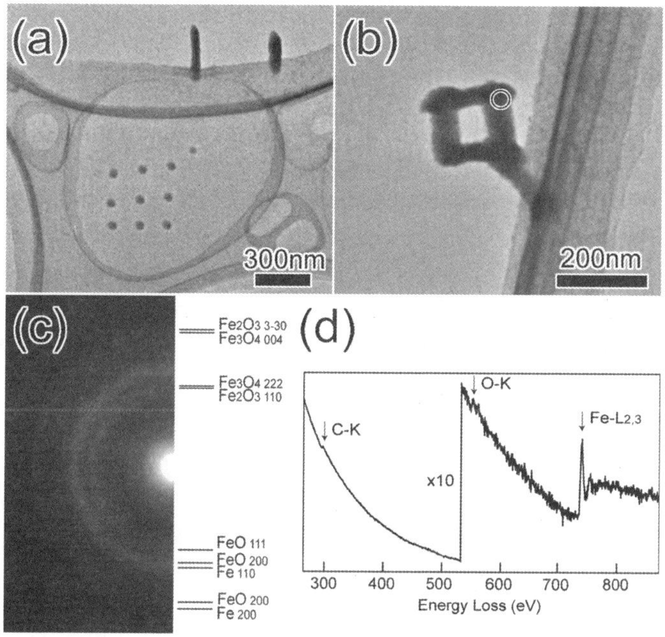

Fig. 3.2. TEM images of (a) nanodots and nanorods and (b) a free-standing square frame fabricated by EBID with $Fe(CO)_5$. They were fabricated on a carbon microgrid film using a 30 keV fine electron probe of 4 nm diameter in FEG-SEM. (c) Electron diffraction pattern taken from area indicated by circle in (b). (d) EELS spectrum from the same area specified by a circle in (b). Reprinted with permission from Ref. 17 by the Japan Society of Applied Physics.

(EELS) result from the region in the circle in Fig. 3.2(b) are indicated in Figs. 3.2(c) and 3.2(d), respectively.

The EELS spectrum indicates that the free-standing rods and frames were composed of iron, carbon and oxygen. Diffraction indices and corresponding ring positions for alpha-iron and possible iron oxides are inserted in Fig. 3.2(c). Their positions are in agreement with the observed ring positions. Shimojo *et al.* have reported that a carbon K-edge showed a characteristic of amorphous carbon but not of iron carbide in their detailed EELS analysis.[16] Hence, it is deduced that some faint polycrystalline-like rings originated from the iron oxide nanocrystals existing near the surface and that broad rings were from the amorphous phases in their body. The existence of surface oxides may be mainly due to the exposure of the specimen to air during the transportation from SEM to TEM machine.

With increasing the scanning speed, the self-standing nanorods became more slender. The deposit shape is determined by secondary electrons emitted from the substrate. After the incident electron beam is moved away from the substrate edge, the self-standing nanorod can be grown by secondary electrons generated from the as-formed nanorod itself. Resultantly, the self-standing nanorod maintains an almost constant width throughout its length. It should be noted that the aspect ratio (e.g., thickness to linewidth) of the nanostructures formed by EBID must be considerably large.[31] In the case of the nanostructures in Fig. 3.2, the thickness was estimated to be 150–250 nm.

It has been reported that the as-deposited nanostructures on a carbon film contained a large amount of carbon (iron and carbon was approximately equivalent) and transformed to iron carbide by the heating process.[16,17] On the other hand, the carbon concentration of the nanostructures on Si substrates was rather smaller than that on the carbon films.[17] Figure 3.3 shows a TEM image of free-standing nanostructures formed on a Si substrate. EELS analysis showed that the carbon concentration in the nanorods is less than 50%.

It is known that electron holography is a powerful tool to observe electric and magnetic field inside and outside of the materials quantitatively with high spatial resolution and applicable to characterization of magnetic fields of the arrays and chains of magnetic nanostructures. Electron holography was carried out to observe magnetic fields around the nanostructure in Fig. 3.3 using a JEOL JEM-3000F

Fig. 3.3. TEM image of the desired-shape self-standing nanostructures on a thin silicon substrate edge formed by EBID with $Fe(CO)_5$ at room temperature. Reprinted with permission from Ref. 11 by IOP Publishing Ltd.

equipped with an electron biprism consisting of a centered wire electrode and grounded electrodes at both the sides. The electron wave passing through the sample and vacuum regions is deflected by the electrostatic potential around the wire electrode so that an interference pattern is formed in an image plane. This interference pattern is called "electron hologram", in which phase information of the object passing through the sample can be recorded. By reconstructing the electron hologram digitally by a computer, a phase image was obtained in the standard manner described in literature.[36] Figure 3.4(a) shows an electron hologram of the nanostructures. Figure 3.4(b) shows a phase image reconstructed from Fig. 3.4(a), displayed as a COS intensity, where the phase was amplified by a factor of 4. This phase image is so called interference micrograph, and one can recognize dark lines as magnetic fluxes. The transition between the adjacent dark lines corresponds to a phase shift of $\pi/2$ (1.57) rad.

The phase difference of the electron wave passing through both sides of the nanorod is described as

$$\Delta\varphi = 2\pi \frac{e}{h} BS \tag{1}$$

where S is a cross section area of the nanorod. The external magnetic stray field above and under the nanorod oriented in opposite direction to the internal magnetic field should be taken into account to estimate the

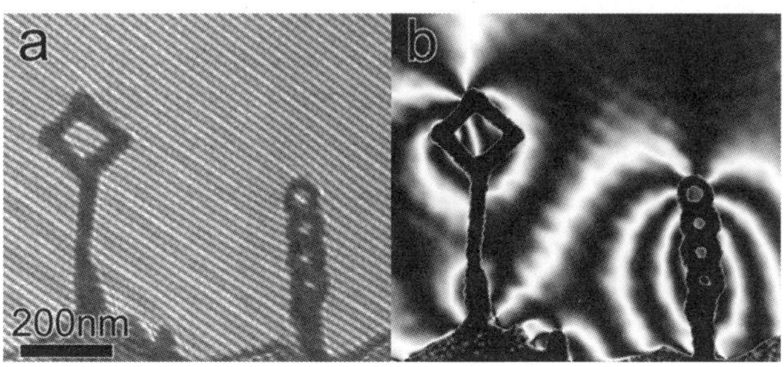

Fig. 3.4. (a) Electron hologram and (b) reconstructed phase image displayed as a COS intensity, so-called "interference micrograph", where the phase was amplified by a factor of 4.

accurate magnetic flux in the nanorod,[37] because the observed phase difference between the both sides of the nanorod is actually the sum of the phase shift due to the internal magnetic field and the vertical integration of the external magnetic stray fields at the nanorod position. The phase shift due to the stray fields at the nanorod position was extrapolated approximately from the phase distribution near the nanorod. As shown in Fig. 3.5, the phase difference between both the sides of the nanorod with a ring-head, indicated by an arrow in Fig. 3.3, was measured to be about 7.3 rad. The phase change due to the external magnetic stray field at the nanorod position was extrapolated approximately from the phase inclination near the nanorod, and found to be about 1.5 rad. Therefore, the phase shift due to the internal magnetic field was estimated to be about 8.8 rad, which corresponds to the magnetic flux of 5.75×10^{-15} Wb. The shape of the cross-section of the nanorod was supposed to be elliptic. The thickness and width of the nanorod was estimated to be about 250 nm and 40 nm, respectively. Thus, the elliptical-shape cross-section of the nanorod was about 7.85×10^{-15} m^{-2}, and finally the remanent magnetic flux density B_r was calculated to be about 0.7 T.

The compositional ratio of metal to carbon in EBID nanostructures significantly influences magnetic properties. Nanostructures with the different compositional ratio of iron to carbon were formed using a mixture gas of $Fe(CO)_5$ and ferrocene, $Fe(C_5H_5)_2$. Stoichiometrically, the ratio of iron to carbon of $Fe(CO)_5$ is twice as many as that of $Fe(C_5H_5)_2$.

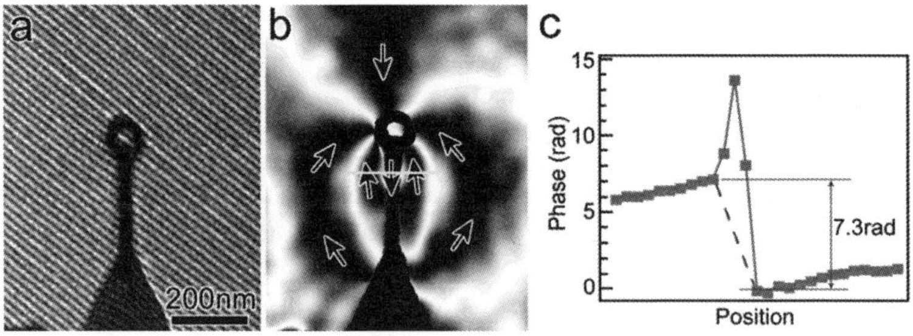

Fig. 3.5. (a) Electron holograms, (b) corresponding COS-intensity-displayed phase images and (c) line profiles of the phase distribution across the nanostructure, which is indicated by an arrow in Fig. 3.3. Reprinted with permission from Ref. 11 by IOP Publishing Ltd.

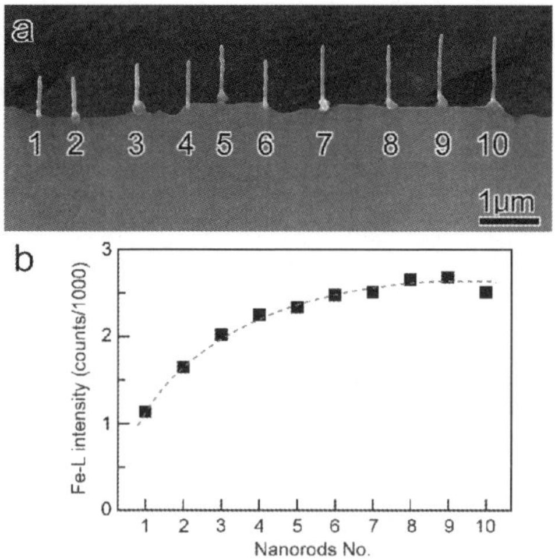

Fig. 3.6. (a) SEM image of the EBID fabricated nanorods formed on an edge of a molybdenum thinned area. (b) The result of EDS analysis, showing Fe-L intensity for each nanorod. Reprinted with permission from Ref. 24 by Springer.

The iron to carbon compositional ratio in the nanostructures was controlled by changing the partial pressure of these gases. Figure 3.6(a) shows a SEM image of nanorods (numbered from 1–10) formed by EBID with various partial pressures of iron carbonyl and ferrocene gases on an edge of a molybdenum thinned area. First, the partial pressure of ferrocene gas was maintained at 1×10^{-5} Pa. Then, carbonyl gas was introduced so that the total pressure was increased from 3×10^{-5} Pa to 12×10^{-5} Pa by 1×10^{-5} Pa for each EBID fabrication of nanorod (totally 10 nanorods were formed). Figure 3.6(b) shows iron-L peak intensity for each nanorod acquired by EDS analysis, indicating that the iron concentration was successfully increased depending on the change of the partial pressure of carbonyl. Carbon-K peak was hardly detected because of insensitivity and poor resolution of the current EDS detector. Previously reported EELS analysis showed that iron, carbon and oxygen contained in the nanorod formed by EBID with only iron carbonyl on the same beam condition as the present experiment was about 70, 10 and 20 at%, respectively.[17] Roughly assuming that the iron concentration of No. 8–10 nanorods is 70 at% and the thickness of all the nanorods is the same, the iron concentration of No. 1 nanorod was approximated to be 30%.

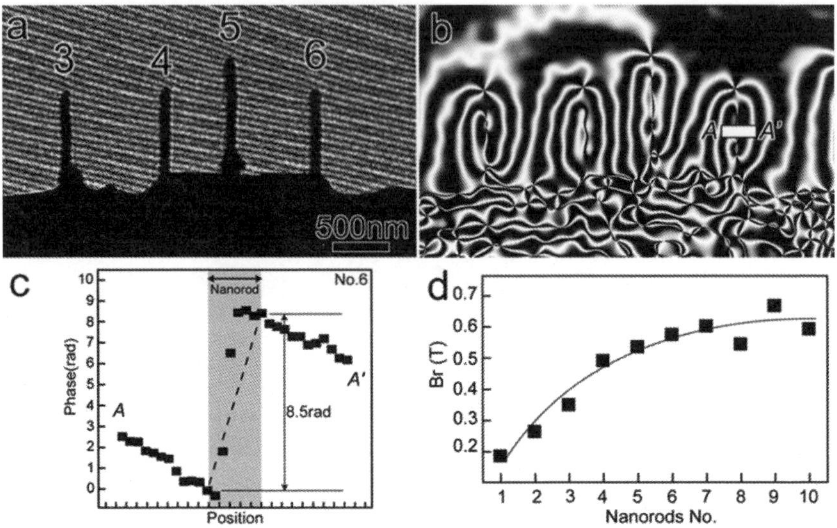

Fig. 3.7. (a) Electron hologram of the No. 3–6 nanorods. (b) Phase image (interference micrograph) reconstructed from (a) with an amplification factor of 4. (c) Lline profile of the phase distribution across the line AA' in the No. 6 nanorod. (d) The remanent magnetic flux density Br for the No. 1–10 nanorods. Reprinted with permission from Ref. 24 by Springer.

The remanent magnetic flux density B_r for each nanorod was evaluated by electron holography. Figures 3.7(a) and 3.7(b) show an electron hologram of the No. 3–6 nanorods and the corresponding reconstructed phase image, respectively. A line profile of the phase across the line *AA'* in the No. 6 nanorod is shown in Fig. 3.7(c). Figure 3.7(d) shows the B_r values for the No. 1–10 nanorods, indicating that a fitting curve (solid line) described in the figure is quite similar to a plotted curve (iron concentration) in Fig. 3.6(b), showing that the remanent magnetic flux density B_r of the EBID nanostructures containing iron and carbon is related to the iron concentration.

3.4. Post-Deposition Heat Treatment: Fabrication of Alpha Iron Nanostructures

It has been reported that substrate heating during deposition results in an enhancement of the decomposition rate, and the nanostructures composed of nanocrystals could be formed (namely, metal concentration and crystallinity could be improved).[20,32,33] Post-deposition heat treatment for chromium-, tungsten- and rhenium-containing nanostructures has been reported; after annealing at about 800°C, chromium- and tungsten-containing nanorods showed the appearance of nanocrystals within them without change in their original shape, while rhenium-containing nanorods transformed to crystalline structures with a large grain size but with change in their shape.[34] Here, we attempt the post-deposition heat treatment to iron-containing nanostructures. The original as-deposited iron and carbon compositional ratio are crucial for the stoichiometry of the productions after the heat treatment. By heating in vacuum, carbon reacts with oxygen to become CO_2 molecules, and the CO_2 molecules are desorbed. If the remaining carbon content is large, carbon reacts with iron to form iron carbides and a suitable phase appears according to a phase diagram. It has been reported that the as-deposited nanostructures on a carbon film contains a large amount of carbon (iron and carbon are approximately equivalent) and therefore transforms to iron carbide during the heating process.[16,17] On the other hand, the carbon

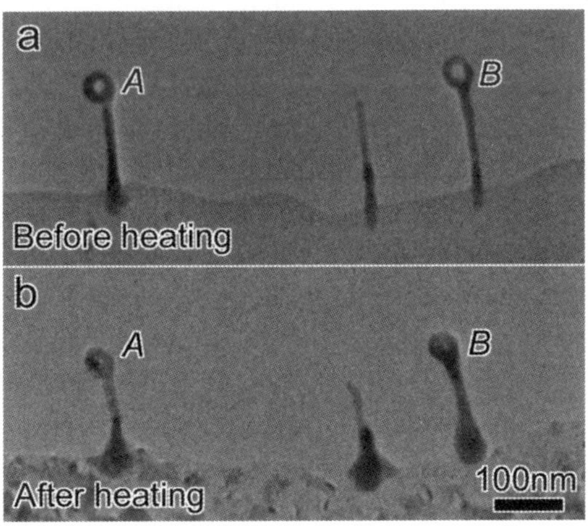

Fig. 3.8. TEM images of the free-standing nanostructures on a silicon substrate before (a) and after (b) a heat treatment at about 600°C for 2 h. Reprinted with permission from Ref. 11 by IOP Publishing Ltd.

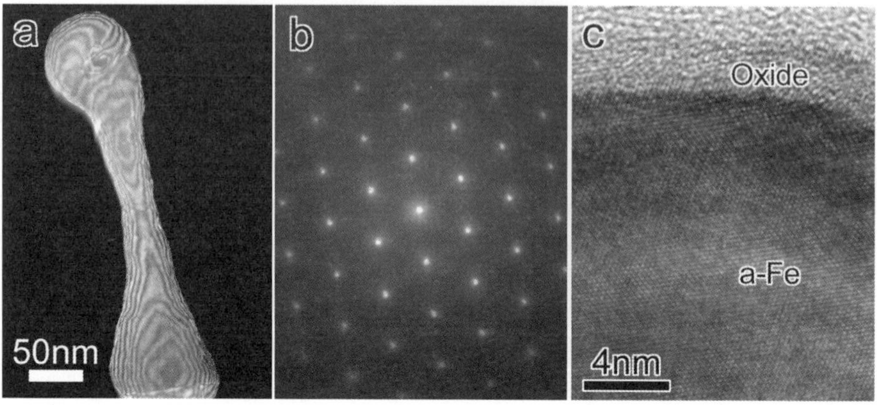

Fig. 3.9. (a) Dark-field TEM image of the nanostructure labeled by B in Fig. 8(b), taken using a (110) diffraction spot at the incidence of <111>a-Fe. (b) Selected area diffraction pattern from the nanostructure in (a). (c) A high-resolution TEM image of the apex part of the nanostructure in (a). Reprinted with permission from Ref. 11 by IOP Publishing Ltd.

concentration of the nanostructures on the Si substrate is rather smaller than that on the carbon film, resulting in the formation of alpha iron.[17]

Figures 3.8(a) and 3.8(b) show TEM images of nanostructures on a silicon film edge before and after the postdeposition heating at about 600°C for 2 h, respectively. On heating, the shape of the nanostructures was maintained except for the following: the linewidth near the root became larger and the inner diameter of the rings was smaller. The tilting of the nanostructures is supposed to be due to the bending of the thin silicon film edge. All the nanostructures in this sample transformed into an alpha-iron phase in their entire body or a part of the body by the heating. Figure 3.9(a) show a dark-field TEM image of the nanostructure labelled by *B* in Fig. 3.8(b), taken using a (110) diffraction spot at the incidence of $<111>_{a-Fe}$. The selected area diffraction pattern from the nanostructure shown in Fig. 3.9(a) is indicated in Fig. 3.9(b), proving that the whole of the body was a single crystal. Figure 3.9(d) shows a high-resolution TEM image of the apex part of the nanostructure.

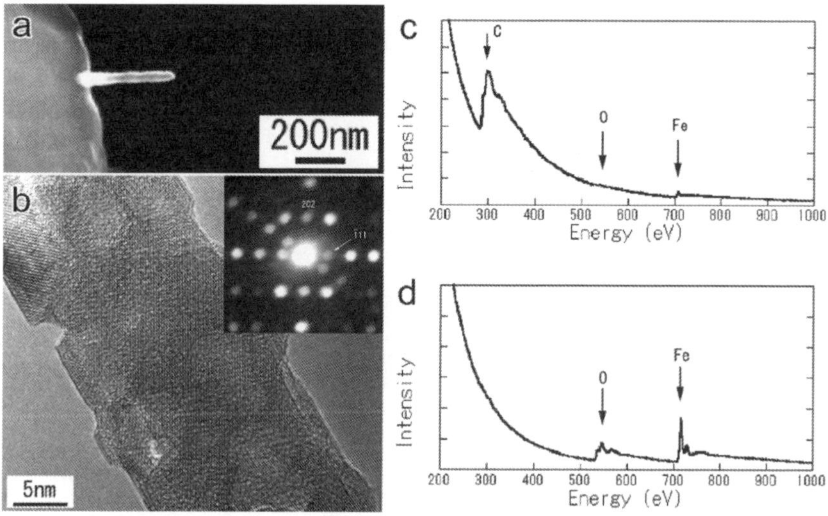

Fig. 3.10. (a) SEM and (b) HRTEM images of the nanorod formed by EBID with a mixture of $Fe(CO)_5$ and water vapor. Nano beam electron diffraction is inserted in (b). (c,d) EELS spectra of nanorods formed withoutand with water vapor, respectively. Reprinted with permission from Ref. 40 by IOP Publishing Ltd.

3.5. EBID with $Fe(CO)_5$ and Water Vapor: Fabrication of Magnetite Nanostructures

Iron oxide, especially magnetite (Fe_3O_4), nanostructures including nanowires and nanodots are also important in spin electronic devices because of their half-metallic ferrimagnetic nature.[39] We have succeeded in the direct formation of Fe_3O_4 nanostructres at room temperature by EBID with a mixture of $Fe(CO)_5$ and water vapor as precursor.[40,41] Water vapor was precisely mixed with an iron carbonyl gas and the mixture was introduced in an EBID SEM chamber as well as the experiment of EBID with a mixture of $Fe(CO)_5$ and ferrocene.

Figure 3.10(a) shows a SEM image of a nanorod formed on an edge of a thinned molybdenum substrate using an $Fe(CO)_5$ precursor mixed with water vapor. The nanorod was about 300 nm in length and 20–30 nm in width. Figure 3.10(b) shows a HRTEM image of the middle part of the nanorod, in which nanobeam electron diffraction pattern is inserted. These indicate the high crystalinity of the deposit. The nanorod

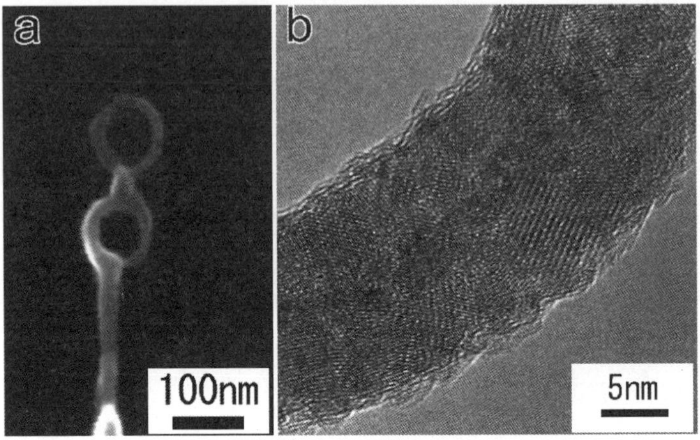

Fig. 3.11. (a) SEM image of a double-ring structure, showing the flexibility in shaping of EBID and (b) HRTEM image of a part of the ring. Reprinted with permission from Ref. 40 by IOP Publishing Ltd.

was polycrystalline and the grain size was relatively large comparing with the width of the nanorod. Figures 3.10(c) and 3.10(d) show EELS spectra of the nanorods with and without mixing water vapor. Iron, oxygen and carbon edges are observed in the spectrum of the nanorod fabricated without water vapor, while no carbon edges are seen in that with water vapor.

EELS analysis for the nanorods formed by various $[H_2O]/[Fe(CO)_5]$ partial pressure ratio showed that oxygen content increases and carbon content decreases with increasing the $[H_2O]/[Fe(CO)_5]$ partial pressure ratio. The microstructure still consisted mainly of amorphous carbon at the partial pressure ratio of 0.4, and carbon content reached substantially naught and atomic ratio of Fe and O was 1: 1.3 ± 0.2 (slightly fluctuated at positions) in the deposit when the partial pressure ratio was larger than 1.0.

Comparing the diffraction pattern shown in Fig. 3.10(b) with those of known iron oxides, it was identified to be Fe_3O_4 (magnetite, lattice constant $a = 0.84$ nm, space group of Fd3m) with an incident beam direction of [1, 2, −1]. (Note that diffraction spots from another crystal are overlapped.) This is consistent with the compositional analysis, described above, which was Fe : O = 1 : 1.30.

Figure 3.11(a) shows a SEM image of a double-ring nanostructure formed at the partial pressure ratio of 1.0. The electron beam position, which was controlled using a computer, was moved from an edge of the substrate into space to make a rod, followed by being circled to make a ring. This procedure was repeated to make a double ring structure. Figure 3.11(b) shows a HRTEM image of a part of the ring. This was of polycrystalline but the grain size was a few nanometers, which was smaller than that in the straight rod shown in Fig. 3.10. This implies that some grains having a preferential growth direction along with the beam moving direction, but no grains can grow larger when the electron beam traces on a curved line.

3.6. Summary

It is demonstrated that magnetic nanostructures with the desired shapes at the desired positions can be fabricated by EBID technique. We believe that the EBID can be one of the most promising methods for nanofabrication tools in point of view of high resolution, abundant types of deposit elements, and capability of three dimensional fabrications.

Acknowledgments

A part of this work was performed with Drs. K. Mitsuishi, M. Tanaka and K. Furuya of National Institute for Materials Science. This work was partly supported by KAKENHI 18510103 and 20560028.

References

1. D. J. Smith, R. E. Dunin-Borkowski, M. R. McCartney, B. Kardynal and M. R. Scheinfein, *J. Appl. Phys.* **87** (2000) 7400.
2. U. Welp, V. K. Vlasko-Vlasov, G. W. Crabtree, J. Hiller and N. Zaluzec, *J. Appl. Phys.* **93** (2003) 7056.
3. S. P. Li, M. Natali, A. Lebib, A. Pepin, Y. Chen and Y. B. Xu, *J. Magn. Magn. Mater.* **241** (2002) 447.
4. Y. Otani, B. Pannetier, J. P. Nozieres and D. Givord, *J. Magn. Magn. Mater.* **126** (1993) 622.

5. A. Sugawara, T. Coyle, G. G. Hembree and M. R. Scheinfein, *Appl. Phys. Lett.* **70** (1997) 1043.
6. K. Liu, J. Nogues, C. Leighton, H. Masuda, K. Nishio, I. V. Roshchin and I. K. Schuller, *Appl. Phys. Lett.* **81** (2002) 4434.
7. H. W. P. Koops, J. Kretz, M. Rudolph, M. Weber, G. Dahm and K. L. Lee, *Jpn. J. Appl. Phys.* **33** (1994) 7099.
8. I. Utke, A. Luisier, P. Hoffmann, D. Laub and P. A. Buffar, *Appl. Phys. Lett.* **81** (2002) 3245.
9. W. F. Van Dorp, B. Van Someron, C. W. Hagen, P. Kruit and P. A. Crozier, *Nano Lett.* **5** (2005) 1303.
10. K. Mitsuishi, M. Shimojo, M. Han and K. Furuya, *Appl. Phys. Lett.* **83** (2003) 2064.
11. M. Takeguchi, M. Shimojo and K. Furuya, *Nanotechnol.* **16** (2005) 1321.
12. M. Shimojo, K. Mitsuishi, A. Tameike and K. Furuya, *J. Vac. Sci. & Technol. B* **22** (2004) 742.
13. K. Mitsuishi, Z. Q. Liu, M. Shimojo, M. Han and K. Furuya, *Ultramicrosc.* **103** (2005) 17.
14. M. Tanaka, M. Shimojo, K. Mitsuishi and K. Furuya, *Appl. Phys. A* **78** (2004) 543.
15. M. Han, K. Mitsuishi, M. Shimojo and K. Furuya, *Philos. Mag.* **84** (2004) 1281.
16. M. Shimojo, M. Takeguchi, M. Tanaka, K. Mitsuishi and K. Furuya, *Appl. Phys. A* **79** (2004) 1869.
17. M. Takeguchi, M. Shimojo and K. Furuya, *Jpn. J. Appl. Phys.* **44** (2005) 5631.
18. H. W. P. Koops, A. Kaya and J. J. Webe, *Vac. Sci. Technol. B* **13** (1995) 2400.
19. I. Utke, P. Hoffmann, B. Dwir, K. Leifer, E. Kapon and P. Doppelt, *J. Vac. Sci. Technol. B* **18** (2000) 3168.
20. R. R. Kunz and T. M. Mayer, *Appl. Phys. Lett.* (1987) 50 962.
21. I. Utke, P. Hoffmann, R. Berger and L. Scandella, *Appl. Phys. Lett.* **80** (2002) 4792.
22. M. Takeguchi, M. Shimojo and K. Furuya, *Nanotechnol.* **16** (2005) 1321.
23. M. Takeguchi, M. Shimojo, R. Che and K. Furuya, *J. Mater. Sci.* **41** (2006) 2627.
24. M. Takeguchi, M. Shimojo, K. Mitsuishi, M. Tanaka, R. Che and K. Furuya, *J. Mater. Sci.* **41** (2006) 4532.
25. M. Takeguchi, M. Shimojo and K. Furuya, *Jpn. J. Appl. Phys.* **44** (2005) 5631.
26. W. W. Pai, J. Zhang and I. F. Wendelken, *J. Vac. Sci. Technol. B* **15** (1997) 785.
27. R. Che, M. Takeguchi, M. Shimojo and K. Furuya *Appl. Phys. Lett.* **87** (2005) 223109.
28. A. Folch, J. Tejada, C. H. Peter and M. S. Wrighton, *Appl. Phys. Lett.* **66** (1995) 2080.
29. K. Molhave, D. N. Madsen, A. M. Rasmussen, A. Carlsson, C. C. Appel, M. Brorson, C. J. H. Jacobsen and P. Boggild, *Nano. Lett.* **3** (2003) 1499.
30. M. Shimojo, M. Takeguchi and K. Furuya, *Nanotechnol.* **17** (2006) 3637.
31. Z. Q. Liu, K. Mitsuishi and K. Furuya, *J. Appl. Phys.* **96** (2004) 619.
32. A. D. Della Ratta, J. Melngailis and C. V. Thompson, *J. Vac. Sci. & Technol. B* **11** (1993) 2195.

33. H. W. P. Koops, C. Schossler, A. Kaya and M. Weber, *J. Vac. Sci. & Technol. B* **14** (1996) 4105.
34. N. A. Kislov, I. I. Khodos, E. D. Ivanov and J. Barthel, *Scanning* **18** (1996) 114.
35. S. Yamamuro, D. F. Farrell and S. A. Majetich, *Phys. Rev. B* **65** (2002) 224431.
36. P. A. Midgley, *Micron* **32** (2001) 167.
37. J. Biskupek, U. Kaiser, H. Lichte, A. Lenk, T. Gemming, G. Pasold and W. Witthuhn, *J. Magn. Magn. Mater.* **293** (2005) 924.
38. M. R. McCartney and M. Gajdardziska-Josifovska, *Ultramicrosc.* **53** (1994) 283.
39. M. P. Nikiforov, A. A. Vertegel, M. G. Shumsky and J. A. Switzer, *Adv. Mater.* **12** (2000) 1351.
40. M. Shimojo, M. Takeguchi and K. Furuya, *Nanotechnol.* **17** (2006) 3637.
41. M. Shimojo, M. Takeguchi, K. Mitsuishi, M. Tanaka and K. Furuya, *Jpn. J. Appl. Phys.* **46** (2007) 6247.

Chapter 4

PREPARATION OF MAGNETIC NANOPARTICLES USING CHEMICAL ROUTE AND FUNCTIONALIZATION FOR MEDICAL APPLICATIONS

Yuko Ichiyanagi

Department of Physics, Yokohama National University
Hodogaya, Yokohama 240-8501, Japan
E-mail: yuko@ynu.ac.jp

Magnetic nanoparticles encapsulated in a silica cage were produced in one step by a wet chemical method. Particles were prepared using a mixture of metal chloride ($M\mathrm{Cl}_2 \cdot n\mathrm{H}_2\mathrm{O}$) aqueous solutions and metasilicate nonahydrate ($\mathrm{Na}_2\mathrm{SiO}_3 \cdot m\mathrm{H}_2\mathrm{O}$) aqueous solutions, in which the metals chosen were Fe, Co, Ni, Mn, Zn, Mg, Ba, Ti and their plural compounds. The obtained particles were monodispersive and the diameters of these particles were estimated in a range of 2 to 34 nm. The magnetic properties observed were ferromagnetic, paramagnetic, antiferromagnetic and superparamagnetic depending on the composition and annealing conditions. We functionalized these magnetic particles by modifiying an amino group. The magnetic nanoparticles present are surrounded by amorphous SiO_2, so that Si ions were located on the outer surface. This characteristic structure enables amino-silane coupling. These functional magnetic nanoparticles are useful in medical applications such as drug-delivery systems, MRI imaging and hyperthermia treatment.

4.1. Introduction

Ever since the National Nanotechnology Initiative (NII) was declared in the United States, interest in nanotechnology has been growing all over the world. In particular, magnetism in nanoscopic systems has drawn a great deal of attention due to the unique magnetic properties, as well as the technological applications, of such systems. Magnetic nanoparticles show peculiar characteristics, such as quantum size effects and quantum magnetic tunneling.[1–6] These interesting phenomena are observed because a particle consists of a very small number of atoms, much smaller than the Avogadro constant N_A ($N_A = 6.02252 \times 10^{23}$). On the other hand, an observation and evaluation system for nano-sized materials remains difficult to achieve and has not been established yet, therefore the evaluation method needs to be improved.

Previously, we obtained monodispersive magnetic nanoparticles (MNPs) with characteristic magnetic properties by an original wet chemical method, and reported their magnetic, structural and thermal properties.[7–27] Magnetic nanoclusters surrounded by the host network of amorphous SiO_2 were produced by mixing aqueous solutions of 3*d* transition-metal chlorides ($M\mathrm{Cl_2}\cdot n\mathrm{H_2O}$) and sodium metasilicate nonahydrate ($Na_2SiO_3\cdot 9H_2O$) to obtain monodisperse nanoparticles with diameters of less than 10 nm in a single step. The particle size can be controlled by adjusting the annealing temperature and pH value of the solutions. The obtained particle sizes were between 2 and 34 nm. Magnetic particles below a critical diameter cannot support more than one domain, and form a single domain. This critical diameter was usually observed for typical materials between 10 and 100 nm.[28–31] A single domain of ferromagnetic particles has a temporary large coercive force at a certain domain size, however, for the smaller particles it is not stable because the magnetic spins of the particle are believed to fluctuate against the thermal energy.

We also prepared MNPs functionalized with surface amino groups by means of an amino-silane coupling procedure to covalently bond target molecules to the particles.[32–36] Due to the extremely small size of the particles, they can be introduced into cells without a cationic coating, and could even have attached to them molecules such as amino group, antibodies or drugs. We proposed that MNPs can be localized in living tissue under the influence of an external magnetic field. In addition, cell selective functional MNPs were also prepared, being modified with folic acid.[34] These MNPs have successfully been internalized in human cancer cells. A cellular recognition system based on functional MNPs could be developed for cell-specific delivery systems.

In this chapter, preparation, characterization and magnetic properties of several kinds of nanoparticles will be introduced and development for biomedical applications will be mentioned.

4.2. Synthesis and Characterization of Magnetic Nanoparticles

We prepared magnetic nanoparticles (MNPs) surrounded by the host network of amorphous SiO_2 by mixing aqueous solutions of 3d transition-metal chlorides ($MCl_2 \cdot nH_2O$) and sodium metasilicate nonahydrate ($Na_2SiO_3 \cdot 9H_2O$). The 3d transition metals M chosen were Fe, Ni, Co, Mn, Cu, Mg, Zn, Ba, Ti and their plural compounds. Metal chlorides aqueous solution mixed with a sodium hydroxide (NaOH) aqueous solution without a metasilicate nonahydrate, were also partially prepared. The resulting precipitates were washed several times with distilled water and dried at around 350 K. The obtained glassy solids were pulverized in a mortar and calcined in an electric furnace between 573–1373 K in air. The particle sizes were controlled by means of the annealing temperature, pH values, and concentration of the silicates. Particle sizes between 2 and 34 nm were obtained.

The prepared samples were examined by means of Cu Kα X-ray ($\lambda = 0.145$ nm) powder diffraction at room temperature (Rigaku, miniflexII). They were additionally examined by transmission electron microscopy (TEM) imaging, X-ray absorption fine-structure (XAFS) experiments, thermogravimetric and differential thermal analysis (TG-DTA, Rigaku TAS-100), and Fourier transform infrared spectroscopy (FT-IR; Horiba FT-720 and Thermo Scientific NICOLET 380). The morphology and diameter distribution of the particles were investigated by means of a transmission electron microscopy (TEM; JEM-1230, JEOL, Japan). Magnetization was measured by a SQUID magnetometer (Quantum Design, MPMS) under a ±50 kOe magnetic field, at temperatures from 2 to 300 K.

4.3. Magnetic Properties of 3d Metal Hydroxide and Metal Oxide Nanoparticles

4.3.1. *Magnetic properties of metal hydroxide nanoparticles*

Nickel hydroxide ($Ni(OH)_2$) nanoparticles encapsulated in amorphous SiO_2 were produced by the wet method described above. $M Cl_2 \cdot n H_2O$ and $Na_2SiO_3 \cdot 9H_2O$ were taken as starting chemical reagents.

Figure 4.1 shows the X-ray powder diffraction patterns of samples measuring 2.4 nm and bulk crystal. The top pattern is the prepared sample and revealed the formation of a monolayer nanocluster (MNC), as the diffraction patterns showed only two-dimensional (2D) broad peaks of (1 0) and (1 1) reflections, corresponding to those of (1 0 *l*) and (1 1 *l*) of the bulk crystal. In addition, noticing the extinction of (0 0 *l*)-reflection, and the splitting of (1 0 *l*)-, (1 1 *l*)-, and (2 0 *l*)-reflections, which appear only in the case of a three-dimensional (3D) structure, the obtained sample was explained as a monolayer. The bottom illustration of Fig. 4.2 shows a typical 3D $Ni(OH)_2$, which has CdI_2-type hexagonal structure, space group $c\bar{3}m$, with lattice constants of $a_{3D} = 0.3117$ nm and $c_{3D} = 0.4593$ nm from our experimental data as illustrated in Fig. 4.2. This is a layered compound composed of a Ni layer sandwiched between two O layers. A curve fitting analysis of the X-ray diffraction patterns based on Warren's formula[37] revealed that the 2D lattice constant a_{2D} of $Ni(OH)_2$-MNC is 0.308 ± 0.003 nm, which is obviously shorter than that of the bulk crystal. Nickel hydroxide 3D crystal can be approximated to the infinite 2D ferromagnetic system, and the monolayer nanocluster can be regarded as a finite ferromagnetic system.

In order to make out the possibility of the monolayer $Ni(OH)_2$, further structural analyses for these species were conducted to understand the correlation between the geometric and magnetic structure. X-ray absorption fine-structure (XAFS) experiments were performed to clarify the microscopic structure of bulk crystal and MNC of $Ni(OH)_2$. Ni K-edge XAFS spectra were taken in the transmission mode at the Beamline 10B[38] of the Photon Factory at the Institute of Materials Structure Science (KEK-PF: typical operation energy: 2.5 GeV, stored current: 450–300 mA), using a channel-cut Si(3 1 1) monochromator. Measured temperatures were 20 K for X-ray absorption near-edge fine-structure (XANES), and room temperature for extended X-ray absorption fine-structure (EXAFS). XANES measurements and DFT calculations[39] suggested that the O-H bonds stand upright on the layer. From the results of EXAFS measurements, the oscillation functions $k^3\chi(k)$ were obtained by standard analysis procedures,[40] then Fourier transforms were operated as shown as Fig. 4.3. From this figure it is obvious that Ni-O_2,O_3 peak intensity and Ni-Ni_3 peak intensity were reduced drastically in MNC in comparison with the bulk curve, in addition the peaks were shifted toward a shorter distance. Analyses of EXAFS spectra exhibited that a significant intralayer lattice contraction takes place with the displacement of oxygen atoms outward along the perpendicular direction.[26]

It is known that $Ni(OH)_2$ is antiferromagnetic, with a Neel temperature of 25.75 K.[41] Magnetization measurements were performed with a SQUID magnetometer (Quantum Design MPMS) in external fields between −50 and 50 kOe at a temperature range from 5 K to 300 K. Figure 4.4 shows typical temperature dependence of both field-cooled (FC) and zero-field-cooled (ZFC) of DC magnetic susceptibilities for $Ni(OH)_2$ MNC samples under a 100 Oe field. The magnetic susceptibilities of FC and ZFC diverge below $T_C \approx 10$ K which agrees with the magnetic transition temperature of $Ni(OH)_2$ MNC determined by AC magnetic susceptibility.[10]

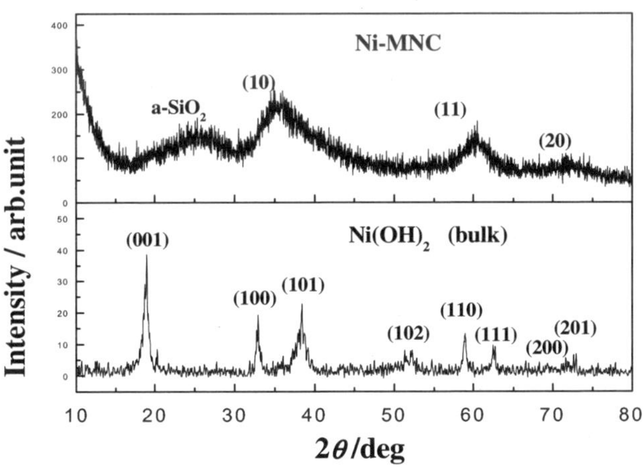

Fig. 4.1. Powder X-ray diffraction patterns of $Ni(OH)_2$ nanoparticles (upper) and bulk crystal (lower). (Reference 24, permission from The Surface Science Society of Japan, Copyright 2008.)

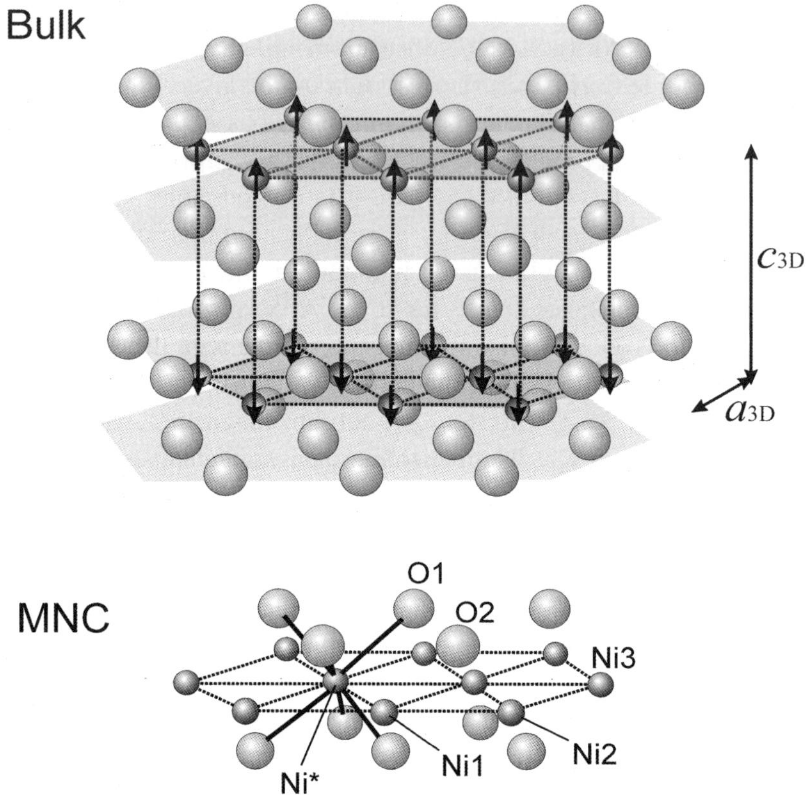

Fig. 4.2. Top: Schematic structure of $Ni(OH)_2$ based on the X-ray diffraction analysis. Large and small spheres represent O and Ni atoms, respectively. Bulk $Ni(OH)_2$ is composed of stacked sheets of the $Ni(OH)_2$ unit. Each arrow on a small sphere indicates direction of spin. Bottom: Schematic view of a monolayer sheet of $Ni(OH)_2$. Coordination shells are indicated with respect to the central Ni atom denoted by Ni*. The third-nearest neighbor oxygen atom (O3) is situated in the adjacent sheet (not shown). (Reference 26, permission from Elsevier, Copyright 2008.)

Particle size dependence of transition temperature of $Ni(OH)_2$ was summarized in Fig. 4.5.[10,41,42] A drastic change of T_C is observed below several nanometer of the particle size, although the current system consists of a two-dimensional system and reflects the effect of fluctuation. This could be one of the finite size effects of extremely small particles.

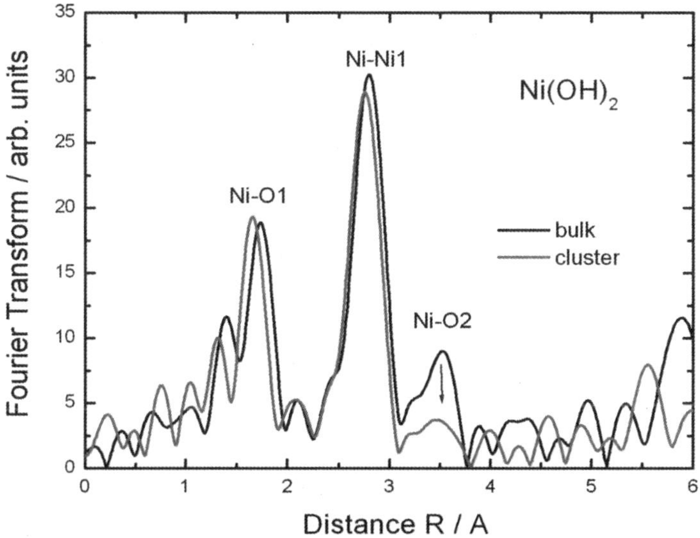

Fig. 4.3. Fourier transforms of Ni K-EXAFS from bulk $Ni(OH)_2$ and $Ni(OH)_2$-MNC. Main contributions are noted above the peaks. See Fig. 4.2 for the numbering of atoms. (Reference 26, permission form Elsevier, Copyright 2008.)

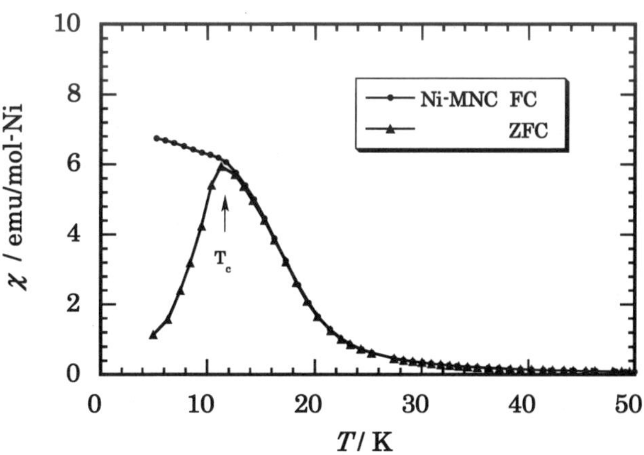

Fig. 4.4. Temperature dependence of DC magnetization of Ni-MNC. (Reference 24, permission from The Surface Science Society of Japan, Copyright 2008.)

By observation of magnetic relaxation time, we can approach a possibility of quantum magnetic tunneling (QMT) of a nanoscopic system. To investigate the magnetic viscosity effect directly, time dependence of the remanent magnetization M_r was measured after switching off the cooling magnetic field of 100 Oe at the lowest temperature, between 5 and 11 K. M_r changes linearly with the logarithm of time t in the above temperature region. From this result, the following expression can be adopted for M_r,[4,5]

$$M_r(t) = M_r(0)[1 - S\ln(t)], \tag{1}$$

where S is the decaying rate of M_r and is called magnetic viscosity. The temperature dependence of S given in Fig. 4.6 changes drastically near $T_C \approx 10$ K, then becomes nearly constant below about 10 K. This phenomenon suggests quantum magnetic tunneling in this system. If we define QMT-temperature as T_Q, magnetic transition temperature T_C is nearly equal to T_Q.

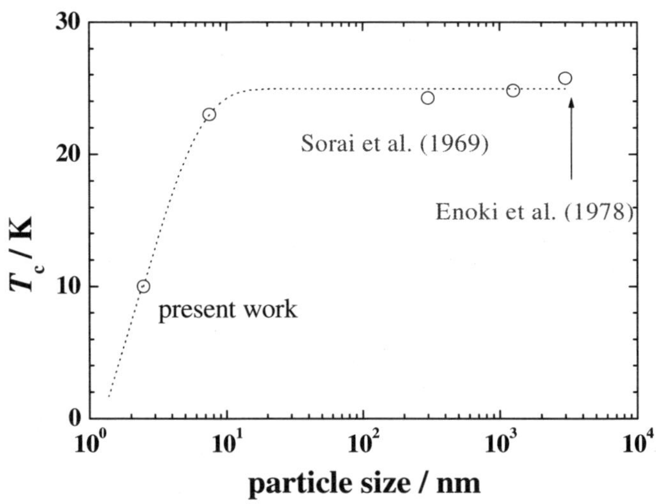

Fig. 4.5. Particle size dependence of transition temperature T_C of $Ni(OH)_2$. (Reference 24, permission from The Surface Science Society of Japan, Copyright 2008.)

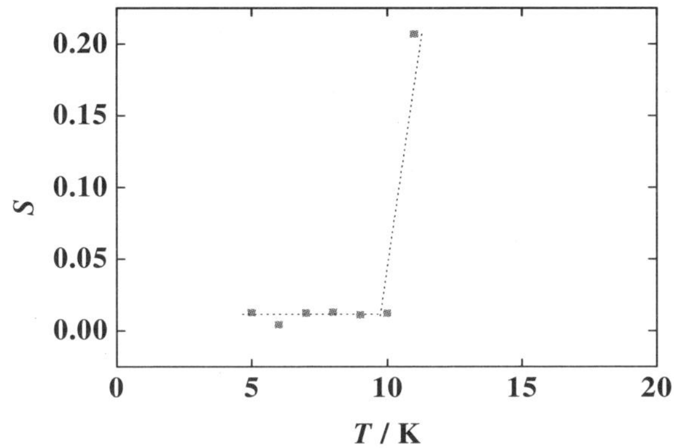

Fig. 4.6. Magnetic viscosity S as a functions of temperature for $Ni(OH)_2$ MNC for measuring the field of 100 Oe. (Reference 13, permission from Elsevier, Copyright 2008.)

As mentioned above, the magnetic structure of bulk $Ni(OH)_2$, the magnetic moments of the Ni atoms in a layer are all directed to the c-axis, but they are antiparallel with those in the adjacent Ni layers, resulting in antiferromagnetism. On the other hand, MNC is composed of a single Ni layer sandwiched by two O layers, and it could exhibit a ferromagnetic character. We observed, in fact, ferromagnetic behavior below 10 K.[11] It is expected that the neighboring of T_C and T_Q in 2D Ising ferromagnetic nanoscopic system can be explained by theoretical studies.

4.3.2. *Metal oxide nanoparticles*

Using $FeCl_2 \cdot 4H_2O$ as a starting reagent precursor of metal chloride, two types, of iron oxides of α-Fe_2O_3 (Hematite) and γ-Fe_2O_3 (Maghemite) nanoparticles ranging from 1.8 to 23 nm in size were obtained depending on the annealing temperature and annealing time. Fe_3O_4 was also obtained by using Na(OH) in place of $Na_2SiO_3 \cdot 9H_2O$. In this Fe-O system, it is very interesting that a relative large coercivity of 1 kOe was observed for 5 nm particles even at room temperature, as shown in Fig. 4.7. These results were considered to be due to the coexistence of γ- and α-Fe_2O_3 phase as evidenced by X-ray diffraction measurements.

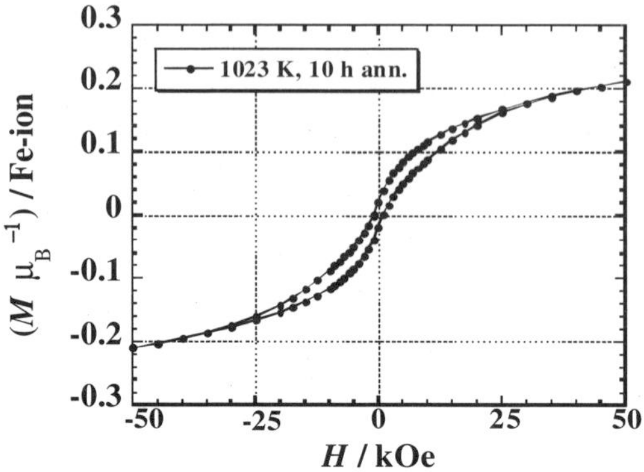

(a)

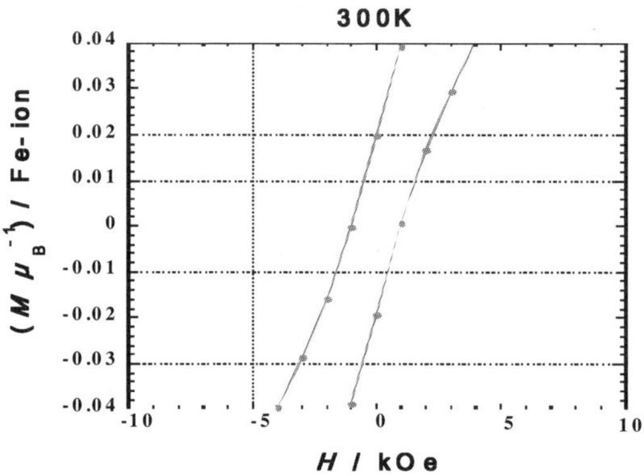

(b)

Fig. 4.7. Magnetization curve of Fe-O system of 5 nm (a-top panel) and enlargement of the origin (b-bottom panel). (Reference 24, permission from The Surface Science Society of Japan, Copyright 2008.)

A gamma phase has an inverse spinel structure and is ferrimagnetic at room temperature, and an alpha phase has a corundum structure and weak ferromagnetism, however, it reorders at 260 K into antiferromagnetism, in the case of bulk crystal. Figure 4.8 shows the temperature dependence of both field-cooled and zero-field-cooled magnetization for various annealing temperature. If we define the bifurcated temperature of FC and ZFC as blocking temperature T_B, above

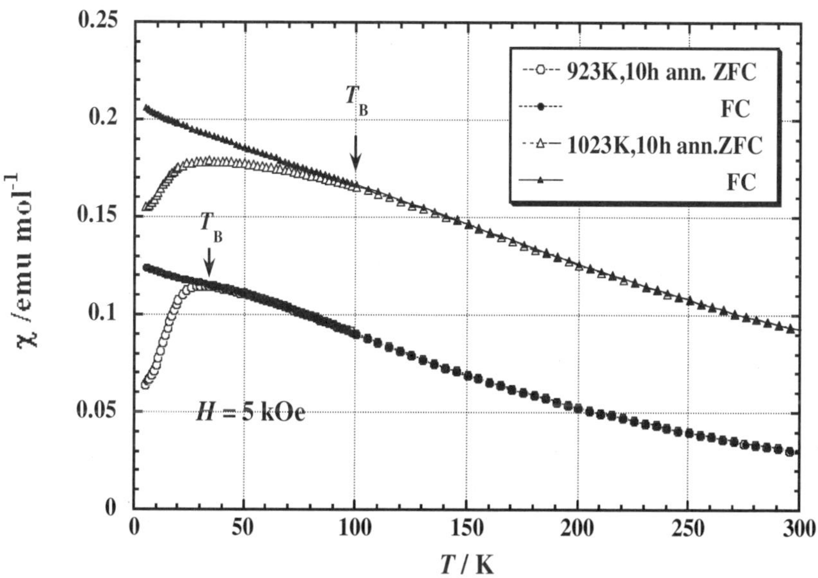

Fig. 4.8. Temperature dependence of magnetization of Fe-O nanoparticles system (original).

the blocking temperature, the magnetic spin of the particles is supposed to fluctuate against thermal energy and behave superparamagnetically. Ferromagnetic behavior can be observed below T_B. The large coercive force described above was observed even in the superparamagnetic region.

In addition to iron oxides, we prepared Co_3O_4, NiO, Mn_3O_4 nanoparticles and reported their magnetic properties.[16,17,20,23]

4.4. Pluralistic Ferrite Nanoparticles

In Secs. 4.3.1 and 4.3.2, I introduced magnetic particles composed of a single atom. In order to approach a practical use, it is important to research the possibilities of ferromagnetic nanoparticles. Here I will summarize pluralistic ferrite nanoparticles.

4.4.1. *Ni-Zn ferrite nanoparticles*

$Ni_{1-x}Zn_xFe_2O_4$ ($0 \leq x \leq 1.0$) (Ni-Zn ferrite) nanoparticles encapsulated in amorphous SiO_2 were prepared by the method mentioned above by mixing solutions of nickel chloride, zinc chloride, iron chloride and sodium metasilicate nonahydrate aqueous solutions. The mole ratio of metal and Si was arranged to be 1:1. The diameters of these particles were estimated from X-ray diffraction patterns as ranging from 2 to 34 nm. The particle size dependence on saturation magnetization is shown in Fig. 4.9. As the particle size decreases, magnetization decreases gradually and abruptly changes at around 6 nm.

Ni ferrite bulk crystal has an inverse spinel structure, where the A sublattice contains half of the Fe^{3+} ions, and the B sublattice contains the other half, together with all the Ni^{2+} ions. Here, A sites are tetrahedral sites, and B sites are octahedral sites. In a mixed Ni-Zn ferrite, the Zn^{2+} ions are expected to remain in the A sites, and Ni^{2+} ion in the B sites. It is a very curious and interesting phenomenon that when Zn ions are added to spinel ferrites, their magnetization increases according to the Zn content, even though Zn ions are nonmagnetic.[43] Such a phenomenon is attributable to the magnetic spin locate counter direction in the A site and the B site. In order to observe the concentration dependence of zinc ion, particle size was controlled at about 6 nm by adjusting the annealing temperature. The particle sizes and lattice constants are summarized in Table 4.1. It is known that the lattice constant of Ni-Zn ferrite nanoparticles are larger than that of bulk crystal,[44] and our data reach agreement on that point.

The magnetic parameters of permeability μ and coercivity H_c, varying with Zn content are summarized in Fig. 4.10. The highest permeability was observed at a Zn content of x = 0.6, where the lowest coercivity was also found. Saturation magnetizations M_s, determined from the *M-H* curves at each Zn concentration, are shown in Fig. 4.11. Reported *M*s values of bulk crystal are also shown in Fig. 4.11.[43] As an overall tendency, the M_s value varies with the Zn concentration, and changes for the bulk crystal in a similar trend, having a maximum value near x = 0.7. But the difference between the nanoparticles and bulk crystal is that for the nanoparticles when x = 0.7, M_s value has the maximum, while for the bulk crystal when x = 0.5, the M_s value is the highest. On the other hand, no difference between nanoparticles and bulk crystal was found at higher concentration of Zn ion above x = 0.7. The maximum value of the saturation magnetization M_s of the nanoparticles tends to be 30% lower than that of bulk crystal. This phenomenon could be due to the canting of the surface spin of nanoparticles at a definite angle. The canting angle θ can be roughly estimated as $\theta = 45^{\circ}$.[45,46] Another possibility is that ion distribution is achieved during the annealing of the particles. The cation distribution in extremely small particles could be different from that in bulk material. For bulk crystal, Ni^{2+} ions are expected to remain in the B sites, however, in the nanoparticles system some Ni ions might to be in A sites, thus resulting in a decrease of the magnetic moment.[47] Spin canting or ion distribution should be determined by Mössbauer spectroscopy in the future.

Table 4.1. Particle size of the samples at various concentrations, annealing temperatures with lattice constants estimated by X-ray diffraction.

x	ann. temp.	d / nm	a / nm
0.0 ($NiFe_2O_4$)	1173 K	5.5	0.834
0.1	1173 K	5.6	0.836
0.2	1173 K	5.6	0.837
0.3	1173 K	5.9	0.837
0.4	1148 K	6.1	0.838
0.5	1173 K	6.2	0.839
0.6	1123 K	5.6	0.841
0.7	1073 K	5.6	0.843
0.8	1073 K	6.2	0.843
0.9	1073 K	6.3	0.845
1.0($ZnFe_2O_4$)	1073 K	6.4	0.845

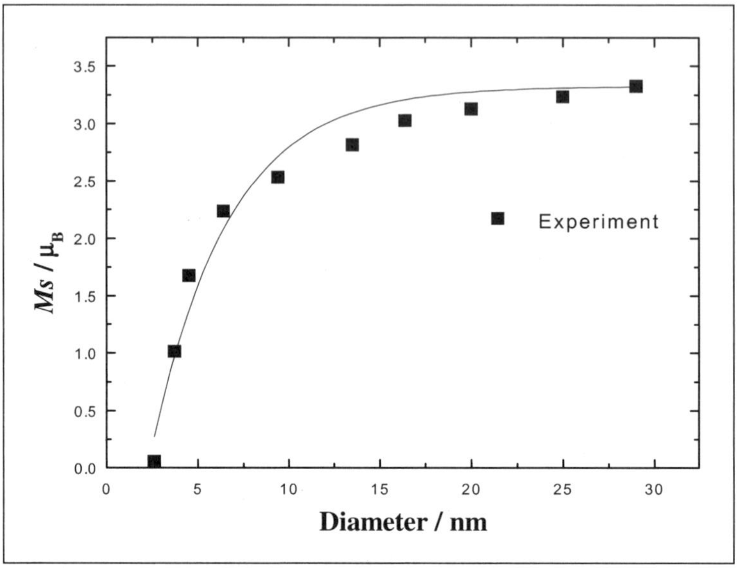

Fig. 4.9. The particle size dependence of the saturation magnetization M_s. (Reference 21, permission from LEXICIA Springer, Copyright 2008.)

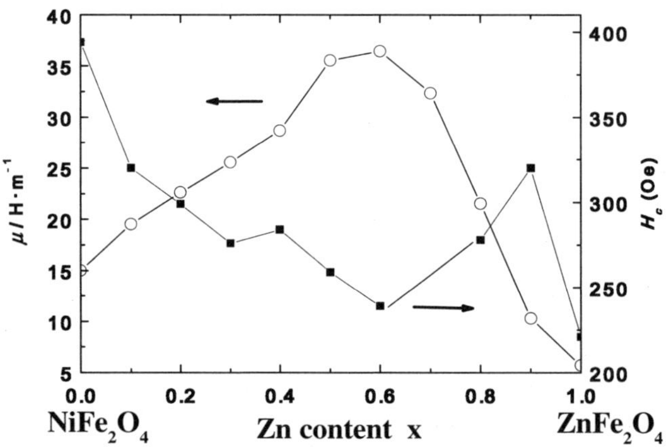

Fig. 4.10. Concentration dependence of permeability (u) and coercivity (Hc) at 5 K under ±50 kOe. (Reference 19, permission from WILEY-VCH, Copyright 2008.)

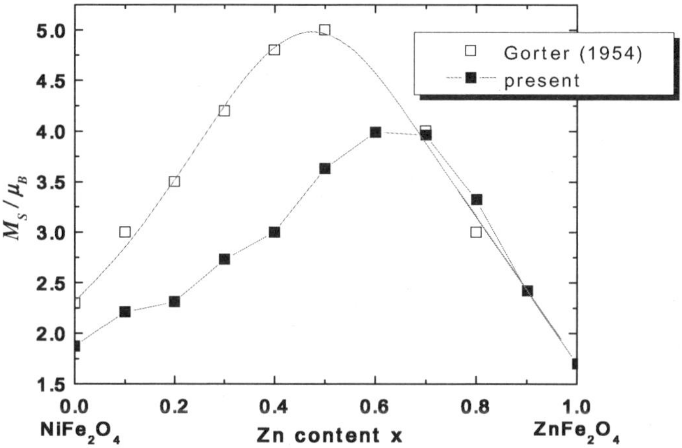

Fig. 4.11. Saturation moment in Bohr magnetons of various Zn concentrations. Closed squares show our experiment and open squares indicate value of bulk crystal reported by Gorter. (Reference 19, permission from WILEY-VCH, Copyright 2008.)

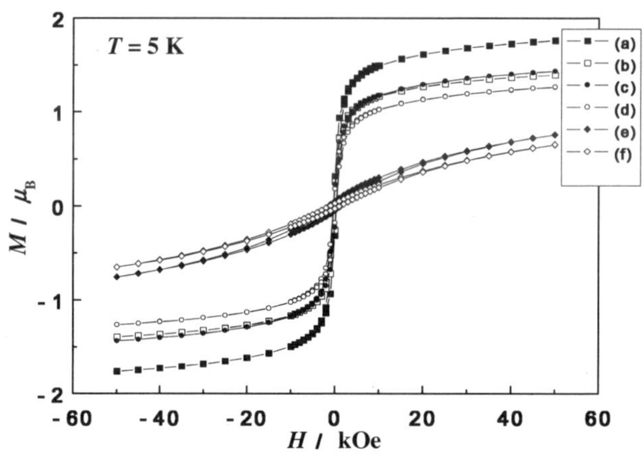

Fig. 4.12. Magnetization curves for particles of 6.1 nm (a, b); 4.9 nm (c, d) and 3.1 nm (e, f) quenched by different methods. Closed marks – Method 1: placing the sample on a copper plate cooled by liquid nitrogen. Open marks – Method 2: quenching to room temperature over a period of 20 h. (Reference 25, permission from LEXICIA Springer, Copyright 2008.)

4.4.2. *Mg ferrite nanoparticles*

$MgFe_2O_4$ (Mg-ferrite) is a very useful soft magnetic material, and is expected to be useful for sensor and catalysis applications.[47–49] It is very interesting that Mg-ferrite shows magnetization depending on the ion distribution, despite the fact that Mg^{2+} ions have nonmagnetic properties. It is known that Mg-ferrite bulk crystal has an inverse spinel structure, where Mg^{2+} ions prefer to exist in the octahedral B sites. When subjected to heat treatment (annealing or quenching), Mg ions tend to position themselves in the tetrahedral A sites in some cases.[50,51] In this section, $MgFe_2O_4$ (Mg-ferrite) nanoparticles were produced, and their magnetic properties and heat-treatment effects were investigated. The starting reagents chosen were $MgCl_2 \cdot 6H_2O$, $FeCl_2 \cdot 4H_2O$ and $Na_2SiO_3 \cdot 9H_2O$.

Magnetization curves were observed after different quenching methods: Method 1: the samples were placed on a copper plate cooled by liquid nitrogen. Method 2: they were quenched to room temperature over a period of 20 hours. Method 3: they were quenched for 4 hours (normally annealed). The *M-H* curves for each particle size measured at a temperature of 5 K under a ±50 kOe field are given in Fig. 4.12. The closed marks (letters a, c, e) correspond to quenching Method 1, and the open marks (letters b, d, f) correspond to quenching Method 2. The magnetization value increased as the quenching time decreased for all the samples, and the maximum saturation magnetization value, M_s, was found to be 1.8 μ_B per molecule at 5 K. These results indicate that some of the Mg ions were distributed more in the A sites as a result of quenching.

In order to clarify the formation process and heat-treatment effect, differential thermal analysis and thermogravimetric (TG-DTA) measurements were carried out using a Rigaku TAS-100 under air. In the TG curve, mass loss by dehydration was observed at temperatures above 370 K, after which the sample mass gradually decreased as the temperature increased up to 1500 K, as shown in Fig. 4.13. Figure 4.13 also shows the DTA curve for the as-prepared samples of Mg-ferrite from room temperature to 1500 K. As can be seen in this figure, there is a deep endothermic peak at around 370 K, corresponding to the above dehydration (Fig. 4.13a). Two exothermic peaks were observed at around 980 and 1280 K. The former small peak could be considered to have been caused by the crystallization, at that temperature, of the amorphous as-prepared sample (Fig. 4.13b), and this explanation is consistent with the powder X-ray diffraction patterns. The slight shoulder in the latter could be due to the sudden aggregation of nanoparticles in this system. The existence of an exothermic peak at 1500 K is unknown at present, however, it is considered that Mg-ferrite was decomposed into other compounds such as α-Fe_2O_3 and $MgSiO_3$ (Fig. 4.13c). The diffraction patterns of those compounds were observed by X-ray measurement after heating to 1500 K.

Thus, it was clarified that some of the Mg ions positioned themselves in the A sites of the spinel structure as a result of quenching. The magnetization values varied with the heat treatment. However, the precise ratio of site distribution was not clear. Mössbauer measurements are needed to determine the ion distribution in this system.

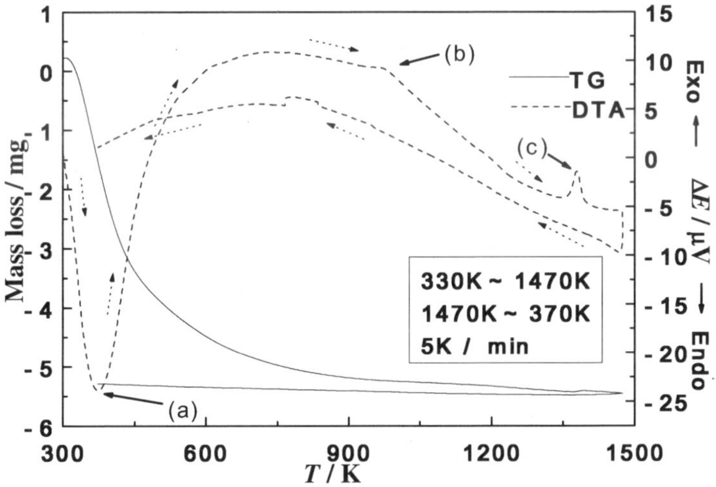

Fig. 4.13. TG-DTA curves of as-prepared $MgFe_2O_4$ nanoparticles under air in the temperature range between room temperature and 1500 K. (Reference 25, permission from LEXICIA Springer, Copyright 2008.)

4.5. Functionalization of Magnetic Nanoparticles

As mentioned above, we have prepared different types of interesting magnetic nanoparticles. They are, of course, useful for high-density magnetic recording media and other magnetic materials. However, recently, we are endeavoring to use the magnetic nanoparticles for biomedical applications. Our magnetic nanoparticles are extremely small, so that the particles can be introduced into cells without a cationic coating, and could even have some molecules attached to them such as amino group, antibodies or drugs. Considering an application for a drug-delivery system or hyperthermia treatment, it is important to be functionalized by modifying amino groups. So we tried to conjugate amino groups with magnetic nanoparticles.[27,32–34]

4.5.1. *Amino-silane coupling*

Functionalized magnetic nanoparticles (f-MNPs; d = 3.5 nm) were prepared by means of a silanization procedure using γ-Fe_2O_3 as introduced in Sec. 4.3.2 and 3-aminopropyl-triethoxysilane (γ-APTES). The mixture of γ-Fe_2O_3 and γ-APTES was stirred for 10 min, heated to 403 K, and then washed several times with pure water. The f-MNPs were labeled with rhodamine, which is a fluorescent material. The labeled MNPs were introduced into PtK2 cells, as well as into the ear of a mouse, wherein they were detected in the ear by use of a button magnet with a surface flux density of 240 mT.

The silanized particles bear surface amino groups, which can be used to covalently attach various materials to the particle surface. The presence of the amino groups on the MNPs was confirmed by Fourier transform infrared spectroscopy (FT-IR). In Fig. 4.14, the typical O–H peak (3660–2991 cm^{-1}), the C–H peak (2972–2842 cm^{-1}), and the C–N peak (1583–1481 cm^{-1}) were detected in the spectrum of the f-MNPs (Fig. 4.14b), whereas only the O–H (3660–2991 cm^{-1}) peak was observed in the spectrum of the unfunctionalized MNPs.

The X-ray diffraction pattern of the f-MNPs indicated that the functionalized particles retained the spinel γ-Fe_2O_3 structure of the unfunctionalized MNPs. The diameters of the amino-MNPs and the

MNPs were independently estimated at 3 nm from the half-width of the X-ray diffraction peaks for both f-MNPs and MNPs. TEM images showed no significant difference between the shape of the f-MNPs and that of the unfunctionalized MNPs. The average diameter of the amino-MNPs was determined to be about 3 nm, which is in good agreement with the X-ray diffraction experimental value.

The magnetization data indicated that the both the f-MNPs and the unfunctionalized MNPs increased linearly with an increasing magnetic field; that is, both types of particles showed paramagnetic behavior. The MNP precipitates could be easily attracted in water solution by means of an external magnetic field of 240 mT. These results indicate that the structure and magnetism were basically unchanged by functionalization with the amino groups.

To evaluate the internalization of the MNPs. the PtK2 cells were cultured after addition of the labeled MNPs. After the cells were cultured for 24 hours, fluorescence was observed from inside the cells even after washing as shown in Fig. 4.15a. On the other hand, no fluorescence was observed from the untreated cells as control. Toxicity was also checked by dividing the cultured cells. The PtK2 cells with MNPs continued to grow even after 5 days.

These results indicate that the PtK2 cell division was not inhibited by the MNPs. We investigated the intracellular space of the living PtK2 cells by TEM to confirm the introduction of the labeled MNPs into the cells. The TEM image of the cells into which MNPs had been introduced showed high-density spots (dashed circles) as shown in Fig. 4.15b. Thus we have successfully introduced magnetic particles into the living cells without cationic coating as usual done. Localization of MNPs in a mouse's ear by means of an external magnetic field was also observed.[32]

4.5.2. *Development for cell selective magnetic nanoparticles*

If functionalized magnetic nanoparticles could be introduced into cancer cells selectively, it would be possible to realize drug-delivery without side effects.

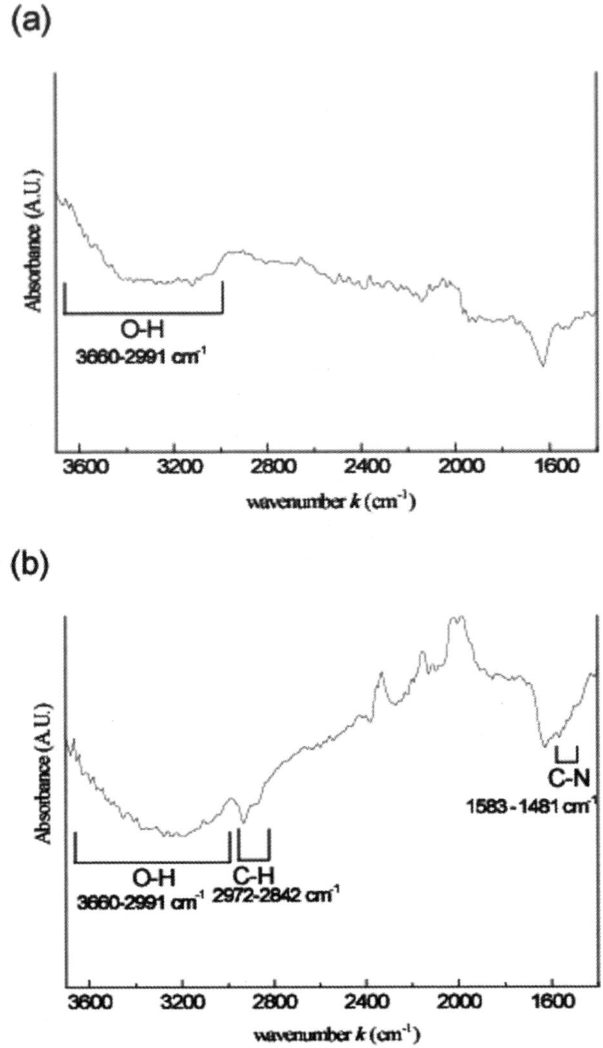

Fig. 4.14. FT-IR spectra of unmodified nanoparticles (a) and amino-nanoparticles (b). (Reference 32, permission form The Surface Science Society of Japan, Copyright 2008.)

In pursuit of this, we developed a cell-specific delivery system that makes use of MNPs (d = 3 nm) conjugated with folic acid (FA) and a coumarin fluorophore (CF) for recognition by folate receptors on the cell surface.[49] The modified MNPs were internalized by human pharyngeal cancer cells (KB cells) after an incubation time that was short compared with the time required for internalization of MNPs without folic acid. Cellular recognition of MNPs may lead to the development of other cell-specific delivery systems.

The amino-MNPs were modified with FA and a CF. The FT-IR spectra of the FA-CF-MNPs showed a peak at 1419 cm^{-1} corresponding to the *p*-amino benzoic acid moiety of FA, and no peaks for the amino-MNPs were observed. The presence of CF moiety on the MNPs was confirmed by fluorescence microscopy. Transmission electron microscopy showed that the shape of the FA-CF-MNPs did not differ markedly from the shape of the unmodified MNPs. The number-average diameter of the FA-CF-MNPs was determined to be about 3 nm.

We evaluated the cellular uptake of the FA-CF-MNPs in three dimensions. After KB cells had been cultured for 3 hours, fluorescence was observed from the cells treated with the MNPs. The lateral image clearly shows that the aggregated particles were present inside the cells. In contrast, no fluorescence was observed from inside the cells treated with CF-MNPs, although weak fluorescence was observed from the cell surface and hollow region, respectively, owing to nonspecific adsorption

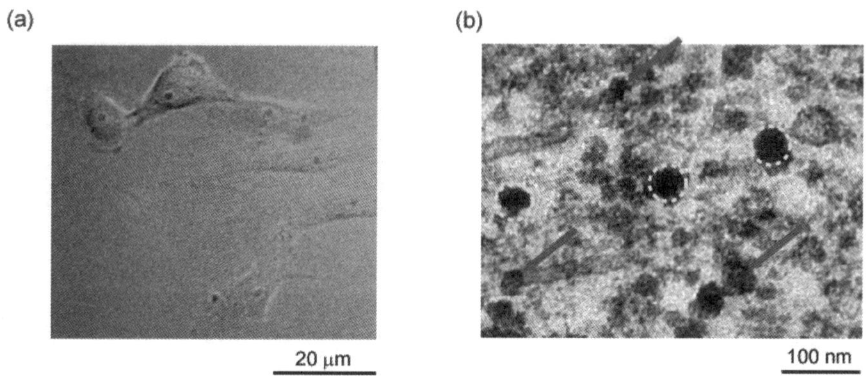

Fig. 4.15. PtK2 cells incubated with Rhodamine labeled MNPs for 24 hours (a) and TEM image of PtK2 cells incubated with labeled MNPs (b). (Reference 32, permission form The Surface Science Society of Japan, Copyright 2008.)

of the particles. Untreated KB cells showed no fluorescence, which indicates that the fluorescence observed from the KB cells treated with the FA-CF-MNPs was not autofluorescence originating from within the cells themselves. Normalized fluorescence intensities (F.I.)/um^2 were estimated. All fluorescence intensities were normalized with respect to the fluorescence intensity of the untreated control cells. The ratio of the fluorescence intensity of the FA-CF-MNPs internalized in the cells to the intensity of the cells treated with CF-MNPs to the intensity of the untreated cells was 158:2:1. In addition, no fluorescence was observed in cells treated with FA-CF-MNPs and excess free FA at the same time. As another control experiment, FA-CF-MNPs were added to rat kangaroo kidney epithelium (PtK2) cells. No fluorescence was observed from either the inside cells or the surficial cells treated with FA-CF-MNPs. These results indicate that FA-CF-MNPs were internalized in the KB cells via folate receptors mediated endocytosis (ligand–receptor interaction).

4.6. Conclusions and Outlook

In summary, preparation and characterization of magnetic nanoparticles were introduced. We found a very interesting 2D Ising ferromagnetic nanoscopic system for $Ni(OH)_2$ monolayered nanocluster. In a Fe-O system, it is very interesting that a relative large coercivity of 1 kOe was observed for 5 nm particles, even at room temperature. Pluralistic ferrite nanoparticles are interesting because their magnetization values vary accordeing to the cation distribution in the spinel structure. In addition, it is a very curious and interesting phenomenon that when Zn ions are added to spinel ferrites, their magnetization increases according to the Zn content, even though Zn ions are nonmagnetic. Quantum size effects and quantum magnetic tunneling were suggested for our magnetic nanoparticles.

For biomedical applications, we studied the feasibility of in vivo introduction of functionalized MNPs. The particles could be incorporated into the subcutaneous tissue of a mouse's ear by means of a magnetic field. We demonstrated that Folic Acid Coumarin Fluorophore MNPs could recognize a specific cell type and be internalized in tumor cells. It therefore would be possible to realize ideal medical treatment, for example, to deliver therapeutic agents to diseased tissue by application of a magnetic field and thus achieve selective delivery at the cellular level using our MNPs. Furthermore, hyperthermia treatment could be also applied after delivery into the cells.

Acknowledgments

This study was partially supported by Japan Society for the Promotion of Science Grant-in Aid for Scientific Research (No. 12650008, No. 15510089, No. 18510090), JST Innovation Bridge, Potentiality Verification Stage, of the Japan Science and Technology Agency and Precursory Research for Embryonic Science and Technology of Japan Science and Technology Agency. I would like to thank the members of the Mitsubishi Kagaku Institute of Life Sciences (MITILS) and Dr S. Taira of JAIST for biological experiments, and I would like to express special thanks to the students of my laboratory.

References

1. A. F. Berkowitz and W. J. Schuele, *J. Appl. Phys.* **30** (1959).
2. R. H. Kodama, A. E. Berkowitz, E. J. Jr McNiff and S. Foner, *Phys. Rev. Lett.* **77** (1996) 394.
3. J. R. Friedmann, M. P. Sarachik, J. Tajada and R. Ziolo, *Phys. Rev. Lett.* **76** (1996) 3830.
4. E. M. Chudnovsky and L. Gunther, *Phys. Rev. Lett.* **60** (1988) 661.
5. M. J. O'Shea and P. Perera, *J. Appl. Phys.* **76** (1994).
6. E. M. Chudnovsky, *Science* **274** (1996).
7. Y. Ichiyanagi, Y. Kimishima, K. Iga and N. Miyata, *Physica B* **194-196** (1994) 297.
8. Y. Ichiyanagi and Y. Kimishima, *J. Magn. Magn. Mater.* **140-144** (1995) 1629.
9. Y. Ichiyanagi and Y. Kimishima, *Jpn. J. Appl. Phys.* **35** (1996) 215.
10. Y. Ichiyanagi and Y. Kimishima, *Mat. Sci. Eng. A* **217** (1996) 358.
11. Y. Ichiyanagi and Y. Kimishima, *J. Magn. Magn. Mater.* **177-181** (1998) 964.
12. Y. Kimishima and Y. Ichiyanagi, *J. Magn. Magn. Mater.* **189** (1998) 62.
13. Y. Ichiyanagi and Y. Kimishima, *J. Magn. Magn. Mater.* **198-199** (1998) 197.
14. Y. Ichiyanagi and Y. Kimishima, *Trans. Mater. Res. Soc. Jpn.* **26** (2001) 1097.
15. Y. Ichiyanagi and Y. Kimishima, *J. Ther. Anal. Calor.* **69** (2002) 919.

16. Y. Ichiyanagi, N. Wakabayashi, J. Yamazaki, S. Yamada, Y. Kimishima, E. Komatsu and H. Tajima, *Physica B* **329-333** (2003) 862.
17. Y. Ichiyanagi, Y. Kimishima and S. Yamada, *J. Magn. Magn. Mater.* **272-276** (2004) 1245.
18. Y. Ichiyanagi, J. Yamazaki, Y. Kimishima and Y. Tachibana, *Trans. Mater. Res. Soc. Jpn.* **29** (2004) 1651.
19. Y. Ichiyanagi, T. Uehashi and S. Yamada, *Phys. Stat. Sol.* (c) **1** (2004) 3485.
20. Y. Ichiyanagi and S. Yamada, *Polyhedron* **24** (2005) 2813.
21. Y. Ichiyanagi, T. Uehashi, S. Yamada, Y. Kanazawa and T. Yamada, *J. Ther. Anal. Calor.* **81** (2005) 541.
22. Y. Ichiyanagi, M. Kubota, S. Moritake, Y. Kanazawa, T. Yamada and T. Uehashi, *J. Magn. Magn. Mater.* **310** (2007) 2378.
23. Y. Ichiyanagi, T. Yamada, Y. Kanazawa and T. Uehashi, *AIP Conference Proceeding* **850** (2006) 1155.
24. Y. Ichiyanagi, *Hyoumen Kagaku* **27-2** (2006) 95.
25. M. Kubota, Y. Kanazawa, K. Nasu, S. Moritake, H. Kawaji, T. Atake and Y. Ichiyanagi, *J. Therm. Anal. Cal.* **92-2** (2008) 461.
26. Y. Ichiyanagi, K. Okamoto, Nagai, H. Kondo, Yokoyama, T. Ohta, *Chem. Phys. Lett.* **379** (2003).
27. Y. Ichiyanagi, S. Moritake, S. Taira and M. Setou, *J. Magn. Magn. Mater.* **310** (2007) 2877.
28. C. KIttel, *Phys. Rev.* **70** (1946) 965.
29. C. Néel, *Compt. Rend.* **224** (1947) 1488.
30. E. C. Stoner and E. P. Wohlfarth, *Nature* **160** (1947) 650.
31. D. J. Dunlop, *Philos. Mag. Ser.* **19** (1969) 329.
32. S. Moritake, S. Taira, Y. Ichiyanagi, N. Morone, S. Song, T. Hatanaka, S. Yuasa and M. Setou, *J. Nanosci. Nanotech.* **7** (2007) 937.
33. S. Moritake, S. Taira, T. Hatanaka, M. Setou and Y. Ichiyanagi, *J. Surf Sci. Nanotech.* **5** (2007) 60.
34. S. Taira, S. Moritake, Y. Kai, T. Hatanaka, Y. Ichiyanagi and M. Setou, *J. Surf. Sci. Nanotech.* **5** (2007) 23-28.
35. S. Moritake, S. Taira, Y. Sugiura, M. Setou and Y. Ichiyanagi, *J. Nanosci. Nanotech.* **8** (2008) 1.
36. S. S. Taira, Y. Sugiura, S. Moritake, Y. Ichiyanagi and M. Setou, *Anal. Chem.* **80** (2008) 4761.
37. B. E. Warren, *Phys. Rev.* **59** (1941) 693.
38. M. Nomura, A. Koyama, *KEK Rep.* **89** (1989) 16.
39. H. Orita, private communication.
40. D. C. Koningsberger, R. Prins (Eds.), X-Ray Absorption: Principles, Applications, Techniques of EXAFS, SEXAFS and XANES, Wiley, New York, 1988.
41. T. Enoki and Tsujikawa, *J. Phys. Soc. Jpn.* **45** (1978) 1515.
42. M. Sorai, A. Kosaki, H. Suga and S. Seki, *J. Chem. Thermodynamics* **1** (1969) 119.

43. E. W. Gorter, *Nature* **165** (1950) 798.
44. L. Wang and F. S. Li, *J. Magn. Magn. Mater.* **223** (2001) 233.
45. J. M. D. Coey, *Phys. Rev. Lett.* **27** (1971) 1140.
46. A. H. Morrish and K. Haneda, *J. Magn. Magn. Mater.* **35** (1983) 105.
47. C. N. Chinnasamy, A. Narayanasamy, N. Pondandian and R. Justin Joseyphus, *J. Magn. Magn. Mater.* **238** (2002) 281.
48. N. Ma, Y. Yue, W. Hua and Z. Gao, *Appl. Catal. A: General* **251** (2003) 39.
49. C. Xiong, Q. Chen, W. Lu, H. Gao and Z. Gao, *Catal. Lett.* **69** (2000) 231.
50. V. Sepelak, D. Schultze, G. Krumeich, U. Steinike and K. D. Becker, *Sol. Stat. Ionics* **141-142** (2001) 677.
51. M. Singh, *J. Magn. Magn. Mater.* **299** (2006) 397.

Chapter 5

ELECTRODEPOSITION AS A FABRICATION METHOD OF MAGNETIC NANOSTRUCTURES

László Péter and Imre Bakonyi

Metals Research Department
Research Institute for Solid State Physics and Optics
Hungarian Academy of Sciences
P.O. Box 49, H-1525 Budapest, Hungary
E-mail: lpeter@szfki.hu

The first part of this chapter gives a general survey on electrochemistry, with a special emphasis on electrodeposition of magnetic metals. The second part comprises a discussion of the electrodeposition of magnetic nanostructures in various forms. This overview is meant to yield an easy-to-read guidance amongst electrodeposited nanostructures even for those who are not involved in this field. Thin magnetic films, nanocrystalline deposits and multilayers are discussed as well as other methods that can be used to prepare a precursor material for magnetic nanostructure formation by a follow-up heat treatment. Template-based methods are listed in a separate section because electrodeposition is a unique method to achieve structures with a very high aspect ratio such as homogeneous and multi-segment nanowires.

5.1. Introduction

Electrodeposition or electroplating has been widely used in many fields of technology. Traditional applications such as hard chromium plating for anti-corrosion and decorative purposes have now been complemented with the incorporation of electrodeposition procedures into micro-electronic procedures. The most important method in the latter group is the formation of Cu interconnects in printed circuit boards by using the so-called damascene process. This is already a part of nanotechnology since the width of Cu interconnects are nowadays in the range of 60 nm.

Electrodeposition of magnetic nanostructures has long been regarded as a possible route of manufacturing individual bit elements for magnetic storage media.

In this chapter, a short overview is given about electrochemistry. The authors' intention was to make the reader familiar with the most important principles in a schematic way but without any overburdening by complicated formulas. The detailed description of the specific methods follows the same line, keeping in mind that the purpose is rather the visualization of the methods than their deep discussion. For subtle details, the reader will be referred to specific reviews.

5.2. Electrodeposition: A General Overview

5.2.1. *Definitions and major principles*

Electrochemistry[1–5] deals with chemical reactions where a *charge transfer* takes place. In *heterogeneous electrochemical reactions*, the charge transfer occurs at a phase boundary, and at least one of the phases involved is an ionic conductor (a melt or an electrolyte solution) whereas the other phase can also be an electron conductor. For electrodeposition,[6–9] one needs a solid metal or a semiconductor substrate on the surface of which the charge transfer takes place and where the deposit is accumulated.

The boundary layer between an ionic and an electron conductor is termed as an *electrode*. The driving force of the electrode reactions is often ranked by the order of *standard electrode potentials* where the standpoint of referencing is the standard hydrogen electrode (SHE), and the more positive standard potential indicates a more noble behaviour. The dependence of the equilibrium potential (E) of a metal electrode on the metal ion concentration [M^{z+}] in the electrolyte is described with the well-renowned Nernst equation:

$$E = E^0 + \frac{RT}{zF}\ln\left[\mathrm{M}^{z+}\right] \tag{1}$$

where E^0 is the standard potential of the M/M^{z+} redox system, F is the Faraday constant, and the other symbols have the usual meaning.

Another thermodynamic approach to electrodeposition is the construction of the potential–pH diagrams[10] in which the relative stability of many possible phases involving the metal, its ion(s), possible oxide phases and the solvent can be displayed simultaneously. This approach describes the thermodynamic prerequisites of the electrodeposition only, while it can account for neither the kinetic conditions of the reactions nor the accessibility of the specific electrode potential in a given system.

Electrochemistry as a fabrication method is unique in the sense that either the driving force of a reaction or the reaction rate can be easily regulated. These operation modes correspond to the *potential control* (potentiostatic mode, P) and the *current control* (galvanostatic mode, G), respectively. While the application of the potential control is indispensable for the investigation of the mechanism of a reaction, large-scale industrial processes take advantage of the simplicity of the current control. In the basic research of electrodeposition of nanostructures, both control methods occur, but they are not always equivalent. For the appropriate choice of the control parameter, the behaviour of the specific system should be known very well.

Instead of relying on the equilibrium potentials, an electrochemical reaction taking place with a nonzero rate should be described with the appropriate kinetic equations. For a one-step reversible electrochemical reaction, O + e = R (O and R are the oxidized and reduced form of the same species), the dependence of the reaction rate (or current density, i) on the electrode potential can be written as

$$i = k_A c_R \exp\left(\frac{\alpha zFE}{RT}\right) - k_C c_O \exp\left(-\frac{(1-\alpha)zFE}{RT}\right) \qquad (2)$$

where k is the rate constant of the anode or cathode reaction as shown by the lower index, c is the concentration of the appropriate reactant, α is the transfer parameter, E is the electrode potential with respect to a reference electrode and z is the number of electron(s) transferred. Equation (2) is known as the Butler–Volmer equation,[11] and it describes the polarization curve when the charge transfer is by far more hindered than any serial reaction step (i.e., diffusion, adsorption, crystallization etc.). *Polarization* is the general term used for the difference between the actual potential of an electrode and its rest potential.

Although Eq. (2) defines an exponential potential dependence of the reaction rate (current density), one has to notice that for complex electrode reactions the actual relationship of the current density and the potential is much more complicated. For practical cases of electrodeposition, at least the transport of the components in the electrolyte/melt must also be taken into account. A key concept is the *limiting current density* (see Fig. 5.1) which means that the electrochemical reaction takes place at a maximum rate allowed by the transport of the reactant in the

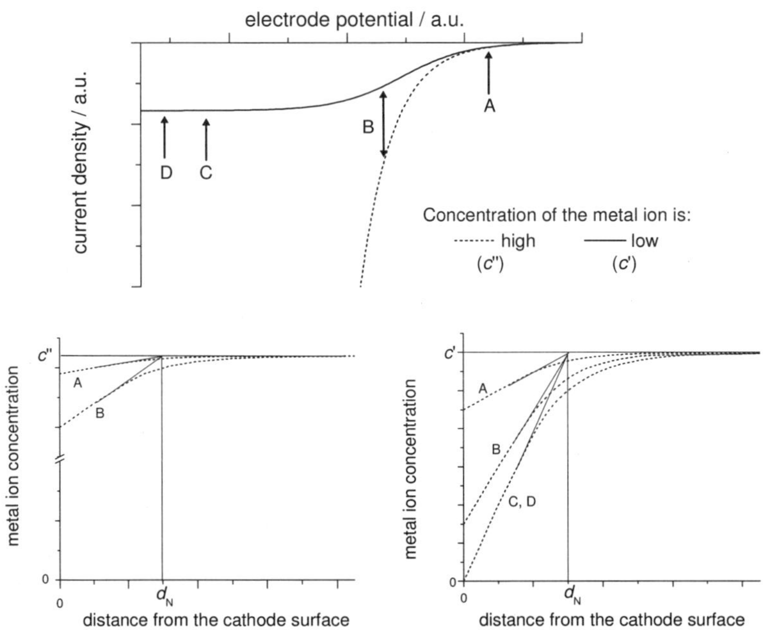

Fig. 5.1. Top: Comparison of the typical polarization curves of metal deposition at low and high metal ion concentrations. At low concentration, the flat section of the polarization curve corresponds to the limiting current region. Bottom: Steady-state metal ion concentrations for high (left) and low (right) bulk metal ion concentrations as a function of the distance from the cathode surface. Capital letters correspond to the potentials shown on the polarization curves. The distance d_N is the so-called *Nernstian diffusion layer thickness*.

electrolyte solution. Under limiting current condition, the electrolyte boundary layer close the electrode surface is completely depleted, and the current can no longer be increased as a result of the increase of the polarization.

Although the theory of electrochemistry was developed with the help of aqueous systems, nonaqueous solvents and ionic liquids have also been used recently. Whereas the basic principles remain valid for the latter two media, too, startpoints of the electrode potential scales are yet to be defined due to the lack of appropriate reference electrodes.

Besides the precursor compounds of the metallic deposits, plating baths almost always have several other components. These can be electrochemically inactive salts to increase the conductivity of the bath (often termed as *supporting electrolytes*), *complexing agents* to increase the solubility of the metal ions, *buffering agents* to regulate the pH, and other (mostly organic) compounds called *additives*.[12] The latter family of bath components exhibiting various functional groups adsorbs more or less reversibly on the cathode and helps to achieve the desired appearance, brightness, low surface roughness, high consistency with reduced porosity, an appropriate crystal size, coatings with low internal stress etc. The formulation of a bath for a specific purpose requires the knowledge of lots of empirical information that can be found in the technical literature.

The dependence of the crystallite size on the parameters of the electrodeposition is of particular interest if nanostructures are aimed at. The general finding is[13,14] that the decrease in polarization (or current density) or the increase of the surface concentration of the metal ions by any means (increase in bulk concentration, bath agitation, temperature elevation) leads to the enlargement of the crystals deposited. Nevertheless, the increase of the crystallite size with the increase of the current density was also reported,[15,16] and the explanation is likely to be the varying intensity of the side reaction (hydrogen evolution). Nano-crystalline deposits can be obtained at highly inhibited deposition conditions, and the application of pulsed current is also useful.[17]

5.2.2. *Electrodeposition of magnetic elements*

Among magnetic elements, cobalt and nickel can be easily deposited in elemental form from aqueous media. The standard electrode potentials for the Ni/Ni^{2+} and Co/Co^{2+} redox systems are −0.257 V and −0.280 V vs. SHE, respectively. These values are out of the stability range of water; hence, hydrogen evolution as a side reaction is unavoidable. Iron exhibits a more negative standard potential (−0.447 V). For iron, the difficulty of obtaining pure deposits stems from various factors such as the sensitivity of iron to water and oxygen, the high corrosion rate as compared to nickel and cobalt, the instability of the electrolyte due to the formation of Fe^{3+}, etc. Therefore, pure iron is seldom obtained via electrodeposition. The standard electrode potential of other magnetic elements (e.g., Gd) is so negative that their deposition from aqueous electrolyte solutions is practically impossible.

5.2.3. *Electrodeposition of magnetic alloys*

The magnetic properties of an alloy can be tuned with the composition of the deposit. Therefore, the codeposition modes of various pairs of metals should be known.

Figure 5.2 serves as a general overview of the most common codeposition modes. *Equilibrium codeposition* means that the composition data all lie on the reference line $y = x$. This codeposition mode is very scarce and cannot be found among magnetic alloys. *Normal codeposition* is obtained if the deposition is regulated mostly by the transport of the ions and is unaffected by any specific interaction of the ions in the electrolyte with the cathode surface. In the normal codeposition, the deposit is composed exclusively from the more noble metal as long as the transport rate of its ions within the electrolyte is sufficient to account for all current passed; in other words, the more noble metal ions have the priority to be discharged. After achieving the limiting current of the more noble metal, the codeposition of the less noble metal takes place to such an extent that the total current passed can be maintained. This deposition mode is observed for many magnetic–non-magnetic pairs like Ni-Cu, Co-Cu, Co-Ru, Co-Ag or Co-Pt. The

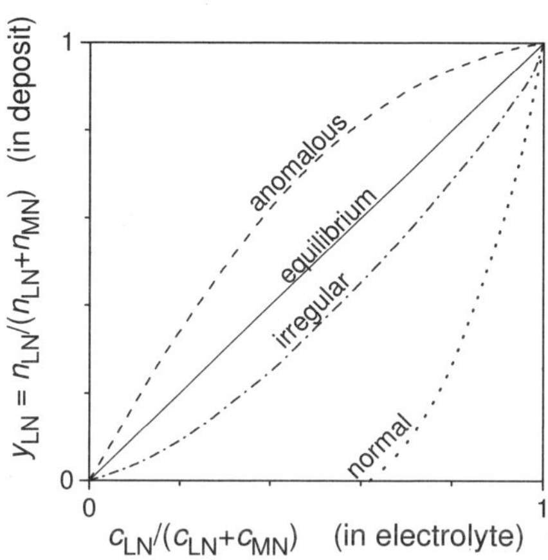

Fig. 5.2. A possible representation of the basic features of the most common codeposition modes. y_{LN} is the molar fraction of the less noble metal in the deposit, while the x axis shows a similar ratio of the elements calculated with the concentration of the ions of the alloy constituent metals in the electrolyte (the index MN refers to the more noble element). All other technical parameters (temperature, total metal ion concentration in the electrolyte, hydrodynamic conditions, current density) are constant.

irregular codeposition means that the ions of the less noble element are discharged with the more noble metal at all concentration ratios, while the deposition of the more noble metal is predominant. This is observed, e.g., for the Zn-Cu pair. The *anomalous codeposition* is very common among magnetic metals since the electrocrystallization of any combination of Fe, Co Ni and Zn from aqueous electrolytes belongs to this group. In this case, the less noble element, which can be present in a minor concentration in the electrolyte, is deposited preferentially so that its molar fraction in the deposit exceeds its electrolyte concentration ratio at all compositions. This is why electrolytes used for Fe-Co-Ni plating employ the precursor metal salts in concentrations increasing with the standard potential of the metals.

Another codeposition mode not displayed in Fig. 5.2 is the so-called *induced codeposition*. In this case, one element (for instance, Ni or Co) can be deposited alone easily. However, another metal (e.g., Mo) cannot be fully reduced from its higher-valency ions (MoO_4^{2-}) in the absence of the former metal. The reason is that the process stops at an intermediate valency oxide (MoO_2) whose further electroreduction is kinetically hindered. However, when the ions of both elements are present, the alloy formation by electrodeposition can take place. This deposition mode is general for any pairs of the (Fe, Co, Ni)-(Mo, W) combination. Although it is not yet described in detail, the codeposition of Gd with Co can also be classified as induced codeposition[18] since the electrodeposition of pure Gd is not feasible from aqueous solutions.

The mutual deposition preferences are often retained when more than two elements are deposited from the same bath to form a multi-component alloy.

5.2.4. *Non-metallic deposits obtained with electrochemistry*[19]

Some *magnetic oxides* can be obtained with a cathode process. Similarly to the production of couprous/coupric oxides and zinc oxide, the key process is the hydrogen evolution and/or the oxoanion reduction at the cathode and the resulting increase in pH, leading to the precipitation of the oxide via a chemical reaction. Here, charge transfer between the magnetic precursor species and the cathode does not always take place. Instead, the electric current changes the media, hence creating the condition of an ordinary chemical reaction.

Some anodic reactions can also be used to produce magnetic oxides. In these reactions, the precursor compound is indeed oxidized. The solubility of the corresponding metal salt is much lower than that of the starting form, and the metal ions of higher valency form an oxide. In this way, higher-valency oxides of iron,[20] cobalt and nickel can be obtained.

The kinetics of both cathode and anode reactions leading to oxide formation is usually complicated and yet to be explored.

5.3. Electrodeposition: A Route Toward Magnetic Nanostructures

5.3.1. *Electrodeposition of ultrathin magnetic films*[21]

For the purpose of this section, a single electrodeposited layer shall be called "ultrathin" if its total thickness does not exceed 2 nm. In this thickness range, the growth of the deposit is often quantified in the relevant literature by the number of equivalent monolayers (ML) instead of nanometers.

The interest in ultrathin magnetic layers[22,23] stems from various features of these layers: (i) Magnetization can be observed only after the accumulation of several atomic layers. The threshold thickness where the ferromagnetic character can be observed depends on the chemical composition of the layers deposited, the morphology of the deposit and the substrate quality (composition, crystallite size and orientation). (ii) There is a thickness range for many magnetic deposits where the strength of the perpendicular magnetic anisotropy (PMA) exceeds the shape anisotropy, and it leads to a perpendicular magnetization. PMA has recently gained particularly strong interest due to the possibility of high-density magnetic data storage. The characteristic thickness where PMA is lost depends on the quality of both interfaces of the ultrathin magnetic layer. Therefore, it may be a function of the electrolyte composition, too, when an in-situ study is performed.

Ultrathin magnetic films can be deposited onto various substrates. In most cases, highly textured noble metal surfaces of well-defined crystal faces are used,[24] Au(111) surface being the most common substrate. The application of a specific crystal face of a noble metal has the advantage that the nucleation pattern of the layer can be studied in-situ by using

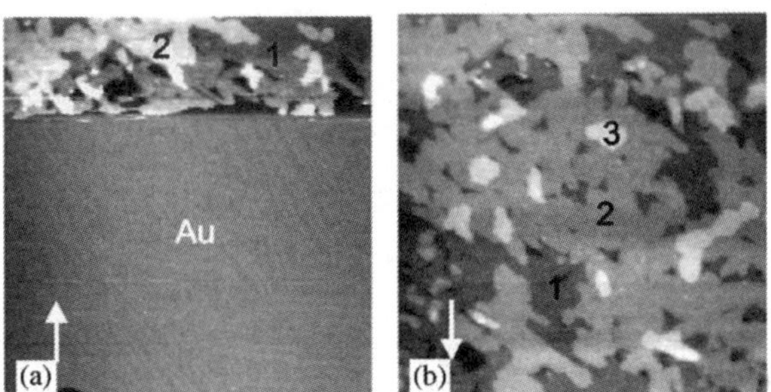

Fig. 5.3. Sequence of in situ STM images (140 nm x 120 nm) showing the initial stages of Fe/Au(111) growth at −1.5 V vs. saturated mercury sulfate electrode.[25] Vertical arrows show in which order the horizontal linescans were recorded, and numbers indicate the local deposit thickness (in ML). Parts (a) and (b) show subsequently recorded pictures for the same area. Reprinted from *Physica B: Condensed Matter* Vol. 354, Gündel *et al.*, Electrodeposition of Fe/Au(111) ultrathin layers with perpendicular magnetic anisotropy, page 243, Copyright (2004), with permission from Elsevier.

scanning probe microscopy (see Fig. 5.3). Surface reconstruction during the deposition of the metal adlayers is commonly observed by STM. The nobility of the substrate makes it possible to study the behaviour of the layer under anodic conditions, too (dissolution).

The concentration of the electroactive species in the electrolyte has to be very low (typically in the range of 1–10 mM) in order to keep the number of atomic layers well controlled with the deposition time at the applied electrode potential. It has to be noted that the nominal thickness of the deposit does not determine the magnetic properties unambiguously, since the growth morphology of the deposit may also depend on the electrode potential applied.

Magnetization of the ultrathin layers can be studied with various methods. Basically, all types of magnetic measurements with high enough sensitivity are viable for thin layers, too. Nevertheless, ex-situ measurements are very cumbersome due to the necessity of the protection of the sample surface from oxidation and the difficulty of the transfer of the sample from the electrolyte to an inert environment. Therefore, in-situ methods like magneto-optical Kerr effect (MOKE) and

Table 1. Summary of the studies of magnetic properties of electrodeposited ultrathin magnetic films. t_{ON}: onset thickness of ferromagnetism; t^*: transient thickness to in-plane magnetization (in contact with the media given).

Substrate	Deposit	t_{ON} (ML)	t^* (ML)	Reference(s)
Au (111)	Ni	6	---	21
	Co	1.6	6.2-7.2 (Cu capped)	21, 26, 27
			1.6 (SO_4^{2-} solution)	
			4 (SCN^- solution)	27
			20 (solution, pH = 8.5)	
		---	function of the deposition potential (Cu capped)	28
		---	>8 (Au capped)	29
	Fe	1	2 (SO_4^{2-} solution)	21, 25, 30
			--- (Cu capped)	
Ag (poly)	Co	<5.5	<15	31
Cu (111)	Co	---	1.5 (SO_4^{2-} solution)	32
		1.5	2 (SO_4^{2-} solution)	33
	Fe	2	8 (SO_4^{2-} solution)	33

alternating gradient field magnetometry are used most often. The in-situ measurement of the Mössbauer spectrum of the deposit is also informative on both cluster formation and the magnetic interaction of the atoms deposited.[31]

The magnetic properties of electrodeposited ultrathin magnetic films are summarized in Table 1.

5.3.2. *Nanocrystalline magnetic deposits*[34,35]

The crystallite size of a deposit is regulated by the relative importance of the nucleation of new crystals and by the growth of existing ones. Nanocrystalline or even amorphous deposits can be obtained if the growth of the existing crystals is suppressed, but the rate of nucleation is high.

The suppression of the growth of the existing crystals can be easily achieved in electrodeposition. If the bath contains additives[12] whose preferred binding site at the metal–electrolyte interface is the same as the entry position of the new metal atoms into the crystal, then the growth is blocked or inhibited. These preferred sites are the step edges and kink positions of the metal. Of course, the adsorption of the additive should be kinetically reversible to avoid the significant incorporation of non-metallic components.

Although the *inhibition intensity*[14] is never measured exactly, it is commonly used for describing bath or additive properties. Roughly speaking, the inhibition intensity increases with surface coverage by the additive and with bond strength between the additive and the metal surface (hence, with additive concentration until a certain saturation level). Electrolytes developed for the deposition of nanocrystalline metals almost always contain some components that are capable of adsorbing at the metal surface.

It is worthwhile to mention that the larger the number of components of a deposit, the more likely is the formation of a nanocrystalline deposit since the non-identical alloying atoms hinder the long-range ordering in the same way as the temporarily adsorbed molecules at the preferred growth sites.

Table 2. Electrodeposited nanocrystalline magnetic metals
(where magnetic properties were of particular interest).

Component(s)	Reference(s)	Component(s)	Reference(s)
Fe	36, 37	Fe-Pd	39, 50
Ni	38–40	Co-Ni	51
Co	41	Co-Fe	51
Ni-Cu	40	Co-Ni-P	52
Ni-Fe	36, 42–46	Co-Fe-P	53
Co-Pd	47, 48	Co-Fe-Ni	51, 54–56
Co-Pt	49		

The rate of nucleation can be enlarged if the surface concentration of the adatoms increases. However, by using a d.c. current, the maximum deposition rate (and hence the maximum surface concentration of the adatoms) is determined by the mass transport limited current density. This limitation can be overcome if pulsed current[17,57–59] is used. If a short but very high cathodic current pulse is applied, the local metal ion concentration in the neighbourhood of the cathode is much larger than under d.c. conditions. Hence, the surface concentration of the adatoms becomes very high, but temporarily only. The high current pulse needs to be followed by an off-time period during which the electrolyte can relax and the bulk metal ion concentration is nearly regenerated in the vicinity of the cathode. The actual crystallite size is determined by the entire deposition process (on-time, off-time, duty cycle, and electrolyte composition). More complicated current waveforms including anodic pulses in order to remove the hydrogen produced have also been reported.[17] From ionic liquids, the application of d.c. current leads to nanocrystalline deposits more often than for aqueous solutions[37] because the chemical properties of the solvent automatically provide some inhibition.

Literature on the electrodeposited nanocrystalline magnetic materials is summarized in Table 2.

5.3.3. *Deposition of metastable precursor alloys and their treatment for obtaining granular magnetic alloys*

A common method to obtain magnetic nanoparticles dispersed in a non-magnetic metallic matrix is the following. A metastable alloy is obtained by the codeposition of a magnetic element with a non-magnetic one, the latter being the majority (matrix-forming) component. Metastability originates from the low miscibility of the constituents. Upon a heat treatment, the activation of the diffusion of the atoms leads to the precipitation of small crystals of the magnetic metal. Figure 5.4 demonstrates how the phase segregation during annealing can be detected by the evolution of the diffraction pattern with time, and the random distribution of the magnetic particles can also be seen in the elemental map.

The electrochemical deposition of such systems is analogous to the physical deposition[60,61] (parallel sputtering or evaporation of the components), and the purpose of the subsequent annealing is exactly the same. The size distribution of the magnetic grains obtained by annealing is usually lognormal, regardless of the preparation method of the precursor

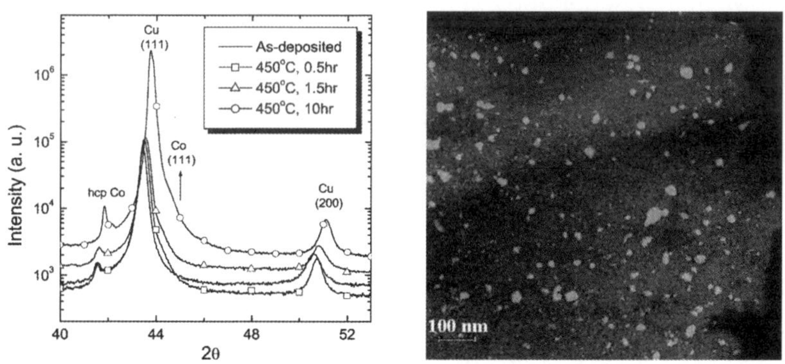

Fig. 5.4. Left: Evolution of the X-ray diffractogram of an electrodeposited metastable Co-Cu alloy with annealing time.[62] The segregation of Co is indicated by the emerging hcp Co line and the shift of the diffraction angle of the Cu peaks. Right: Elemental map of a Co-Cu alloy annealed for 1.5 hours. The light spots are the small Co crystals in the Cu matrix. Reproduced from *Journal of Materials Science* **39**, 5701–5709 (2004); T. Cohen-Hyams *et al.*, Microstructural dependence of giant-magnetoresistance in electrodeposited Cu-Co alloys, Figs. 5.1 and 5.4.d. With kind permission of Springer Science and Business Media.

alloy, and the mean particle size and interparticle distance can be regulated by the initial concentration and the conditions of the heat treatment (temperature, duration). Several reports are available on magnetic and/or magnetotransport properties of electrodeposited Co-Cu,[62–76] Co-Ni-Cu[77] and Co-Ag[65] systems.

From the point of view of the electrolysis, it is necessary to promote the alloy formation, despite the constituent elements cannot form a stable alloy. In this case, the selection of the bath constituents needs to follow the general rule that the deposition potential of the metals should be brought close to each other. This usually makes it necessary to use some kind of complexing agent that shifts the deposition potential of the more noble element to the cathodic direction to a larger extent than the deposition potential of the less noble one. The application of the complexing agent is likely to lead to the formation of a nanocrystalline precursor alloy, although the crystallite size is usually not a subject of investigation prior to the heat treatment. The recrystallization certainly leads to a grain coarsening of the matrix forming non-magnetic metal.

If the alloy formation is not forced by the appropriate choice of the bath formulation, the codeposition of immiscible metals may lead to the formation of a granular alloy in one step.

5.3.4. *Electrodeposition of magnetic/non-magnetic multilayer films with nanometer-scale periodicity*

Electrodeposition of metallic multilayers has been the subject of research for a long time.[78] However, the significance of the magnetic/non-magnetic multilayers comes from the giant magnetoresistance (GMR) effect which was discovered only in the late 1980's[79,80] and was also awarded with the Nobel Prize in Physics two decades later. A few years after the discovery of the GMR it was shown[81] that such nanostructured materials can be produced by electrochemical deposition, too. An example for a multilayer structure obtained by electrodeposition is presented in Fig. 5.5, and typical magnetoresistance curves are shown in Fig. 5.6. The details of the deposition method[82] and literature data on electrodeposited multilayers with GMR[83–85] are summarized in various reviews (see also Part III- Chapter 7 of this book).

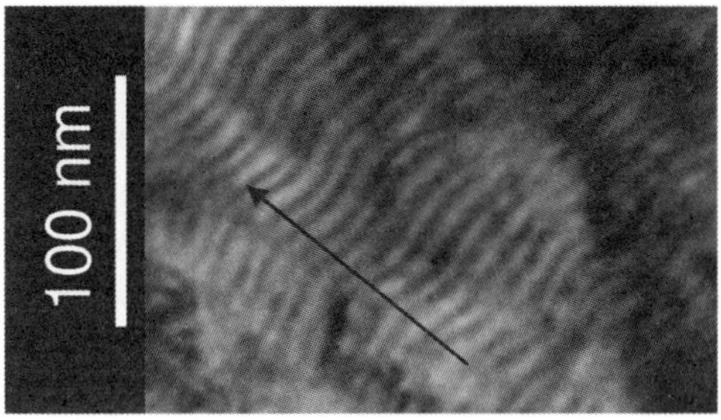

Fig. 5.5. Cross-sectional TEM image of an electrodeposited magnetic/non-magnetic multilayer sample with alternating thick and thin magnetic layers[86] (pseudo-spin valve structure). The Cu layers appear in dark. The repeating structure elements are: NiCo(6.0 nm)/Cu(3.6 nm)/NiCo(2.0 nm)/Cu(3.6 nm). The arrow indicates the growth direction of the sample.

Metallic multilayers can be deposited by contacting the substrate alternatingly with two different electrolytes,[87] each of them capable of yielding one type of layer only. This method is rather cumbersome for producing nanolayers, and the deterioration of the sample surface during the electrolyte change is difficult to eliminate. Hence, nanoscale multilayers are usually obtained with the *single bath method*.[78]

The key feature of the single bath method is that the salt of the more noble non-magnetic metal is applied in a fairly low concentration, while the concentration of the salt of the magnetic metal is often close to the limit defined by its solubility. The normal codeposition mode is a prerequisite; otherwise the non-magnetic metal cannot be obtained in the pure form. For pairs like (Ni,Co,Fe)/(Cu,Ag) this condition is fulfilled. At low cathodic current density or at moderately negative potential the non-magnetic "spacer" layer is deposited with no magnetic impurity. At much larger cathodic current density or at a more negative potential, the deposit obtained contains the magnetic metal as the major component with some non-magnetic impurity. If the molar fraction of the non-magnetic component in the magnetic layer is not very high, it does not impact significantly the magnetic properties of the magnetic layer.

It is clear from the above train of thoughts that the deposition potentials of the magnetic and non-magnetic metals should not be too close, and there must be a potential window where the non-magnetic metal can be deposited alone. Hence, for multilayer deposition, one should not strive for a good alloy formation, but rather a sufficient separation of the potential regimes of the deposition of various metals is favourable. Multilayers with high magnetoresistance are usually obtained from baths containing no additives, brighteners and complexing agents because such bath components are deleterious for GMR.[88,89]

Both the magnetic and the non-magnetic layers can be obtained by either a current pulse or a potential pulse. However, the potential control for the deposition of the non-magnetic layer is much better to obtain a good layer structure since the spontaneous exchange reaction between the metals (e.g., $Co + Cu^{2+} = Cu + Co^{2+}$) can then be excluded.

The deposition potential of the more noble metal has to be tuned precisely so that neither the dissolution nor the codeposition of the

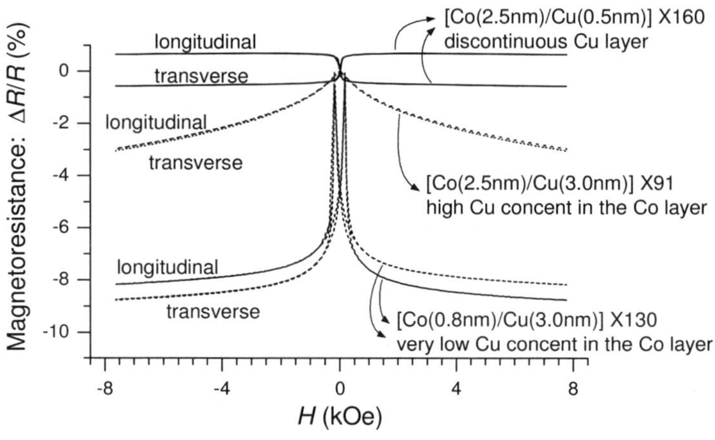

Fig. 5.6. Typical magnetoresistance curves of various electrodeposited Co/Cu multilayers.[90] Pinholes in the magnetic layer result in anisotropic magnetoresistance due to the direct ferromagnetic coupling of the magnetic layers (top curves). A high Cu content in the Co layer leads to a superparamagnetic behaviour because of the magnetic fragmentation of the Co layer (middle). A sample with a satisfactory separation of the Co layers with low Cu content exhibits giant magnetoresistance with a dominant ferromagnetic contribution (bottom). Longitudinal and transverse magnetoresistance means that the external magnetic field was parallel and perpendicular to the measuring current, respectively.

magnetic metals can take place. One of the optimization methods is based on the analysis of the current transients during the less negative deposition pulse,[91] while another one relies on the weight change of the substrate by using an electrochemical quartz crystal microbalance.[92] The current transient analysis can be regarded as being more sensitive since the data acquisition frequency can be higher and it measures an actual reaction rate instead of an integral parameter determined from the total weight change.

Besides the appropriate technical realization, a general requirement of the formation of subsequent layers at fairly large areas is that the nucleation of the alternating layers onto each other should take place easily. The smallest possible difference in the lattice parameters in the pure forms of the individual layers favours the growth of a deposit whose composition is modulated in the growth direction only (this is the case for Ni/Cu and Co/Cu pairs). However, a large difference in the lattice parameters prevents the continuous coverage of the previous layer and may result in an island-like growth, yielding a magnetic structure similar to the granular alloys.

Apart from metal/metal type magnetic multilayers, some attempts were also made to deposit metal/insulator type nanoscale iron/iron-oxide multilayers by using a cathodic/anodic pulse sequence.[93]

5.3.5. *Deposition of nanostructures at preferred nucleation sites*

When a metal ion is discharged on the surface of the parent metal, the adatom tends to occupy the so-called *half crystal position*,[94] especially if the deposition conditions are not very far from the equilibrium. The special feature of the half-crystal or kink position is that either the removal or the incorporation of an atom creates a position with identical properties as the initial one.[94] Hence, these kink positions are the locations where the step edge can propagate in the case of either the deposition or the dissolution of a metal.

A single adatom on the single-crystal surface of the parent metal has higher energy than that of an atom in the kink position. However, this energy difference is quite small when it is compared with the energy difference of a metal adatom at similar crystal positions of a non-metallic surface. The energy of a metallic adatom on a carbon or silicon surface is so high that the nucleation of a metallic deposit cannot take place anywhere else but on an active site or along a step edge of the surface. After the nucleation, which is mostly instantaneous, further electrodeposition takes place on these grains only, and the deposit remains discontinuous until the nominal (average) coverage achieves a few hundred nanometers.

The deposit formation at the step edges of highly oriented pyrolitic graphite (commonly known as HOPG) is often termed as *electrochemical step-edge decoration.*[95–97] After the initial nucleation and grain growth, the coalescence of the grains results in nanowires, this is why the appearance of the wire is often similar to a string of pearls, as seen in Fig. 5.7. The diameter can be finely tuned by applying an either cathodic current (increasing wire diameter) or anodic current (wire thinning) after the wire formation. The deposition of Ni[96] and Ni-Pd[98] wires has been reported so far, though the successful deposition was not accompanied with the study of magnetic properties.

If the nucleation of the metal takes place at a high-energy single crystal surface (e.g., Si), the nucleation is also instantaneous,[99] which leads to the growth of particles with a narrow size distribution if the deposition does not last until the coalescence of the grains. Interestingly, this deposition mode often leads to the formation of core-shell magnetic particles.[100,101]

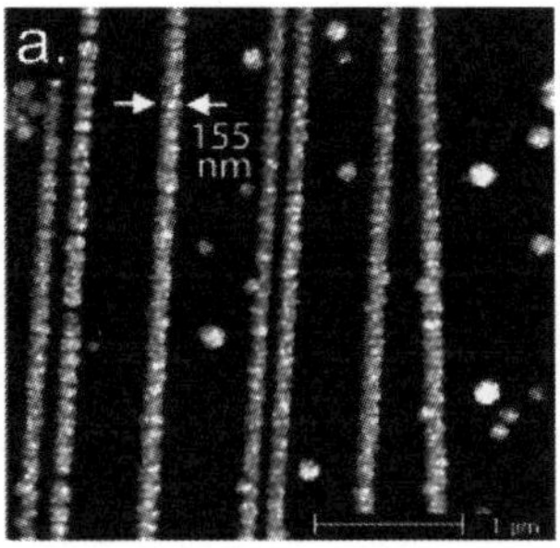

Fig. 5.7. Scanning electron micrograph of a Ni nanowire system obtained by electrodeposition onto HOPG. The wires were formed along the step edges of the HOPG surface. Reprinted in part with permission from Ref. 96. Copyright (2002) American Chemical Society.

From technical point of view, the preparation of nanoparticles or nanowires usually comprises several steps. A preliminary current pulse is often applied to achieve appropriate conditions for the nucleation, and then a short high-potential cathodic nucleation pulse is applied, followed by the growth pulse with a low or moderate overpotential. The concentration of the electroactive species is usually much lower than in normal plating electrolytes.

5.3.6. *Electrodeposition into templates*

The word "template" is used here as the general term for the solid spatial confinement at the electrode surface that regulates the shape of the deposit being formed. Hence, the deposit will not grow in a bulk form but will fill up the empty space defined by the template. The templates prepared with lithographic methods[102] have now been replaced with etched self-organized templates. The emerging importance of this field is well explained by the fact that objects with an aspect ratio of as high as 10^3–10^4 and with an areal pore density of around 10^{11} cm^{-2} can be obtained with a fairly simple method, making electrodeposition an indispensable tool of nanotechnology.

A very simple template can be a piece of sintered glass disk covered with a conducting coating at one side and filled up with the electrolyte. Therefore, the deposit fills up the pores within the sintered glass disk. Such pore structures are random or irregular.

A more regular template is a plastic (mostly polycarbonate) membrane that is bombarded with heavy ions perpendicular to the surface.[103] Along the ion tracks the bond structure of the polymer is destroyed, and the track can be etched, typically with a strong alkaline solution to widen the columnar pores. This type of template is regular in the sense that the pores are perpendicular to the surface at least to the same extent as the trajectory of the bombarding ions is unchanged by the subsequent collisions. The pore density is defined by the ion dose, while the pore diameter can be tuned by the etching time. Hence, the relative position of the pores is random, so this system is still laterally disordered. However, this type of template is suitable for the preparation of a single-ion track pore and for the study of an individual wire deposited within.[104]

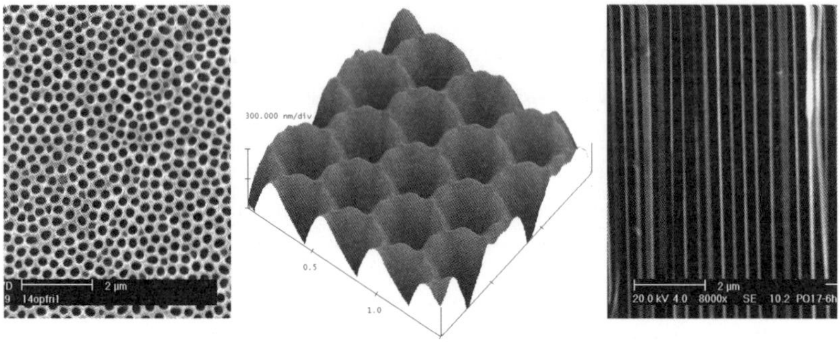

Fig. 5.8. Left: Scanning electron micrograph of an etched porous alumina template (top view). Middle: AFM picture of the top of the pore structure. Right: Cross-sectional view of the AAO template. Anodization conditions: 1 M phosphoric acid at 195 V. Anodization was repeated 3 times by dissolving the oxide layer produced after the 1st and 2nd step with 0.2 M chromic acid. By courtesy of J. Gong and G. Zangari.[105]

The most ordered type of template up to now is the anodic aluminum oxide (AAO).[106,107] The preparation method is that the aluminum foil is immersed in an acidic solution (phosphoric, oxalic or sulfuric acid), then it is anodized at a high voltage (20–195 V). This anodization leads to the growth of a porous oxide whose surface is still disordered, but the bottom part (i.e., the metal/oxide boundary) becomes ordered upon extended anodization. Then the initial oxide is dissolved, and the anodization is repeated. Therefore, a laterally ordered hexagonal pore system can be produced (see Fig. 5.8) in which the ordered areas form domains with some disorder within the domain walls. The disadvantage of this method is the lengthy multistep preparation of the template.

The available pore size for AAO templates is 50–500 nm, but the ratio of the pore diameter to the pore axis distance is practically fixed, and it can be modified only by a pore widening during a subsequent chemical etching. The degree of ordering can be increased by a mechanical initialization of the pore formation by using an imprinting method. If the aluminum is fully oxidized, one side of the resulting alumina honeycomb structure can be coated with a metal by evaporation to ensure the conductivity, or the deposition can be started by making use of the remaining aluminum.

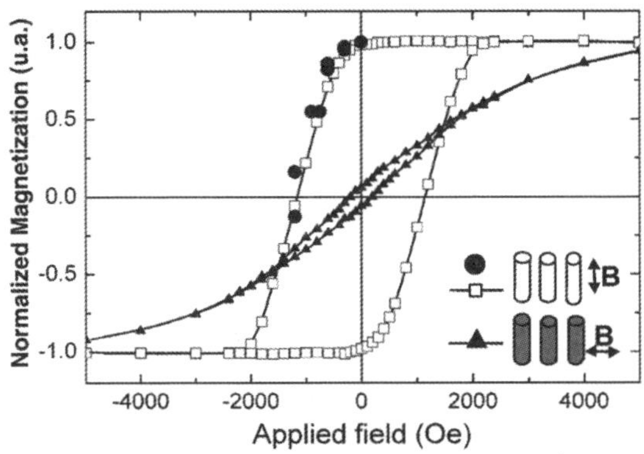

Fig. 5.9. Typical magnetization curves of metallic nanowires electrodeposited into the pores of AAO templates.[111] Deposit structure: 30 nm diameter and 700 nm long Ni wires. Continuous curves show SQUID magnetization data for the field direction shown in the figure, and filled circles refer to magnetic force microscopy measurements. The hysteresis loops are determined by the shape anisotropy of the wires. Reprinted from *Journal of Magnetism and Magnetic Materials*, Vol. 249, K. Nielsch *et al.*, High density hexagonal nickel nanowire array, page 236, Copyright (2002), with permission from Elsevier.

Recently, porous templates based on anodized titania[108] or porous silicon[109] have also become available. The deposition on silicon substrates patterned with submicrometer holes by using focused ion beam[110] can be very significant since any regular (and hence not only hexagonal) surface structure is easily available by this method.

There are two basic growth modes in a template. One of them is the growth of a wire. In this case, the growing column fills the pore entirely. However, it may happen that the plating in the vicinity of the walls is accelerated as compared to the center of the cavity. This leads to a fast depletion of the electrolyte within the metal channel being formed, and the growth mode becomes tubular. It was found[112] that a low current efficiency and the resulting electrolyte flow pattern induced by the release of hydrogen bubbles in the center of the template channel can promote the tubular growth mode.

There is a great variety of magnetic nanostructures obtained by electrodeposition in templates (see, e.g., Fig. 5.9). Besides homogeneous magnetic nano-wires, multisectioned nanowires can be obtained; here, the length of each section in the wire is much larger than the wire diameter. The structure is, however, called a multilayered nanowire when the layer thickness is much smaller than wire diameter. The guidelines of the electrodeposition of each type of structure are identical to those discussed for their deposition without any template. In very special (and unpredictable) cases a core-shell structure can also form spontaneously.[113–115]

After the removal of the template, a second electrolysis step can provide a coated nanowire system[116] in which either the internal wire core or the coating can be a magnetic metal. The field of template-based electrodeposited magnetic nanostructures is expanding very fast, and hence the reader is advised to perform an own literature search for the topic of interest.

5.3.7. *Electrodeposition on surfaces modified by self-assembly of colloids*

Self-assembly of adsorbing molecules at a metal surface has been well known for a long time. However, entities much larger than a single molecule are also capable of self-assembly, forming a structure that can be used similarly to solid templates for electrodeposition.

The self-ordering of non-conducting (e.g., polystyrene, poly(methyl-metacrylate) etc.) colloidal particles on the surface of a metal is similar to the crystal growth that leads to ordered domains with some domain boundaries in between. Depending on the amount of colloid particles applied, a single adlayer structure or a multilayer structure can be produced. The empty space between the ordered particles is suitable for a fill-up process by electrodeposition. In this case, the magnetic nanostructure will be the inverse of the non-conducting particle system, exhibiting a typical periodicity of some hundreds of nanometers. After the electrolysis, the nanostructure produced can be studied as it is or after the removal of the colloids by dissolving them in an appropriate organic solvent. Besides structural properties[117] and magnetoresistance,[118] coercivity behaviour of the system as a function of the colloid particle size[119–121] has deserved particular attention so far.

Similarly to the nanotube formation in the porous solid templates, the specific interaction of the metal to be deposited with the structure-forming colloid can lead to an incomplete void filling, hence leading to a foam-like magnetic nanostructure (see Fig. 5.10).[122] If the empty space between the colloids is first filled up with a conductive polymer, then the colloids are removed, the inversion of the ordinary template is suitable

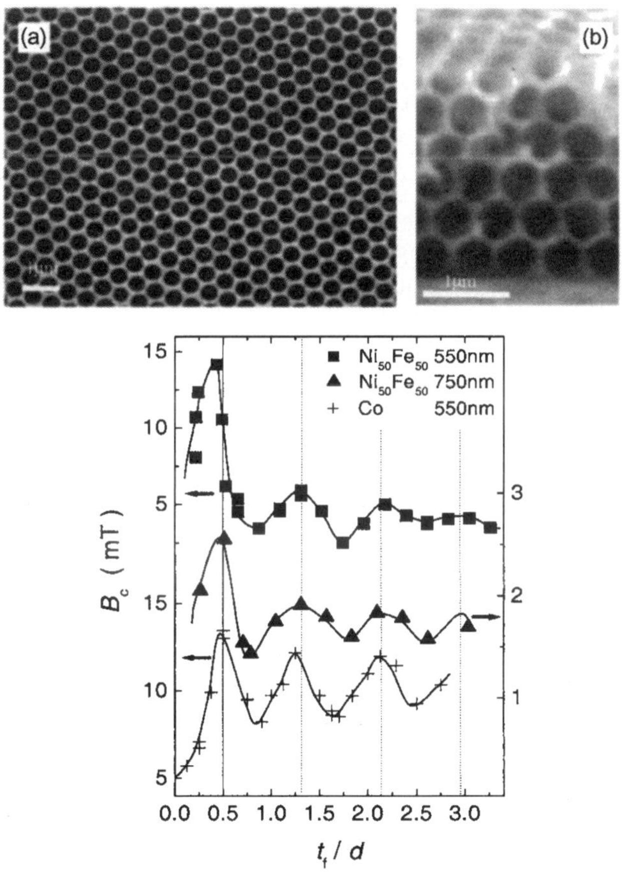

Fig. 5.10. Top: SEM image of structured Ni film prepared using 0.5 μm polystyrene spheres (a) and cross-sectional view of a thick film after cleaving (b). In both cases the white marker corresponds to 1 μm. Bottom: Coercivity of various magnetic hollow structures as a function of the deposit thickness normalized to the diameter of the colloid particles. Reprinted with permission from Ref. 119. Copyright (2006), American Institute of Physics.

for the electrodeposition of an ordered system of magnetic nanospheres.[123] A similar method can be used with silica spher es and a polymer filling, and the silica spheres can be removed with hydrogen fluoride.[124] The key factor is the chemical dissimilarity of the self-ordered nanospheres and the material filling the empty space between them.

Another mean of the self-assembly is the spontaneous structure formation of the two-phase mixture of a lyotropic liquid crystal with water. Here, the hydrophobic and hydrophilic regions form separated zones with cylindrical hydrophilic columns within a practically insulating matrix. The electrolyte formation by dissolving various salts in the water-rich phase usually does not modify significantly the miscibility properties of the system. The columnar regions with the aqueous electrolyte solution behave similarly to the pores in a solid template while the non-conducting liquid crystal matrix serves as a wall. After the removal of the liquid crystal, one can obtain an ordered system of nanorods attached to the substrate. Nanostructured cobalt films have been obtained with this method.[125]

5.3.8. *Suspension plating with magnetic particles*

Suspension plating[126–128] was originally developed for the codeposition of inert particles in order to enhance tribological, wear resistance and corrosion properties. The host metal and the particles embedded (Al_2O_3, ZrO_2, SiO_2, SiC, Si_3N_4, diamond etc.) provide the desired properties together. While the particle to be incorporated is never involved in an electrochemical reaction, its surface charge and adsorption properties can strongly influence the entire codeposition process.

Reports on codeposited magnetic particles in electroplating are scarce. Incorporation of Ni particles in Zn deposit[129] and barium ferrite particles in various magnetic metals,[130] even combining with a lithographic surface patterning,[131] has been reported. However, the particle size was in both cases in the order of micrometers. With the application of nanosized particles, this field is expected to develop significantly in the near future.

5.3.9. *Formation of suspended magnetic particles by electrochemistry*

Particles produced by electrodeposition sometimes do not attach to the substrate surface well enough to make a continuous layer or an ordered system. This statement is particularly valid for non-metallic particles. Hence, the particles can leave the vicinity of the electrode, and in the case when the particles within the electrolyte are not stabilized by any mean, the product can be precipitated.[132] However, the stabilization of the particles as a suspension can be promoted by the application of various surfactants.[133–135] The surfactant also has a role in the regulation of the particle size. Magnetic particles with a typical diameter of 5–50 nm can be synthesized in this manner.

5.4. Summary

The sections above presented a variety of electrodeposited nanostructures following the logical order of the deposition modes and conditions. Figure 5.11 gives another type of overview, showing that principally different deposition modes may result in similar structures, although there is very little commonality in the methods themselves.

Electrodeposition of nanostructures is a fast-expanding field, and the versatility of available electrodeposited nanostructures is almost unlimited. From the point of view of technical application, electro-deposition of magnetic nanostructures is very promising in fields where it can contribute to the development of information storage technology (magnetic nanodots, deposition in regular templates). It can be anticipated that electrodeposition in templates may gain some important role in manufacturing of magnetic storage media in the near future.

Acknowledgment

This work was supported by the Hungarian Scientific Research Fund (OTKA) via the grant # K-60821.

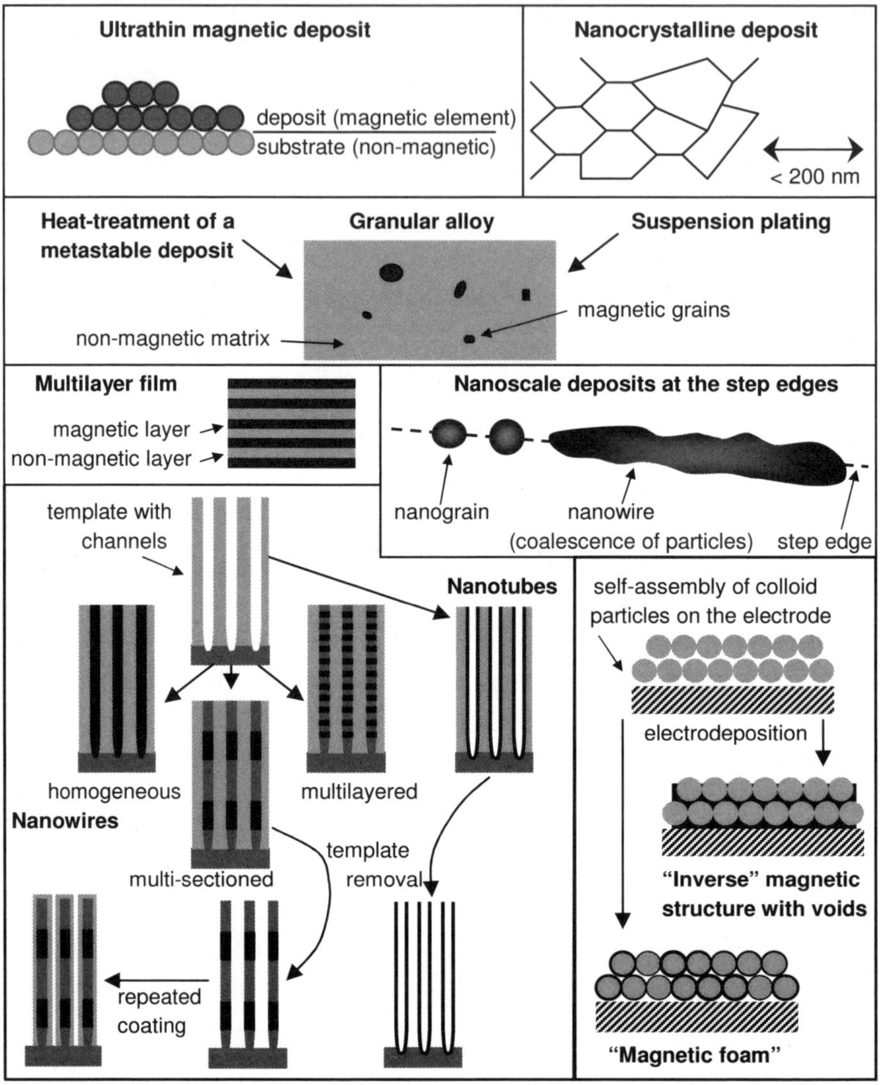

Fig. 5.11. A schematic summary of the main types of magnetic nanostructures that can be obtained with electrodeposition.

References

1. A. J. Bard, L. M. Faulkner, *Electrochemical Methods. Fundamentals and Application, 2nd Edition* (John Wiley & Sons, Inc., New York, 2001).
2. C. H. Hamann, A. Hamnet, W. Vielstich, *Electrochemistry* (WILEY-VCH, Weinheim, Gemany, 2007).
3. C. M. A. Brett, A. M. O. Brett, *Electrochemistry. Principles, Methods, and Application* (Oxford University Press, Oxford, UK, 1993).
4. J. O'M. Bockris, A. K. N. Reddy, *Modern Electrochemistry, Second Edition* (Kluwer Academic Publishers, New York, USA, 1998).
5. W. Plieth, *Electrochemistry for Materials Science* (Elsevier, Amsterdam, 2007).
6. A. Brenner, *Electrodeposition of Alloys I-II* (Academic Press, New York, USA 1963).
7. W. H. Safranek, *The Properties of Electrodeposited Metals and Alloys – A Handbook* (American Elsevier Publishing, New York, 1974).
8. J. W. Dini, *Electrodeposition* (Noyes Publications, Park Ridge, New Jersey, USA, 1993).
9. M. Paunovich, M. Schlesinger, *Fundamentals of Electrochemical Deposition* (John Wiley & Sons, Inc., New York, USA, 1998).
10. M. Pourbaix, *Atlas of Electrochemical Equilibra in Aqueous Solutions* (National Association of Corrosion Engineers, Houston, Texas, USA, 1974).
11. T. Erdey-Grúz, M. Volmer, *Z. Phys. Chem.* A **150** (1930) 203.
12. L. Oniciu, L. Muresan, *J. Appl. Electrochem.* **21** (1991) 565.
13. J. W. Dini, *Plat. Surf. Fin.* **75** (1988) 10.
14. R. Winand, *Electrochim. Acta* **39** (1994) 1091.
15. I. Bakonyi, E. Tóth-Kádár, L. Pogány, Á. Cziráki, I. Gerőcs, K. Varga-Josepovits, B. Arnold, K. Wetzig, *Surf. Coat. Technol.* **78** (1996) 124.
16. E. Tóth-Kádár, I. Bakonyi, L. Pogány, Á. Cziráki, *Surf. Coat. Technol.* **88** (1997) 57.
17. J. C. Puippe, F. Leaman eds., *Theory and Practice of Pulse Plating* (American Electroplaters and Surface Finishers Society, Orlando, Florida, 1986).
18. R. Mishra, E. J. Podlaha, *Electrochem. Solid-State Lett.* **9** (2006) C199.
19. G. H. A. Therese, P. V. Kamath, *Chem. Mater.* **12** (2000) 1195.
20. T. A. Sorenson, S. A. Morton, G. D. Waddill, J. A. Switzer, *J. Am. Chem. Soc.* **124** (2002) 7604.
21. P. Allongue, F. Maroun, in *Electrocrystallization in Nanotechnology*, ed. G. Staikov (Wiley-VCH, Weinheim, Germany, 2007), p. 217.
22. J. A. C. Bland, B. Heinrich, *Ultrathin Magnetic Structures I-IV* (Springer-Verlag, Berlin, 2005).
23. C. A. F. Vaz, J. A. C. Bland, G. Lauhoff, *Rep. Prog. Phys.* **71** (2008) 056501.
24. P. Allongue, F. Maroun, *Curr. Opin. Solid State Mater. Sci.* **10** (2006) 173.

25. A. Gündel, T. Devolder, C. Chappert, J. E. Schmidt, R. Cortès, P. Allongue, *Physica B* **354** (2004) 282.
26. L. Cagnon, A. Gündel, T. Devolder, A. Morrone, C. Chappert, J. E. Schmidt, P. Allongue, *Appl. Surf. Sci.* **164** (2000) 22.
27. J. G. Borges, P. Prod'homme, F. Maroun, R. Cortès, J. Geshev, J. E. Schmidt, P. Allongue, *Physica B* **384** (2006) 138.
28. J. L. Bubendorff, E. Beaurepaire, C. Mény, J. P. Bucher, *J. Appl. Phys.* **83** (1998) 7043.
29. P. Prod'homme, F. Maroun, R. Cortès, P. Allongue, J. Hamrle, J. Ferré, J. P. Jamet, N. Vernier, *J. Magn. Magn. Mater.* **315** (2007) 26.
30. A. Gundel, A. Morrone, J. E. Schmidt, L. Cagnon, P. Allongue, *J. Magn. Magn. Mater.* **226–230** (2001) 1616.
31. I. Dézsi, Cs. Fetzer, *Electrochem. Commun.* **9** (2007) 1846.
32. W. Schindler, J. Kirschner, *Phys. Rev. B* **55** (1997) R1989.
33. W. Schindler, O. Schneider, J. Kirschner, *J. Appl. Phys.* **81** (1997) 3915.
34. T. Watanabe, *Nano-Plating* (Elsevier, Amsterdam, 2004).
35. U. Erb, K. T. Aust, G. Palumbo, in: *Nanostruct. Mater.*, ed. C. C. Koch (Noyes Publications/William Andrew Publishing, Norwich, New York, USA, 2002), p. 179.
36. M. L. Trudeau, *Nanostruct. Mater.* **12** (1999) 55.
37. H. Natter, M. Bukowski, R. Hempelmann, S. Zein El Abedin, E. M. Moustafa, F. Endres, *Z. Phys. Chem.* **220** (2006) 1275.
38. M. J. Aus, B. Szpunar, A. M. El-Sherik, U. Erb, G. Palumbo, K. T. Aus, *Scr. Met. Mater.* **27** (1992) 1639.
39. S. Lauer, Z. Guan, H. Wolf, H. Natter, M. Schmelzer, R. Hempelmann, T. Wichert, *Nanostruct. Mater.* **12** (1999) 955.
40. H. Wolf, Z. Guan, X. Li, T. Wichert, *Hyperfine Interact.* **136–137** (2001) 281.
41. M. J. Aus, C. Cheung, B. Szpunar, U. Erb, *J. Mater. Sci. Lett.* **17** (1998) 1952.
42. H. L. Seet, X. P. Li, Z. J. Zhao, Y. K. Kong, H. M. Zheng, W. C. Ng, *J. Appl. Phys.* **97** (2005) 10N304.
43. L. J. Gao, G. W. Anderson, P. R. Norton, Z. H. Lu, J. P. McCaffrey, M. J. Graham, *J. Appl. Phys.* **78** (1995) 5795.
44. C. Cheung, G. Palumbo, U. Erb, *Scr. Met. Mater.* **31** (1994) 735.
45. F. Czerwinski, J. A. Szpunar, U. Erb, *J. Mater. Sci. – Mater. El.* **11** (2000) 243.
46. H. Li, F. Ebrahimi, *Mater. Sci. Eng. A* **347** (2003) 93.
47. R. D. Noce, N. Barelli, P. T. A. Sumodjo, D. R. Cornejo, A. V. Benedetti, *J. Magn. Magn. Mater.* **306** (2006) 199.
48. R. D. Noce, N. Barelli, R. F. C. Marques, P. T. A. Sumodjo, A. V. Benedetti, *Surf. Coat. Technol.* **202** (2007) 107.
49. F. Wang, K. Hosoiri, S. Doi, N. Okamoto, T. Kuzushima, T. Totsuka, T. Watanabe, *Electrochem. Commun.* **6** (2004) 1149.

50. F. M. Takata, G. Pattanaik, W. A. Soffa, P. T. A. Sumodjo, G. Zangari, *Electrochem. Commun.* **10** (2008) 568.
51. D. Kim, D. Y. Park, B. Y. Yoo, P. T. A. Sumodjo, N. V. Myung, *Electrochim. Acta* **48** (2003) 819.
52. D. Y. Park, N. V. Myung, M. Schwartz, K. Nobe, *Electrochim. Acta* **47** (2002) 2893.
53. E. E. Kalu, *J. Solid State Electrochem.* **11** (2007) 1145.
54. T. Nakanishi, M. Ozaki, H. S. Nam, T. Yokoshima, T. Osaka, *J. Electrochem. Soc.* **148** (2001) C627.
55. E. H. du Marchie van Voorthuysen, F. T. ten Broek, N. G. Chechenin, D. O. Boerma, *J. Magn. Magn. Mater.* **266** (2003) 251.
56. Y. Zhang, D. G. Ivey, *Chem. Mater.* **16** (2004) 1189.
57. G. Devaraj, S. Guruviah, S. K. Seshdari, *Mater. Chem. Phys.* **25** (1990) 439.
58. N. V. Mandrich, *Met. Fin.* **100** (2002) 359.
59. M. S. Chandrasekar, M. Pushpavanam, *Electrochim. Acta* **53** (2008) 3313.
60. J. Q. Xiao, J. S. Jiang, C. L. Chien, *Phys. Rev. Lett.* **68** (1992) 3749.
61. A. E. Berkowitz, J. R. Mitchell, M. J. Carey, A. P. Young, S. Zhang, F. E. Spada, F. T. Parker, A. Hutten, G. Thomas, *Phys. Rev. Lett.* **68** (1992) 3745.
62. T. Cohen-Hyams, J. M. Plitzko, C. J. D. Hetherington, J. L. Hutchinson, J. Yahalom, W. D. Kaplan, *J. Mater. Sci.* **39** (2004) 5701.
63. Y. Ueda, M. Ito, *Jpn. J. Appl. Phys.* **33** (1994) L1403.
64. V. M. Fedosyuk, O. I. Kasyutich, D. Ravinder, H. J. Blythe, *J. Magn. Magn. Mater.* **156** (1996) 345.
65. H. Zaman, A. Yamada, H. Fukuda, Y. Ueda, *J. Electrochem. Soc.* **145** (1998) 565.
66. G. R. Pattanaik, D. K. Pandya, S. C. Kashyap, *J. Magn. Magn. Mater.* **219** (2000) 309.
67. S. Ge, H. Li, C. Li, L. Xi, W. Li, J. Chi, *J. Phys.: Condens. Matter* **12** (2000) 5905.
68. G. R. Pattanaik, D. K. Pandya, S. C. Kashyap, *J. Magn. Magn. Mater.* **234** (2001) 294.
69. G. R. Pattanaik, D. K. Pandya, S. C. Kashyap, *J. Alloy. Comp.* **326** (2001) 260.
70. T. A. Tochitskii, G. A. Jones, H. J. Blythe, V. M. Fedosyuk, J. Castro, *J. Magn. Magn. Mater.* **224** (2001) 221.
71. G. R. Pattanaik, D. K. Pandya, S. C. Kashyap, *J. Electrochem. Soc.* **149** (2002) C363.
72. S. Kainuma, K. Takayanagi, K. Hisatake, T. Watanabe, *J. Magn. Magn. Mater.* **246** (2002) 207.
73. G. R. Pattanaik, D. K. Pandya, S. C. Kashyap, *Thin Solid Films* **433** (2003) 247.
74. R. López Anton, M. L. Fdez-Gubieda, G. Kurlyandskaya, A. García-Arribas, *J. Magn. Magn. Mater.* **254–255** (2003) 85.
75. E. Gómez, A. Labarta, A. Llorente, E. Vallés, *J. Electrochem. Soc.* **151** (2004) C731.
76. S. Yao, H. Wu, W. Zhang, H. Wang, *J. Mater. Sci. Technol. (China)* **21** (2005) 307.

77. S. Pané, E. Gómez, E. Vallés, *J. Electroanal. Chem.* **596** (2006) 87.
78. C. A. Ross, *Annu. Rev. Mater. Sci.* **24** (1994) 159.
79. M. N. Baibich, J. M. Broto, A. Fert, F. Nguyen Van Dau, F. Petroff, P. Etienne, G. Creuzet, A. Friederich, J. Chazelas, *Phys. Rev. Lett.* **61** (1988) 2472.
80. G. Binasch, P. Grünberg, F. Saurenbach, W. Zinn, *Phys. Rev. B* **39** (1989) 4828.
81. M. Alper, K. Attenborough, R. Hart, S. J. Lane, D. S. Lashmore, C. Younes, W. Schwarzacher, *Appl. Phys. Lett.* **63** (1993) 2144.
82. L. Péter, I. Bakonyi, in *Electrocrystallization in Nanotechnology*, ed. G. Staikov (Wiley-VCH, Weinheim, Germany, 2007), p. 242.
83. W. Schwarzacher, D. S. Lashmore, *IEEE Trans. Magn.* **32** (1996) 3133.
84. I. Bakonyi, L. Péter, in: *Magnetic Materials, Processes and Devices VIII, The Electrochemical Society PV 2004-23*, Eds. S. Krongelb *et al.* (The Electrochemical Society, Pennington, New Jersey, USA, 2006) p. 227.
85. I. Bakonyi, L. Péter, *Progr. Mater. Sci.*, in preparation.
86. A. Bartók, L. Péter, I. Bakonyi, J. Pádár, P. B. Barna, unpublished work.
87. L. M. Goldman, C. A. Ross, W. Ohasi, D. Wu, F. Spaepen, *Appl. Phys. Lett.* **55** (1989) 2182.
88. E. Chassaing, *J. Electrochem. Soc.* **148** (2001) C690.
89. S. K. J. Lenczowski, C. Schönenberger, M. A. M. Gijs, W. J. M. de Jonge, *J. Magn. Magn. Mater.* **148** (1995) 455.
90. L. Péter, I. Bakonyi, unpublished results.
91. L. Péter, Q. X. Liu, Zs. Kerner, I. Bakonyi, *Electrochim. Acta* **49** (2004) 1513.
92. S. K. Gosh, A. K. Grover, P. Chowdhury, S. K. Gupta, G. Ravikumar, D. K. Aswal, M. Senthil Kumar, R. O. Dusane, *Appl. Phys. Lett.* **89** (2006) 132507.
93. M. Lakatos-Varsányi, A. Mikó, L. K. Varga, E. Kálmán, *Corr. Sci.* **47** (2005) 681.
94. E. Budewski, G. Staikov, W. J. Lorentz, *Electrochemical Phase Formation and Growth* (VCH, Weinheim, 1996).
95. R. M. Penner, in: *Electrocrystallization in Nanotechnology*, ed. G. Staikov (Wiley-VCH, Weinheim, Germany, 2007), p. 195.
96. E. C. Walter, B. J. Murray, F. Favier, G. Kaltenpoth, M. Grunze, R. M. Penner, *J Phys. Chem B* **106** (2002) 11407.
97. R. M. Penner, *J. Phys. Chem. B* **106** (2002) 3339.
98. Y. Xiao, G. Yu, J. Yuan, J. Wang, Z. Chen, *Electrochim. Acta* **51** (2006) 4218.
99. K. H. Kim, J. D. Lee, J. J. Lee, B. Y. Ahn, H. S. Kim, Y. W. Shin, *Thin Solid Films* **483** (2005) 74.
100. H. W. Wang, H. C. Lin, Y. C. Yeh, C. H. Kuo, *Appl. Phys. Lett.* **88** (2006) 033111.
101. L. Y. Zhao, H. Jalili, N. Panjwani, T. Chan, Z. H. He, N. F. Heinig, K. T. Leung, *Electrochem. Solid-State Lett.* **10** (2007) K47.
102. C. A. Ross, M. Hwang, M. Shima, H. I. Smith, M. Farhoud, T. A. Savas, W. Schwarzacher, J. Parrochon, W. Escoffier, H. Neal Bertram, F. B. Humphrey, M. Redjdal, *J. Magn. Magn. Mater.* **249** (2002) 200.
103. P. Apel, *Radiat. Meas.* **34** (2001) 559.

104. M. Daub, I. Enculescu, R. Neuman, R. Spohr, *J. Optoel. Adv. Mater.* **7** (2005) 865.
105. J. Gong, G. Zangari, unpublished results.
106. H. Masuda, K. Fukuda, *Science* **268** (1995) 1466.
107. H. Asoh, S. Ono, in *Electrocrystallization in Nanotechnology*, ed. G. Staikov (Wiley-VCH, Weinheim, Germany, 2007), p. 138.
108. Y. Kang, J. Zhao, J. Tao, X. Wang, Y. Li, *Appl. Sur. Sci.* **254** (2008) 3935.
109. S. Aravamudhan, K. Luongo, P. Poddar, H. Srikanth, S. Bhansali, *Appl. Phys. A* **87** (2007) 773.
110. M. V. Rastei, S. Colis, R. Meckenstock, O. Ersen, J. P. Bucher, *Surf. Sci.* **600** (2006) 2178.
111. K. Nielsch, R. B. Wehrspohn, J. Barthel, J. Kirschner, S. F. Fischer, H. Kronmüuller, T. Schweinböck, D. Weiss, U. Gösele, *J. Magn. Magn. Mater.* **249** (2002) 234.
112. D. M. Davis, E. J. Podlaha, *Electrochem. Solid-State Lett.* **8** (2005) D1.
113. Q. Wang, G. Wang, X. Han, X. Wang, J. Hou, *J. Phys. Chem B* **109** (2005) 23326.
114. Z. Liu, L. Guo, C. L. Chien, P. C. Searson, *J. Electrochem. Soc.* **155** (2008) D569.
115. Z. Liu, G. Xia, F. Zhu, S. Kim, N. Markovic, C. L. Chien, P. C. Searson, *J. Appl. Phys.* **103** (2008) 064313.
116. P. R. Evans, W. R. Hendren, R. Atkinson, R. J. Pollard, *J. Electrochem. Soc.* **154** (2007) K79.
117. Y. W. Chung, I. C. Leu, J. H. Lee, J. H. Yen, M. H. Hon, *J. Electrochem. Soc.* **154** (2007) E77.
118. M. E. Kiziroglou, X. Li, D. C. Gonzalez, C. H. de Groot, A. A. Zhukov, P. A. J. de Groot, P. N. Bartlett, *J. Appl. Phys.* **100** (2006) 113720.
119. A. A. Zhukov, A. V. Goncharov, P. A. J. de Groot, M. A. Ghanem, P. N. Bartlett, R. Boardman, H. Fangohr, V. Novosad, G. Karapetrov, *Appl. Phys. Lett.* **88** (2006) 062511.
120. A. A. Zhukov, A. V. Goncharov, P. A. J. de Groot, M. A. Ghanem, I. S. El-Hallag, P. N. Bartlett, R. Boardman, H. Fangohr, V. Novosad, G. Karapetrov, *J. Appl. Phys.* **97** (2005) 10J701.
121. A. A. Zhukov, M. A. Ghanem, A. V. Goncharov, P. A. J. de Groot, I. S. El-Hallag, P. N. Bartlett, R. Boardman, H. Fangohr, *J. Magn. Magn. Mater.* **272–276** (2004) 1621.
122. Y. W. Chung, I. C. Leu, J. H. Lee, J. H. Yen, M. H. Hon, *Electrochim. Acta* **53**, (2007) 1703.
123. M. A. Ghanem, P. N. Bartlett, P. de Groot, A. Zhukov, *Electrochem. Commun.* **6**, (2004) 447.
124. L. Xu, L. D. Tung, L. Spinu, *Adv. Mater.* **15** (2003) 1562.
125. P. N. Bartlett, P. N. Birkin, M. A. Ghanem, P. de Groot, M. Sawicki, *J. Electrochem. Soc.* **148** (2001) C119.
126. A. Hovestad, L. J. J. Janssen, *J. Appl. Electrochem.* **25** (1995) 519.

127. J. L. Stojak, J. Fransaer, J. B. Talbot, in *Advances in Electrochemical Science and Engineering, Vol. 7*, eds. R. C. Alkire, D. M. Kolb (Wiley-VCH, Weinheim, Germany, 2002), p. 193.
128. C. T. J. Low, R. G. A. Wills, F. C. Walsh, *Surf. Coat. Technol.* **201** (2006) 371.
129. R. A. Tacken, P. Jiskoot, L. J. J. Janssen, *J. Appl. Electrochem.* **26** (1996) 129.
130. S. Guan, B. J. Nelson, K. Vollmers, *J. Electrochem. Soc.* **151** (2004) C545.
131. S. Guan, B. J. Nelson, *J. Microelectromech. Syst.* **15** (2006) 330.
132. R. Amigó, J. Asenjo, E. Krotenko, F. Torres, J. Tejada, *Chem. Mater.* **12** (2000) 573.
133. C. Pascal, J. L. Pascal, F. Favier, *Chem. Mater.* **11** (1999) 141.
134. F. J. Santos, L. C. Varanda, L. C. Ferracin, M. Jafelicci, Jr., *J. Phys. Chem. C* **112** (2008) 5301.
135. L. Cabrera, S. Gutierrez, N. Menendez, M. P. Morales, P. Herrasti, *Electrochim. Acta* **53** (2008) 3436.

Chapter 6

MAGNETOELECTRIC MATERIALS FOR SPINTRONICS

Faik Mikailzade

Department of Physics, Gebze Institute of Technology
P.O. Box 141,41400 Gebze, Kocaeli, Turkey
E-mail: faik@gyte.edu.tr

Recent developments in the fields of multiferroic and magnetoelectric materials are reviewed. The investigations of magnetoelectric effects in multiferroic crystals and magnetoelectric composites are considered in a frame of their possible applications in spintronics and magnetoelectronics.

6.1. GMR and Spintronics

Intensive investigations in the field of the nanoscale materials promote to the great progress in various kinds of the storage media and memory applications. The technical progress of last years in the preparation of multilayer thin films promote to discovering the Giant Magnetoresistance (GMR) phenomena,[1,2] consisting in extraordinary changing of resistivity/impedance of the material while applying external magnetic field. Nobel Prize in Physics in 2007 is awarded to Albert Fert and Peter Grünberg for their discovery of GMR. Applications of this phenomenon have revolutionized techniques for retrieving data from hard disks. The discovery also plays a major role in various magnetic sensors as well as for the development of a new generation of electronics. The use of Giant Magnetoresistance can be regarded as one of the first major applications of nanotechnology. The GMR- materials have already found applications as sensors of low magnetic field, computer hard disk heads, magnetoresistive RAM chips etc. The "read" heads for magnetic hard disk drives (HDD) have allowed to increase the storage density on a disk drive from 1 to 20 Gbit per square inch, merely by the incorporation of the new GMR materials.[3,4] The market only in the field of GMR-nanotechnology is estimated over $100 billion annually. In addition to hard disk drives, the technologies developed have also been applied to magnetic random access memories (MRAMs).[5–7] Further advance in these fields is the key to realizing terabits/in^2 hard disk drives and gigabit nonvolatile memories within this decade.

Magnetic recording has dominated the area of peripheral information storage ever since the beginning of the computer era a half century ago, with tapes and disks representing the two main application areas. The embodiment of the technology involves a relatively thin magnetic layer supported by a flexible or rigid substrate, which can be magnetized by an external magnetic field and which retains its magnetization after the field is removed. Information is recorded in the form of oppositely magnetized regions in the surface layer of the medium utilizing the fringing field of an inductive transducer. In this frame, half-metallic dioxide films are expected to be the key element of Magnetic Tunnel Junction Magnetoresistive Random Access Memories (MTJ MRAM), prototypes of which, based on conventional metallic ferromagnets, have already been tested under laboratory conditions. Use of these materials with two orders of magnitude discrimination between the “zero”-state and the “one”-state current essentially simplifies the control electronics and increases the operational speed. They are also considered as spin aligner/filters in FET, LED, RTD devices and optical switches. Another class of the oxide-based spintronics materials is magnetically doped transition metal oxides (TiO_2, SnO_2, ZnO, etc.). Among other techniques, laser ablation, co-sputtering and sol-gel methods have so far been primarily used to synthesize these unique oxides doped with iron-group metals.

Evolution beyond passive magneto-electronic components is envisioned in the next generation of spintronics devices, which should combine memory and logical functions in order to set new standards in future information technology.[8] Recently, there has been growing interest in studying magnetization reversal involving spin transfer from a spin-polarized current injected into the device as an alternative to stray magnetic fields for switching the magnetic configurations in GMR or TMR devices.[9,10] The required huge current densities hamper the technical realization of this attractive concept. One can use the electric field as alternative means for controlling the magnetic configuration of magnetoresistive systems. So, the quest for higher data density in information storage is motivating investigations into approaches for manipulating magnetization without using magnetic field. This is also evidenced by the recent boom in magnetoelectronics and spintronics,[11] where phenomena such as carrier effects in magnetic semiconductors[12] and high-correlation effects in colossal magnetoresistive compounds[13] are studied for their device potential.

6.2. History and Invention of Magnetoelectricity

The magnetoelectric effect – the induction of polarization by a magnetic field and of magnetization by an electric field – provides another route for linking magnetic and electric properties. Hence, one expects the direct coupling between the magnetic and dielectric properties and their control by the application of magnetic and/or electric fields.

Although the ME effect was prophetically predicted by Curie early in 1894 on the basis of crystal symmetry consideration[14] no further work was done until 1958 when Landau and Lifshitz proved the feasibility of the ME effect in certain crystals. Subsequently, the symmetry argument was applied by Dzyaloshinskii[15] to antiferromagnetic Cr_2O_3 and it was suggested that the ME effect could appear in Cr_2O_3. This was followed by experimental confirmation.[16] The ME effect was observed in antiferromagnetic Cr_2O_3 in 1961[16,17] and later some single phase crystal families, e.g., perovskite-type $BiFeO_3$ $BiMnO_3$, and $TbMnO_3$, hexagonal $(RE)MnO_3$ (RE = rare earths), and the rare-earth molybdates, were found to have the ME effect.[18–22] Note that most of these compounds display an antiferromagnetic behavior.

As it has just been mentioned, these materials can display a magnetoelectric effect in which change in magnetization is induced by an electric field and change in electric polarization is induced by an applied magnetic field. Hence, one expects the direct coupling between the magnetic and dielectric properties and their control by the application of magnetic and/or electric fields. In multiferroic materials, in which ferroelectric and ferromagnetic ordering occur simultaneously, magnetic domains can be tuned by the application of an external electric field, and likewise electric domains are switched by magnetic field, which is unachievable separately in either ferroelectric or magnetic materials. So, multiferroics, characterized by simultaneous ferroelectric and magnetic ordering, may exhibit a larger magnetoelectricity exhibiting potential in a wide range of applications, including the emerging field of spintronics, data-storage media, and multiple-state memories, information storage, sensors, transducers, actuators, storage devices etc.[23]

Besides application potential, the fundamental physics of multiferroic materials is rich and fascinating. This nontrivial spin-lattice coupling in the magnetoelectrics has been manifested through various forms, such as linear and bilinear magnetoelectric effects, polarization change through field-induced phase transition, magneto-dielectric effect, and dielectric anomalies at magnetic transition temperatures.

6.3. Linear Magnetoelectric Effect

The magnetoelectric (ME) effect in its most general definition denominates the coupling between electric and magnetic fields in matter. A systematic progression of contributions to the ME effect is obtained from the expansion of the free energy (F) of a material, i.e.[24]

$$F\left(\vec{E},\vec{H}\right) = F_0 - P_i^s E_i - M_i^s H_i - \frac{1}{2}\varepsilon_0\varepsilon_{ij}E_iE_j - \frac{1}{2}\mu_0\mu_{ij}H_iH_j$$
$$- \alpha_{ij}E_iH_j - \frac{1}{2}\beta_{ijk}E_iH_jH_k - \frac{1}{2}\gamma_{ijk}H_iE_jE_k - \ldots,$$

where E_i and H_i denote the components of electric and magnetic fields. Differentiation leads to the polarization

$$P_i\left(\vec{E},\vec{H}\right) = -\frac{\partial F}{\partial E_i}$$
$$= P_i^s + \varepsilon_0\varepsilon_{ij}E_j + \alpha_{ij}H_j + \frac{1}{2}\beta_{ijk}H_jH_k + \gamma_{ijk}H_jE_k + \ldots$$

and to the magnetization

$$M_i\left(\vec{E},\vec{H}\right) = -\frac{\partial F}{\partial H_i}$$
$$= M_i^s + \mu_0\mu_{ij}E_j + \alpha_{ij}E_j + \beta_{ijk}E_jH_k + \frac{1}{2}\gamma_{ijk}E_jE_k + \dots,$$

where P_i^s and M_i^s denote the spontaneous polarization and magnetization, whereas ε and μ are the electric and magnetic susceptibilities. The tensor α corresponds to induction of polarization by a magnetic field or of magnetization by an electric field which is designated as the linear ME effect. It is supplemented by higher-order ME effects like those parametrized by the tensors β and γ. The vast majority of research on the ME effect is devoted to the linear ME effect and it is generally acceptable to omit the prefix 'linear' and simply to refer to the linear manifestation as the 'ME effect'.

As it was mentioned by Fiebig,[24] in order to estimate the magnitude of the coupling coefficients, it is necessary to understand the microscopic mechanism driving ME behaviour. Hornreich and Shtrikman[25] presented a model which explains the temperature dependence of both the parallel and the perpendicular ME effects in Cr_2O_3. The generalization of this approach with a compilation of magnetic interaction mechanisms, their modification by an electric field and the consequences for different types of magnetic ordering are published in Ref. 26. According to these works, the microscopic mechanism of ME behavior in Cr_2O_3 can be explained as described in Fig. 6.1. As it is seen from the figure, the movement of the Cr^{3+} ions in an electric field shows that the $A_{1,2}$ ions move closer to the small O^{2-} triangle while the $B_{1,2}$ ions move away from it. This breaks the equivalence of the ferromagnetic Cr^{3+} sublattices, introducing a ME magnetization. Thus, the figure shows that in differently magnetized sublattices the sign or magnitude of the described effects may be different. This is a very common source for emergence of a ME net magnetization in a compensated antiferromagnet subjected to an electric field.

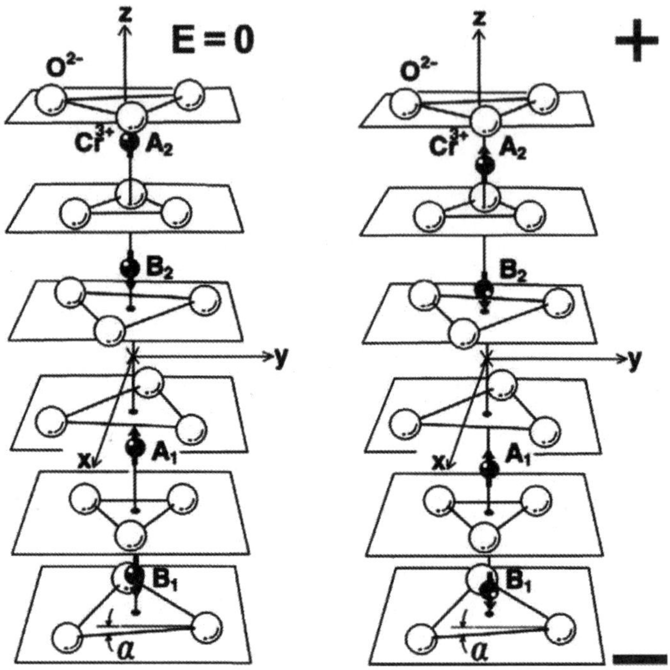

Fig. 6.1. Microscopic sources of ME behaviour in Cr_2O_3.[25,26]

It was shown that the ME response is limited by the relation[27]

$$\alpha_{ij}^2 < \varepsilon_{ii}\mu_{jj}\,.$$

So, one can conclude that the ME effect can only be large in ferroelectric and/or ferromagnetic materials, which takes place in multiferroics.

6.4. Multiferroics

The term multiferroic was first used by H. Schmid[28] in 1994. His definition referred to multiferroics as single phase materials which simultaneously posses two or more primary ferroic properties. Today the term multiferroic has been expanded to include materials, which exhibit any type of long range magnetic ordering, spontaneous electric polarization, and/or ferroelasticity. Working under this expanded definition the history of magnetoelectric multiferroics can be traced back to the 1960's.[29] In the most general sense the field of multiferroics was born from studies of magnetoelectric systems.

Multiferroics were grown for the first time in 1958, when magnetically active 3d ions were used to substitute ions with a noble gas shell in ferroelectrically distorted perovskite lattices.[28,29] This leads us to ferroelectric antiferromagnetic compositions like $PbFe_{0.5}Nb_{0.5}O3$ and $PbFe_{0.5}Ta_{0.5}O_3$.[30] Up to now more than 80 single-phase multiferroics were grown either as a discrete composition or as a solid solution. Multiferroics have been the topic of various review papers with a focus on structure and materials science,[31,32] compounds,[33] phase diagrams,[34] symmetries[29,35,36] and theory.[28]

Currently four major crystallographic types of multiferroics can be distinguished as: 1) compounds with perovskite structure – $BiFeO_3$, $BiMnO_3$, $TbMnO_3$ and others;[37–39] 2) compounds with hexagonal structure, which include ferroelectric antiferromagnetic rare-earth manganites RMnO$_3$ with R = Sc, Y, In, Ho, Er, Tm, Yb, Lu;[40–48] 3) Boracite compounds with the general formula $M_3B_7O_{13}X$ (M = Cr, Mn, Fe, Co, Cu, Ni) which are ferroelectric ferroelastic antiferromagnets;[49–51] 4) Orthorhombic $BaMF_4$ compounds (M = Mg, Mn, Fe, Co, Ni, Zn).[52,53]

Aside from these major types a large number of multiferroics with different structures are known. Specific examples are discussed in the before mentioned review articles.[28,31,33] A systematic classification of symmetries, related types of ferroic ordering and compounds can be found in Refs. 35, 36.

Generally, magnetically driven multiferroics are insulating materials, mostly oxides, in which macroscopic electric polarization is induced by magnetic long-range order. A necessary but not sufficient condition for the appearance of spontaneous electric polarization is the absence of inversion symmetry. In these materials inversion symmetry is broken by magnetic ordering. Such a symmetry breaking often occurs in so-called frustrated magnets, where competing interactions between spins favor unconventional magnetic orders. The microscopic mechanisms of magnetically induced ferroelectricity involve the polarization of electronic orbitals and relative displacement of ions in response to magnetic ordering.

However, finding a choice for multiferroic material is very limited because a few materials exhibit coexistence of ferroelectric and ferromagnetic properties at room temperature. There are very few natural multiferroic magnetoelectrics with both magnetic and electric polarizations existing or present in nature or synthesized in the laboratory, on the other hand, the observed ME effect for most such materials is too weak to be applicable.[54] The mechanisms and reasons for the existence of a very few magnetoelectric systems has been discussed in a number of works.[18,24,35,55–57] According to these works, in perovskite compounds the transition metal ion can trigger two types of Jahn–Teller distortion, which is responsible for appearing ferroelectric polarization. A first-order Jahn–Teller distortion occurs in the case of partially filled 3d orbitals, which retains the centre of symmetry, e.g. by elongation of the octahedron of ligands (as in the case of $LaMnO_3$ and $YTiO_3$). A second-order Jahn–Teller distortion, which is weaker than the first order one, requires an empty 3d shell for which a first-order Jahn–Teller distortion cannot occur, and breaks the centrosymmetry by off-centre movement of the transition-metal ion (examples are $BaTiO_3$ and PZT). Since only a partially filled 3d orbital can lead to magnetic ordering, whereas the breaking of centrosymmetry is a necessary condition for the formation of a spontaneous polarization, the conditions for ferroelectricity and (anti-) ferromagnetism are mutually exclusive. In an alternative approach[56] the incompatibility was explained by Hund's rule coupling that tends to keep the 3d spins parallel to one another. This mechanism breaks the strong covalent bonds that are necessary for ferroelectricity. In turn magnetic ferroelectrics must be materials in which the double well potential provoking the ferroelectric distortion is not caused by the hybridization of transition-metal ions in a noble gas configuration.

So, all these reasons and mechanisms led to presence of little number of single phase multiferroic materials and a low value of magnetoelectric coefficient in these materials, which made them unsuitable for possible applications.

6.5. Magnetoelectric Composites

Researchers, therefore, attempt to bring the desired effects by growing heterostructures of ferroelectric and ferromagnetic materials.[58] However, it is a difficult task to arrive at suitable compositions, which would give desired properties. Consequently, an additional electronic or structural driving force must be present for magnetic and ferroelectric ordering to occur simultaneously. A strong ME effect, however, could be realized in the composite consisting of magnetorestrictive and piezoelectric effects.[59] A much higher ME effect has been identified in specially designed composites in which the magnetostrictive phase is combined with the piezoelectric one so that an efficient magnetomechanical–piezoelectric coupling between the two phases is achieved.[59] The ME effect in composite materials of magnetostrictive and piezoelectric phases originates from the elastic interaction between the magnetostrictive and piezoelectric subsystems. In a magnetic field, the magnetostriction in the magnetostrictive phase gives rise to mechanical stresses that are transferred into the piezoelectric phase, owing to the piezoelectric effect, resulting in an electric polarization of the piezoelectric phase.

Van Suchtelen and co-workers[59–62] made the first artificial magnetoelectric material by combining a ferroelectric (piezoelectric) material $BaTiO_3$ (BTO) and a ferromagnetic (piezomagnetic) material $CoFe_2O_4$ (CFO) in an eutectic sintered composite. Most studies in the past focused exclusively on ferrite - $PbZrTiO_3/BaTiO_3$ composites,[60–64] although the experimental ME yield was only about 1%–2% of the theoretical prediction.

There also have been reports of magnetoelectric effects in layered composites such as ($PbZr_{1-x}Ti_xO_3$ (PZT)–$Tb_{0.3}Dy_{0.7}Fe_{1.92}$ (Terfenol-D), PZT-$NiFeO_4$, polyvinylidenefluoride–Terfenol-D and laminate PbMgNbO–PbTiO–Terfenol-D. (PZT)–$Tb_{0.3}Dy_{0.7}Fe_{1.92}$ (Terfenol-D), polyvinylidenefluoride-Terfenol-D and $PbMg_{1/3}Nb_{2/3}O_3$–$PbTiO_3$–Terfenol-D).[65–70] All these composite materials have been exclusively studied in bulk form. Interestingly, the materials made in the form of superlattice structure yielded unusual transport properties that cannot be obtained by classical solid-state chemistry route. Thus it is possible to construct superlattices whose structure consists of alternating FE and FM layers, and one can investigate the coupling between the two properties.

The CFO-PZT ($Pb(Zr_{0.52}Ti_{0.48})O_3$) multilayered structures prepared by Harshe *et al.*[64] show a ME-voltage coefficient α_E of 75 mV/cm Oe, also much lower than the predicted value. The α_E for a NFO ~nickel ferrite - PZT multilayered structure prepared using the tape casting technique was; 1500 mV/cm Oe.[65,66] A giant ME effect was measured in a sandwiched composite structure stacked alternatively by large magnetostrictive material - Terfenol-D and PZT disks,[67,68] and in the Terfenol-D/PVDF laminate.[69] Since then, in particular, Dong *et al.*[70–90] have reported the giant ME effect in a number of laminate composites of Terfenol-D and various piezoelectric materials including PZT ceramics, $Pb(Mg_{1/3}Nb_{2/3}O_3)$–$PbTiO_3$ (PMN-PT) or $Pb(Zn_{1/3}Nb_{2/3}O)_3$–TiO_3 (PZN-PT) single crystal, or electroactive PVDF copolymers.

Previous investigations of ME laminates have focused on piezoelectric and magnetostrictive layers that were, respectively, poled and magnetized along their thickness (transverse geometry – T-T mode) directions. For these previous laminate designs, relatively large ME coefficients were only observed under a high dc magnetic bias. Relatively large ME voltage coefficients of $\alpha_E = 4.8$ V/cm Oe was reported by Ryu *et al.*[68] under dc magnetic field bias of $H_{dc} \geq 4000$ Oe; although later, other investigators repeated the actual value to be $\alpha_E = 1.3$ V/cm Oe.[91,92] Figure 6.2 shows these later experimental results for laminate in T-T mode, where a maximum ME voltage of 66 mV/Oe or 1.32 V/cm Oe was observed under $H_{dc} \approx 4000$ Oe.

The main problem with the transverse (T-T) mode laminates is that a quite high dc magnetic bias H_{dc} is required to obtain a maximum value of α_E. This high H_{dc} is caused by a large demagnetization factor in transversely magnetized Terfenol-D layers. To reduce the demagnetization factor effect, a long-type configuration that uses a longitudinal magnetization was designed.[70–72] This dramatic decrease in the demagnetization factor results in a large reduction in the H_{dc}, required to achieve the maximum ME coefficient.[51] Long rectangular shaped Terfenol-D/PZT/Terfenol-D and Terfenol-D/PMNPT/Terfenol-D three layer laminates with a longitudinal magnetization and transverse polarization (L-T mode) were then reported[70–72] based on this consideration. Experimental results confirmed that at low magnetic biases of $H_{dc} < 500$ Oe much large values of α_E could be obtained for L-T laminates relative to T-T ones. It has been established that the induced ME voltage for the L-T Terfenol-D/PZT laminates under $H_{dc} = 500$ Oe was 0.085 V/Oe (or α_E = 1.7 V/cm Oe and for L-T Terfenol-D/PMN-PT ones ~0.11 ṼOe (α_E = 2.2 V/cm Oe); whereas that for the T-T mode of Terfenol-D/PZT laminate under $H_{dc} = 500$ Oe was only ~0.015 V/Oe (α_E = 0.3 V/cm Oe). Clearly, long type L-T laminates have significantly higher ME voltage coefficients than T-T ones under modest magnetic biases.

To achieve a high output voltage, a L-L mode (longitudinal magnetization and longitudinal polarization) of Terfenol-D/PZT/Terfenol-D or Terfenol-D/PMN-PT/Terfenol-D laminate is a good choice due to its large dielectric displacement along the length (longitudinal) direction.[73,88,92] For a long-type ME laminate, the length of the piezoelectric layer is much larger than its thickness, in addition, the longitudinal electromechanical coupling coefficient and piezoelectric voltage constant are higher than the corresponding transverse quantities. So, the Terfenol-D/PZT laminate operated in an L-L mode should have a

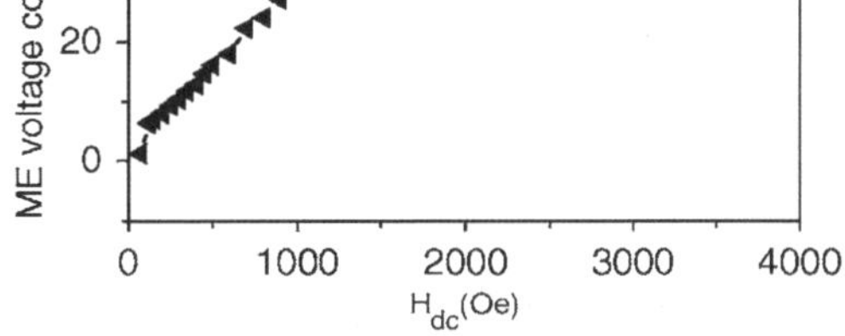

Fig. 6.2. ME voltage coefficient of Terfenol-D/PZT laminate (T-T mode) as a function of the applied dc magnetic field bias H_{dc} (Dong *et al.*[92]).

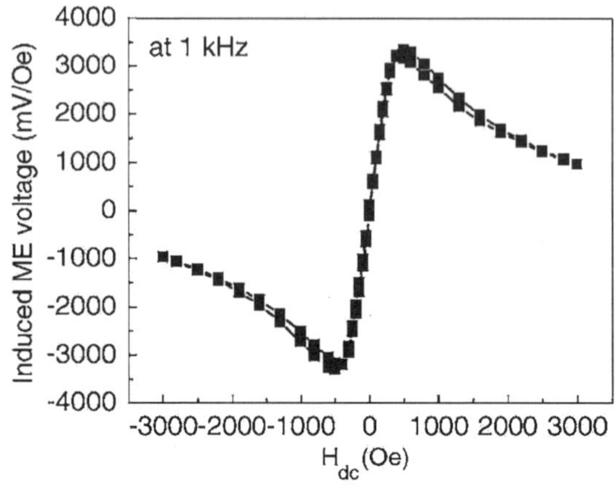

Fig. 6.3. Induced ME voltage in the L-L Terfenol-D/PZT/Terfenol-D laminate (Dong *et al.*[88]).

much higher induced voltage under magnetic field excitation. For example, Fig. 6.3 shows measurements for an L-L mode terfenol-D/PZT laminate, which exhibits a maximum $V_{ME} = 3.5$ V/Oe at 1 kHz (or $\alpha_E = 2.4$ V/cm Oe) at $H_{dc} = 500$ Oe.

Thus, magnetostrictive-piezoelectric laminate composites have been found to have higher ME coefficients than that of single-phase materials or particle composites. A very comprehensive review on magnetoelectric composites have been presented recently in Ref. 93.

6.6. Conclusions and Outlook

Hence, in normal practice, desired magnetoelectric effect is achieved by growing heterostructures of ferroelectric and magnetic materials. Realization of heterostructures with desired properties is not only difficult but also involves complicated lengthy procedures. Recently, attention to ME materials has been gradually drawn toward composite thin films. Compared to bulk composites, ME composite thin films exhibit unique advantages. Their composition and connectivity could be modulated at the microscopic scale, and the artificial thin film heterostructures can thus be achieved, which have potential applications in all kinds of microdevices and integrated units such as microsensors, MEMS devices, and high density information storage devices. It has been identified that a number of magnetostrictive-piezoelectric composite structures show significant ME output qualified for potential applications.

Recently, efforts have been made to fabricate the artificial layers and tailor their structures for the suitability of multiferroics.[94–98] These materials are in the form of either composites, superlattices, or multilayers. Interestingly, the material made in the form of superlattices, whose structure consists of alternating ferroelectric and ferromagnetic layers, yielded unusual electrical and magnetic transport properties that cannot be obtained in either of their constituents. As a result of these studies of sandwich devices, practical applications of magnetoelectric effects now appear more feasible. Nevertheless, sandwich devices are intrinsically limited in the feature size and difficult to miniaturize. Therefore, further advances in new magnetoelectric composites are still desirable.

Acknowledgments

The author is indebted to Research Projects Commission of Gebze Institute of Technology for supporting this work by the Grant No. 2007-A-14 and to The Scientific & Technological Research Council of Turkey (TÜBİTAK) for supporting by Project. No. 106M540.

References

1. M. N. Baibich, J. M. Broto, A. Fert, F. Nguyen van Dau, F. Petroff, P. Etienne, G. Creuzet, A. Friederich, and J. Chazelas, *Phys.Rev. Lett.* **61**, 2472 (1988).
2. G. Binasch, P. Grünberg, F. Saurenbach, and W. Zinn, *Phys. Rev. B* **39** 4828 (1989).
3. G. A. Prinz, *Sci.* **282**, 1660 (1998).
4. P. Grünberg, *Acta Mater.* **48**, 239 (2000).
5. J. M. Dauton, *Thin Solid Films* **216**, 162 (1992); J. M. Dauton, *J. Appl. Phys.* **81**, 3758 (1997).

6. S. S. P. Parkin, K. P. Roche, M. G. Samant, P. M. Rice, R. B. Beyers, R. E. Scheurlein, E. J. O'Sullivan, S. L. Brown, J. Bucchigana, D. W. Abraham, Y. Lu, M. Rooks, P. L. Trouilloud, R. A. Wanner, and W. J. Gallagher, *J. Appl. Phys.* **85**, 5828 (1999).
7. S. Tehrani, E. Chen, M. Durlam, M. Deherrera, J. M. Slaughter, J. Shi, and G. Kerszkowski, *J. Appl. Phys.* **85**, 5822 (1999).
8. A. Ney, C. Pampuch, R. Koch, and K. H. Ploog, *Nature Londond* **425**, 485 (2003).
9. J. Slonczewski, *J. Magn. Magn. Mater.* **159**, L1 (1996); **195**, 261 (1999).
10. L. Berger, *Phys. Rev. B* **54**, 9353 (1996).
11. G. A. Prinz, *J. Magn. Magn. Mater.* **200**, 57–68 (1999).
12. H. Ohno *et al.*, *Nature* **408**, 944–946 (2000).
13. A. Asamitsu, Y. Tomioka, H. Kuwahara, and Y. Tokura, *Nature* **388**, 50–52 (1997).
14. P. Curie, *J. Physique* **3**, 393 (1894).
15. I. E. Dzyaloshinskii, *Zh. Eksp. Teor. Fiz.* **37**, 881 (1959), *Sov. Phys. JETP* **10**, 628 (1960).
16. D. N. Astrov, *Sov. Phys. JETP* **13**, 729 (1961).
17. G. T. Rado and V. J. Folen, *Phys. Rev. Lett.* **7**, 310 (1961).
18. N. A. Hill, *J. Phys. Chem. B* **104**, 6694 (2000).
19. N. Hur, S. Park, P. A. Sharma, J. S. Ahn, S. Guha, and S.-W. Cheong, *Nature* **429**, 392 (2004).
20. T. Kimura, S. Kawamoto, I. Yamada, M. Azuma, M. Takano, and Y. Tokura, *Phys. Rev. B* **67**, 180401R (2003); T. Kimura, T. Goto, K. Thizaka, T. Arima, and Y. Tokura, *Nature* **426**, 55 (2003).
21. H. Zheng, J. Wang, S. E. Lofland, Z. Ma, L. M. Ardabili, T. Zhao, L. S. Riba, S. R. Shinde, S. B. Ogale, F. Bai, D. Viehland, Y. Jia, D. G. Schlom, M. Wutting, A. Roytburd, and R. Ramesh, *Sci.* **303**, 661 (2004).
22. P. Murugavel, D. Saurel, W. Prellier, Ch. Simon, and B. Raveau, *Appl. Phys. Lett.* **85**, 4424 (2004).
23. G. R. Slemon, Magnetoelectric Devices Transducers, Transformers, and Machines, Wiley, New York, 1966.
24. M. Fiebig, *J. Phys. D: Appl. Phys.* **38**, R123 (2005).
25. R. M. Hornreich and S. Shtrikman, *Phys. Rev.* **161**, 506 (1967).
26. G. A. Gehring, *Ferroelectrics* **161**, 275 (1994).
27. T. H. O'Dell, *Phil. Mag.* **8**, 411 (1963).
28. H. Schmid, *Ferroelectrics* **162**, 317 (1994).
29. G. A. Smolenskii and I. E. Chupis, *Sov. Phys.—Usp.* **25**, 475 (1982).
30. G. A. Smolenskii, A. I. Agranovskaya, and V. A. Isupov, *Sov. Phys. Solid State* 1, **149**, (1959).
31. Y. N. Venevtsev, V. V. Gagulin, and I. D. Zhitomirsky, *Ferroelectric*s **73**, 221 (1987).
32. G. A. Smolenskii, V. A. Bokov, V. A. Isupov, N. N. Krainik, and G. H. Nedlin, *Helv. Phys. Acta* **41**, 1187 (1968).

33. Y. N. Venevtsev and V. V. Gagulin, *Ferroelectrics* **162**, 23 (1994).
34. S. M. Skinner, *IEEE Trans. Parts, Mater. Packaging* **6**, 68 (1970).
35. H. Schmid, *Int. J. Magn.* **4**, 337 (1973).
36. K. Aizu, *Phys. Rev. B* **2**, 754 (1970).
37. S. M. Skinner, *IEEE Trans. Parts, Mater. Packaging* **6**, 68 (1970).
38. F. Kubel and H. Schmid, *Acta Crystallogr. B* **46**, 698 (1990).
39. S. V. Kizelev, R. P. Ozerov, and G. S. Zhdanov, *Sov. Phys.—Dokl.* **145**, 1255 (1962).
40. A. V. Kovalev and G. T. Andreeva, *C. R. Acad. Sci.* **256**, 1958 (1963).
41. K. H. Hellwege and A. M. Hellwege (ed), Numerical Data and Functional Relationships (Landolt-Bornstein, New Series) Group III, vol 16a (Berlin, Springer, 1981).
42. B. B. van Aken, Structural Response to Electronic Transitions in Hexagonal and Ortho-Manganites PhD Thesis University of Groningen, 2001.
43. E. F. Bertaut and M. Mercier, *Phys. Lett.* **5**, 27 (1963).
44. H. Sugie, N. Iwata, and K. Kohn, *J. Phys. Soc. Japan* **71**, 1558 (2002).
45. C. Moure, M. Villegas, J. F. Fernandez, J. Tartaj, and P. Duran, *J. Mater. Sci.* **34**, 2565 (1999).
46. B. B. van Aken, J. W. G. Bos, R. A. de Groot, and T. T. M. Palstra, *Phys. Rev. B* **63**, 125127 (2001).
47. M. Bieringer, J. E. Greedan, and A. S. Wills, *Appl. Phys. A* **74**, S601 (2002).
48. N. Fujimura, H. Sakata, D. Ito, T. Yoshimura, T. Yokota, and T. Ito, *J. Appl. Phys.* **93**, 6990 (2003).
49. Y. Y. Tomashpol'ski, Y. N. Venevtsev, and V. N. Beznozdrev, *Fiz. Tverd. Tela* **7**, 2763 (1965).
50. E. Ascher, H. Schmid, and D. Tar, *Solid State Commun.* **2**, 45 (1964).
51. H. Schmid, H. Rieder, and E. Ascher, *Solid State Commun.* **3**, 327 (1965).
52. J. F. Scott, *Ferroelectrics* **24**, 127 (1980).
53. M. Eibschütz and H. J. Guggenheim, *Solid State Commun.* **6**, 737 (1968).
54. J. Wang, J. B. Neaton, H. Zheng, V. Nagarajan, S. B. Ogale, B. Liu, D. Viehland, V. Vaithyanathan, D. G. Schlom, U. V. Waghmare, N. A. Spaldin, K. M. Rade, M. Wuttig, and R. Ramesh, *Sci.* **299**, 1719 (2003).
55. B. B. van Aken, T. T. M. Palstra, A. Filippetti, and N. A. Spaldin, *Nat. Mater.* **3**, 164 (2004).
56. D. I. Khomskii, *Bull. Am. Phys. Soc. C* 21.002 (2001).
57. N. A. Hill and A. Filipetti, *J. Magn. Magn. Mater.* **242–245**, 976 (2002).
58. K. Ueda, H. Tabata, and T. Kawai, *Appl. Phys. Lett.* **75**, 555 (1999).
59. J. van Suchtelen, *Philips Res. Rep.* **27**, 28 (1972).
60. J. van den Boomgaard, D. R. Terrell, and R. A. J. Born, *J. Mater. Sci.* **9**, 1705 (1974).
61. J. van den Boomgaard, A. M. J. G. van Run, and J. van Suchtelen, *Ferroelectrics* **14**, 727 (1976).

62. J. van den Boomgaard, A. M. J. G. van Run, and J. van Suchtelen, *Ferroelectrics* **10**, 295 (1976).
63. J. van den Boomgaard and R. A. J. Born, *J. Mater. Sci.* **13**, 1538 (1978).
64. G. Harshe, J. P. Dougherty, and R. E. Newnham, *Int. J. Appl. Electro-magn. Mater.* **4**, 161 (1993).
65. G. Srinivasan, E. T. Rasmussen, J. Gallegos, R. Srinivasan, Yu. I. Bokhan, and V. M. Laletin, *Phys. Rev. B* **64**, 214408 (2001).
66. G. Srinivasan, E. T. Rasmussen, B. J. Levin, and R. Hayer, *Phys. Rev. B* **65**, 134402 (2002).
67. J. Ryu, S. Priya, A. V. Carazo, and K. Uchino, *J. Am. Ceram. Soc.* **84**, 2905 (2001).
68. J. Ryu, A. V. Carazo, K. Uchino, and H. E. Kim, *Jpn. J. Appl. Phys. Part 1*, **40**, 4948 (2001).
69. K. Mori and M. Wuttig, *Appl. Phys. Lett.* **81**, 100 (2002).
70. S. X. Dong, J. F. Li, and D. Viehland, *Appl. Phys. Lett.* **83**, 2265 (2003).
71. S. X. Dong, J. F. Li, and D. Viehland, *IEEE Trans. Ultrason. Ferroelectr. Freq. Control* **50**, 1236 (2003).
72. S. X. Dong, J. F. Li, and D. Viehland, *J. Appl. Phys.* **95**, 2625 (2004).
73. S. X. Dong, J. F. Li, and D. Viehland, *Appl. Phys. Lett.* **85**, 2307 (2004).
74. S. X. Dong, J. F. Li, and D. Viehland, *J. Appl. Phys.* **96**, 3382 (2004).
75. S. X. Dong, J. F. Li, and D. Viehland, *Appl. Phys. Lett.* **85**, 2307 (2004).
76. S. X. Dong, J. F. Li, and D. Viehland, *Appl. Phys. Lett.* **85**, 3534 (2004).
77. S. X. Dong, J. F. Li, and D. Viehland, *Appl. Phys. Lett.* **84**, 4188 (2004).
78. S. X. Dong, J. Y. Zhai, Z. P. Xing, J. F. Li, and D. Viehland, *Appl. Phys. Lett.* **86**, 102901 (2005).
79. S. X. Dong, J. G. Bai, J. Y. Zhai, J. F. Li, G. Q. Lu, D. Viehland, S. J. Zhang, and T. R. Shrout, *Appl. Phys. Lett.* **86**, 182506 (2005).
80. S. X. Dong, J. Y. Zhai, F. M. Bai, J. F. Li, D. Viehland, and T. A. Lograsso, *J. Appl. Phys.* **97**, 103902 (2005).
81. S. X. Dong, J. Y. Zhai, N. G. Wang, F. M. Bai, J. F. Li, D. Viehland, and T. A. Lograsso, *Appl. Phys. Lett.* **87**, 222504 (2005).
82. S. X. Dong, J. Zhai, J. F. Li, and D. Viehland, *Appl. Phys. Lett.* **88**, 082907 (2006).
83. S. X. Dong, J. Y. Zhai, J. F. Li, and D. Viehland, *Appl. Phys. Lett.* **89**, 252904 (2006).
84. S. X. Dong, J. Y. Zhai, J. F. Li, and D. Viehland, *Appl. Phys. Lett.* **89**, 122903 (2006).
85. S. X. Dong, J. F. Li, and D. Viehland, *J. Appl. Phys.* **100**, 124108 (2006).
86. S. X. Dong, J. Y. Zhai, J. F. Li, D. Viehland, and M. I. Bichurin, *Appl. Phys. Lett.* **89**, 243512 (2006).
87. J. Y. Zhai, J. F. Li, S. X. Dong, D. Viehland, and M. I. Bichurin, *J. Appl. Phys.* **100**, 124509 (2006).
88. S. X. Dong, J. F. Li, and D. Viehland, *IEEE Trans. Ultrason. Ferroelectr. Freq. Control* **50**, 1253 (2003).

89. S. X. Dong, J. F. Li, and D. Viehland, *J. Mater. Sci.* **41**, 97 (2006).
90. S. X. Dong, J. R. Cheng, J. F. Li, and D. Viehland, *Appl. Phys. Lett.* **83**, 4812 (2003).
91. J. Y. Zhai, Z. Xing, S. X. Dong, J. F. Li, and D. Viehland, *Appl. Phys. Lett.* **88**, 062510 (2006).
92. S. X. Dong, J. F. Li, and D. Viehland, *IEEE Trans. Ultrason. Ferroelectr. Freq. Control* **51**, 794 (2004).
93. C.-W. Nan, M. I. Bichurin, S. Dong, D. Viehland, and G. Srinivasan, *J. Appl. Phys.* **103**, 031101 (2008).
94. K. S. Wang *et al.*, *Appl. Phys. Lett.* **84**, 3091 (2004).
95. H. Zheng *et al.*, *Sci.* **303**, 661 (2004).
96. C.-W. Nan, N. Cai, L. Liu, J. Zhai, Y. Ye, and Y. Lin, *J. Appl. Phys.* **94**, 5930 (2003).
97. G. Liu, C.-W. Nan, N. Cai, and Y. Lin, *ibid.* **95**, 2660 (2004).
98. P. Murugavel, D. Saurel, W. Prellier, Ch. Simon, and B. Raveau, *Appl. Phys. Lett.* **85**, 4424 (2004).

Chapter 7

GMR IN ELECTRODEPOSITED SUPERLATTICES

Gholamreza Nabiyouni

Department of Physics, University of Arak
Shahid Beheshti Avenue, P.O. Box 38156, Arak, Iran
E-mail: g-nabiyouni@araku.ac.ir

Following the discovery of GMR in 1988, and since the very first observation of GMR in electrodeposited superlattices, there has been an extensive interest in electrodeposition of magnetic thin films and multilayers. Among the different multilayer systems, electrodeposition of magnetic/nonmagnetic superlattices attracted an especial interest because of their magnetoresistive properties and their applications in the modern data storage and information technology. After giving a brief introduction we introduce electrodeposition as a versatile technique to fabricate GMR materials. The chapter then deals with origin of magnetoresistance, and ends with dealing the parameters involved in the GMR in electrodeposited superlattices.

7.1. Introduction

Technological applicability of metallic nano-structures has generated tremendous interest in recent years because of their wide range of applications, especially in computer and microelectronic industries. Progress in the growth of high quality thin films, be they metals, semiconductors or insulators, has allowed the realisation of many modern electronic, magnetic and optical devices. The study of magnetic films, consisting of alternating magnetic and non-magnetic layers is of interest as they exhibit new magnetic phenomena with the potential for application in data storage technology, and also for obtaining a better understanding of the magnetic properties of materials. The properties of multilayers strongly depend on the layer thicknesses and effects of the boundaries between magnetic and non-magnetic layers. Multilayer thin films usually consist of layers with thicknesses from a few angstroms up to ~100Å. The mechanical, transport and magnetic properties of multilayer thin films are often radically different from the bulk properties of their constituents. With the introduction of magnetoresistance (MR) heads, the data reading function of the head is now performed by a MR sensor. Thus, with an increase in performance parameters, a demand for materials with improved properties, both magnetic properties and corrosion resistance, is desired for the recording head.

Co-rich FeCoNi thin film alloys are good candidates as soft magnetic materials.[1,2] Layering the Co-rich alloy with copper can increase the alloy resistivity, and avoiding eddy currents and make them alternative for the traditional permalloy NiFe write materials.

Much attention has been paid to metallic superlattices consisting of alternating layers of a ferromagnetic metal e.g. Ni, Co, Fe or their alloys and a non-magnetic transition or noble metal e.g. Cu, Ag, Pt, Au and Ru since they can exhibit special properties such as the magneto-optic Kerr effect (MOKE), anisotropic magnetoresistance (AMR) or giant magnetoresistance (GMR).[3–14]

A range of methods are available for the fabrication of thin films. Multilayer and superlattice structures are most commonly prepared by vacuum based techniques such as evaporation, sputtering and molecular beam epitaxy (MBE). In these techniques a metal is evaporated or sputtered and deposited on a substrate to form a thin film. In MBE under certain conditions, it is possible to grow one atomic layer after another, forming a nearly perfect single crystal. However the required equipment is complicated and expensive. Sputtering which is less complicated and less expensive has been widely used to produce thin films and superlattices although the produced materials are not perfect single crystals and do not always have very sharp interfaces.

Electrodeposition, or electroplating, as an alternative deposition technique to grow thin films has been little used compared to vacuum deposition till last two decades. This may be because electrodeposition was not expected to produce pure and high quality thin films. Nevertheless Brenner[15] reported that in the right conditions, it was possible to electrodeposit material that had impurity levels as low as 100 ppm. Electrodeposition has long been used for depositing decorative and protective coatings and is now hugely popular as it is the method of choice for metal deposition in the fabrication of integrated circuits[16] and magnetic recording heads for computer hard discs.[2]

Electrodeposition has some advantages over vacuum deposition techniques including a high deposition rate, simplicity in the experimental set-up, deposition at nearly room temperature and low cost. It is also a very useful technique to fabricate metal deposits into recessed and curved areas.

Moreover the properties of the electrodeposited thin films can be easily controlled by changing the deposition parameters such as applying the voltage and current density, electrolyte concentration and pH, and temperature. Electrodeposited alloys are used widely in the microelectronics field, due to the cost efficiency, and the applicability to irregular geometries. It is also an advantageous method compare to other vapour technique when depositing alloys of elements with extremely different vapour pressures.

Electrodeposition, usually refers to deposition of a metal or an alloy from an electrolyte by passing a charge between two electrodes located in the electrolyte. It dates back to Volta's discovery of the production of electricity by chemical reactions. It is a method by which a relatively pure metal or alloy may be deposited from a solution. Electrodeposition is based on Faraday's law i.e. the amount of chemical reaction caused by the flow of current is proportional to the amount of electricity passed.

The first electrodeposition of alloys, brass and bronze, probably took place at the same time. However up to 1900 the application of alloy plating was very limited and quite empirical. The first systematic work on alloy deposition was that of Fritz Spitzer[17] published in 1905, which discussed the electrodeposition of brass. To date the electrodeposition of many metals and alloys has been reported, and the conditions for their deposition may vary considerably. Modern electrodeposition in the sense of the deposition of thin and dense metal films for the purposes of decoration and protection took place after the discovery of the value of cyanides for plating baths.

Artificially layered materials consisting of alternating layers of two different metals or alloys are widely used to study their mechanical, optical and magnetic properties which might be different from those of bulk materials. Electrodeposition was possibly the earliest deposition technique for metal multilayers, as Blum deposited Cu/Ni and other multilayer systems by alternate deposition from two different electrolytes as early as 1921.[18] Electrodeposition of multilayers was then developed and reported on by a number of groups in different metal (alloy)/metal systems.[13,19–22] A few years after the discovery of GMR in MBE grown Fe/Cr superlattices,[23] GMR was observed in electrodeposited Co-Ni-Cu/Cu superlattices.[24]

We first deal with the electrodeposition of metallic thin films and superlattices, which has not been studied as widely as vacuum-based deposition techniques, despite its capability for making superlattices or precision microstructures. Some fundamental concepts of electrochemistry are described in this section before dealing with the electrodeposition of metals and superlattices. The electrodeposition of nanostructure superlattice thin films and study their magnetic properties will then be described. We then focus of the magnetic nanostructures with GMR properties.

7.2. Electrodeposition

Electrodeposition usually takes place in a single or double cell containing a solution (electrolyte) and two or three electrodes. The electrode is a conducting material, usually a metal or a semiconductor, in which charge is carried by electron movement and it may be a solid or a liquid. The electrolyte is a conducting medium, usually a liquid or a fused salt solution, in which charge is carried by the movement of ions.

An electrode reaction is a chemical process involving the transfer of one or more electrons to or from the electrode surface, for example

$$\begin{aligned} &M_{(aq)}{}^{n+} + ne^{-} \longrightarrow M_{(s)} \\ &M \longrightarrow M^{n+} + ne^{-} \end{aligned} \tag{1}$$

where M indicates a metal atom. The former reaction occurs at the cathode while the latter occurs at the anode.

A cathode is an electrode at which reduction occurs and positive ions are discharged or negative ions formed. An anode is an electrode at which oxidation occurs. Current which consists of electrons flowing from an electrode to the electrolyte solution is called a reduction or cathodic current. When there is no current flowing through the cell, the cathode potential eventually reaches a value which is known as the equilibrium potential (V_{eq}), given by the Nernst equation:

$$V_{eq} = V_s + \frac{R'T}{nF} \ln \frac{C_O}{C_R} \qquad (2)$$

where V_s is the standard potential of O and R ions, (O reduced to R in the solution), R' is the ideal gas constant ($J\ K^{-1}\ mol^{-1}$), T is the absolute temperature, F is Faraday's constant (96485 coulomb/mol) and C_O and C_R are the concentrations of the species O and R respectively.

The standard electrode potential is defined as the potential of the metal in equilibrium with a one molar solution of its ions at 25°C, measured relative to the standard hydrogen electrode (SHE). From a practical point of view the SHE is not convenient; so in routine experiments, an alternative reference electrode is used.

Often it is necessary to apply an external potential (V_{ext}) to drive a reaction. Consider a reaction at the cathode:

$$V_{ext} = V_{eq} + \eta + iR\,. \qquad (3)$$

In this equation V_{eq} is the equilibrium potential of the cathode and i is the current flowing between cathode and anode in the solution, so that iR is the potential drop between the two electrodes due to the solution resistance R. η is the overpotential which is defined as the deviation of the potential from its equilibrium value.

$$\eta = V - V_{eq}\,. \qquad (4)$$

Both iR and η vary with the current but in different ways. In order to achieve a situation in which the variation of i with V is characteristic of the processes at only one electrode (known as the working electrode), a reference electrode (RE), should be used in addition to an electrode used to complete the circuit (known as the secondary electrode). The reference electrode should be placed in the cell quite close to the working electrode (WE). A potentiostat applies a potential difference between the WE and the secondary electrode (SE) such that the potential difference between the WE and RE is kept constant at a specific voltage. This can be done using a DC power supply and a feedback circuit. Since no current is passed through the RE, its potential is constant and it is not polarised during the experiment.

The more positive the standard electrode potential of a metal, the more easily it is reduced and the more noble the metal. For example Cu^{2+} has a more positive standard electrode potential than Co^{2+} or Ni^{2+} so it has a greater tendency to absorb two electrons, be reduced and deposit on the cathode.

For electrodeposition the electrolyte is a conducting solution containing ions of the metals which are to be deposited on the working electrode. It can also contain other ions to decrease the resistance of the solution between the working and secondary electrodes and thereby decrease the potential drop iR, reduce unwanted Joule heating and make the potential distribution between working and secondary electrodes more uniform.

7.3. Electrodeposition of Metals and Alloys

In order to understand the multilayer electrodeposition process it is easier to start by describing the deposition of a single metal from a solution under potentiostatic control. Figure 7.1 shows a cell containing an electrolyte and three electrodes connected to a potentiostat.

The potentiostat applies a potential difference between the WE and the RE. Since the RE is at fixed potential relative to the solution, any change in the potential applied to the working electrode changes the WE potential relative to the solution by the same amount.

The working electrode is the substrate (a metal or a semiconductor) on which the desired material is subsequently deposited. The choice of

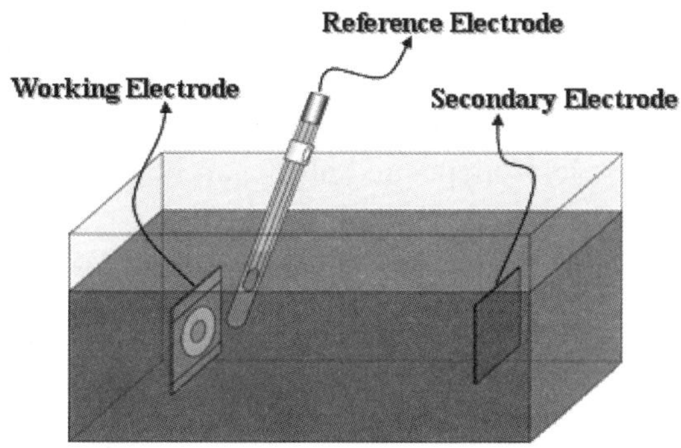

Fig. 7.1. Schematic of a simple electrodeposition cell.

electrode material is determined by the nature of the material deposited. In order to obtain good electrodeposited films, it is essential that the WE is polished (mechanically, chemically or electrochemically) and cleaned to avoid absorbed impurities.

The secondary electrode supplies the current required by the WE. It can be any nonpolarisable metal or other conductor that has no chemical reaction with the electrolyte. It is located far from the WE to gives as uniform a current distribution as possible across the cell in order to give uniform deposition.

The metal ions are reduced to metal atoms and deposit on the WE. Noble metals are reduced at less negative electrode potentials than other metals. Metal atoms in the working electrode may be oxidised and can dissolve in the solution.

The mass of metal (m) deposited on or dissolved from the WE can be calculated using Faraday's law:

$$m = \frac{qM}{nF} \tag{5}$$

where q is the charge passed between WE and SE, F is Faraday's constant and n and M are the valence and atomic mass of the metal respectively.

It is quite possible to electrodeposit more than one metal at a time in the form of an alloy by selecting an appropriate electrolyte containing the different metal ions and applying a suitable potential. The concentrations of the metals in the electrodeposited alloy can be controlled by selecting different ion concentrations in the solution.

The quality of the electrodeposited material is affected by factors including the pH, temperature, current density and external voltage applied between the electrodes.

7.4. Electrodeposition of Multilayers and Superlattices

One approach in developing materials with new physical properties is to artificially structure the materials; and an easy way to do this is by layering them and making a sandwich or multilayer structure rather than the homogenous alloys.

Multilayer thin films consist of successive layers of two or more different materials deposited on a substrate. The thickness of each layer is typically between a few angstrom and hundreds of angstrom and the multilayer may contain up to several hundred layers. A superlattice is a periodic multilayer. Superlattices have been widely studied because their properties may be different from bulk materials.

MBE, sputtering and evaporation are currently the main deposition techniques for making superlattices. Electrodeposition also offers advantages however, including cheapness, a high deposition rate and the simplicity of the apparatus. Electrodeposition was used as a method for fabricating multilayers with thin layers by Brenner in 1939 who grew Cu/Bi multilayers with a bilayer thickness of about 100 nm from a single bath by varying the current between two suitable values.[13] So far most of the research on electrodeposited multilayers has concentrated on metallic multilayers in which magnetic metal layers are sandwiched between two non-magnetic layers, because they exhibit interesting magnetic properties as well as interesting mechanical properties.

Tench and White[18] electrodeposited Ni/Cu superlattices from a single bath, with a period of tens of nanometers which was one order of magnitude less than Brenner's films. Lashmore[20] grew Ni/Cu superlattices from a single bath on (100), (111) and (110) oriented Cu single crystal substrates, and they were highly textured. They also had relatively sharp interfaces, and magnetic measurements confirmed that the quality of these electrodeposited superlattices compared well with films grown by vacuum deposition methods. However there has also been a significant amount of work on other metallic and ceramic multilayers. For example Cohen *et al.*[19] electrodeposited Ag/Pd multilayers and Switzer *et al.*[25] electrodeposited Pb-Ti oxide ceramic superlattices.

Electrodeposition of superlattices can be divided in two major categories, using either the dual bath technique or the single bath technique. Each technique has its own advantages and suffers from some limitations.

7.4.1. *Dual bath electrodeposition*

In dual bath electrodeposition the substrate is transferred between two different baths containing different ions in solution. The idea is to deposit only one metal, either a single element or a particular alloy composition from each electrolyte.

Clearly the substrate should be cleaned before the deposition of each layer in order to avoid contamination of one electrolyte by the other. This can be done by rinsing the substrate in distilled water. However it is hard to keep the substrate absolutely clean and to protect it from undesirable surface reactions occurring during the substrate transfer from one bath to the other. For instance, the thickness of any oxide layers formed during transfer could be significant if the multilayer consists of very thin layers.

Dual bath electrodeposition also requires relatively complex mechanical apparatus which could limit its application. Apart from these restrictions, the dual bath deposition method may be used more in the future because of its flexibility and ability to deposit two different pure metals.

7.4.2. *Single bath electrodeposition*

In the single bath deposition technique, a superlattice structure is formed by periodically varying either the deposition voltage (potentiostatic control) or deposition current (galvanostatic control) between two suitable values. All of the metal ions to be deposited are present in the single electrolyte used. At a less negative potential the more noble metal present in the solution (say metal A) is deposited on the working electrode, forming layer A, while at a more negative potential (depending on the equilibrium potentials of the metal ions present in the solution) the other metal (say metal B) is deposited and layer B formed. Thus it is possible to electrodeposit metallic superlattices by switching the potential between two suitable values. It should be noted that the more noble metal will be deposited as well as the less noble metal at the more negative potential, so that the real composition of layer B is A_xB_{1-x} instead of B and the superlattice structure will consisting alternative layer of A, A_xB_{1-x}, A, A_xB_{1-x} .... Although it is not possible to deposit a layer of pure metal B, the concentration of metal A in the A_xB_{1-x} layer may be kept low if the concentration of metal A in the solution is sufficiently low. Figure 7.2 show a schematic diagram for a superlattice structure.

The single bath technique has been widely used to electrodeposit superlattices containing different metals, but most work has been done on the Ni-Cu/Cu system.[20,22] This systems used for mechanical, electrical and magnetic studies, since Ni and Cu, one a ferromagnet and the other non-magnetic, have the same crystal structure (fcc) and very similar lattice parameters. Electrodeposition of Ni-Cu/Cu multilayers is straightforward, Ni often becoming passive in the electrolyte so that it does not dissolve during the Cu deposition. Electrodeposited Ni-Cu/Cu multilayers have very well-defined layer structures. Nevertheless, this system is not expected to have a GMR as large as that observed in the Co/Cu systems.

Other superlattice systems prepared using single bath method include Co-Cu/Cu and Co-Ni-Cu/Cu. Figure 7.3 shows the high angle X-ray diffraction pattern of an electrodeposited Ni-Cu/Cu superlattice.[26]

Fig. 7.2. A schematic diagram for a superlattice structure.

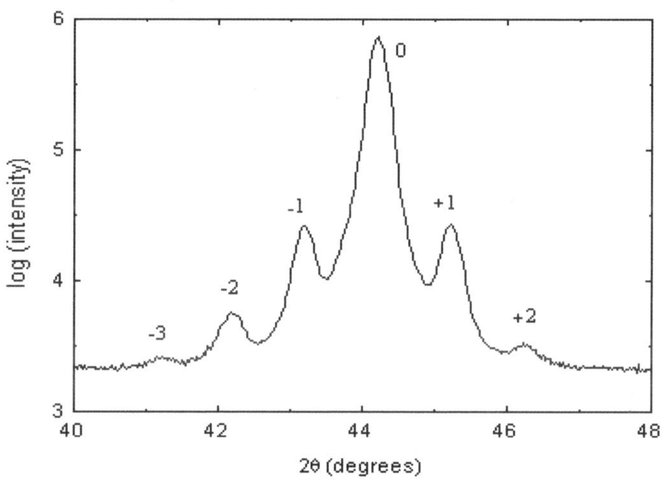

Fig. 7.3. High angle X-ray diffraction pattern measured using Cu Kα radiation for a 100x($Cu_{0.19}Ni_{0.81}$ 6nm/$Cu_{0.79}Ni_{0.21}$ 2nm) electrodeposited multilayer. The position of the central peak labelled '0' is determined by the average lattice spacing in the <111> direction of the film, while the positions of the satellite peaks labelled –3 (third order) to +2 (second order) are determined by the period of the multilayer.[26] [Reproduced by kind permision of I. Kazeminezhad].

The satellite peaks surrounding the central <111> peak correspond to the period of the composition modulation. A low angle X-ray diffraction pattern (using synchrotron radiation and choosing a wavelength of

1.610Å) of an electrodeposited Co-Ni-Cu/Cu superlattice grown on Cu(100) single crystal[27] is shown in Fig. 7.4. Observation of up to two orders of superlattice peaks indicates the existence of well defined superlattice structure.

Dissolution of the less noble metal during the deposition of the more noble metal is a problem which can restrict the applicability of the single bath method and make it difficult to control the thickness of each layer. This can be minimised by choice of solution and deposition waveform.

One major reason for growing the magnetic/nonmagnetic superlattices is to observe the GMR effect. While GMR in Ni/Cu or Ni-Cu/Cu superlattices is small (only a few percent, and hardly reaches to 5%), in Co/Cu or Co-Cu/Cu superlattices, the effect could be considerably larger. By adding Co ions to the Ni-Cu electrolyte and electrodepositing of Ni-Co-Cu/Cu superlattices, a GMR magnitude of 25% at room temperature was achieved by Nabiyouni *et al.*,[28] the result is shown in Fig. 7.5.

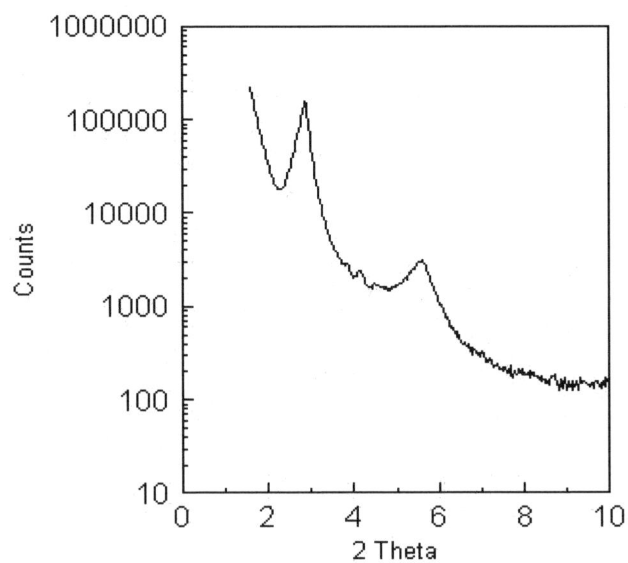

Fig. 7.4. Low angle X-ray diffraction from an electrodeposited super lattice.[28]

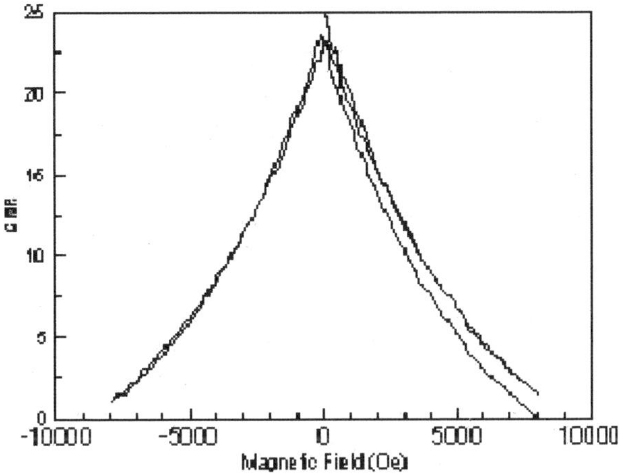

Fig. 7.5. Transverse MR for a superlattice consisting of 100 repeats of nominally Co-Ni-Cu(30Å)/Cu(7Å) grown on a polycrystalline Cu substrate.[28]

The problem with growing pure Co/Cu superlattices by electrodeposition is that the deposition of Co is a more reversible reaction than that of Ni. The consequence of this is that at the potential for the deposition of the Cu layer the Co in the Co-Cu layer will dissolve back into the electrolyte giving a rough and uneven interface between layers. By depositing the Co as a Ni-Co-Cu alloy this dissolution reduces and has less effect on the quality of the layers. This effect of one metal depositing preferentially to another even though they have similar standard potentials is known as anomalous co-deposition and is thought to be related to an increase in hydroxide concentration at the electrode.

7.4.3. *Electrodeposition of metallic thin films onto semiconductor substrates*

A problem that may have limited the utilisation of electrodeposition in modern devices is the requirement that the substrate has to be conducting which in most cases means metallic. Metals, are not, however the generally preferred substrate for thin film growth due to their physical and electrical properties. Physically they are soft which makes it very difficult to prepare a high quality defect free flat surface for deposition, while if the electronic transport properties of the film are of interest these will be short-circuited by the low resistance of the substrate.

A semiconducting substrate has the advantage of allowing resistivity measurements to be made without having to remove the film, as the semiconductor has a much higher resistivity than the metal as well as being partly isolated from the film by the Schottky barrier formed at the semiconductor/metal interface. Semiconductors can also be prepared with flatter surfaces than metals and results achieved would be directly comparable to those for vacuum techniques which also tend to employ semiconducting substrates.

The choice for choosing a semiconductor as the substrate is usually between GaAs and Si, both of which can be readily obtained as high quality polished wafers as they are commonly used in the electronics industry. Si has the advantage of lower cost but the problem of removing the insulating SiO_2 layer from the surface before deposition made it unattractive. GaAs also has native oxides but these are easier to remove and it had been used in the majority of previous studies of metal electrodeposition on semiconductors. It is also a commonly used substrate in vacuum deposition of multilayers.

The requirement that the substrate must conduct does not preclude intrinsic semiconductors. By using doped semiconductors it is possible to pass enough current through the bulk of the material to allow electrodeposition, although the mechanisms for charge transfer at the solid/electrolyte interface are different between semiconductors and metals. This is especially the case for p-type material where illumination is often required to generate enough minority carrier electrons at the surface to reduce the ions in solution. A detailed discussion of the consequences of using a semiconductor instead of a metal as a substrate for electrodeposition is given by R. Hart.[29]

The growth of epitaxial single crystal thin films on matching semiconductor substrates has been an area of great interest especially where the thin film is ferromagnetic. A number of devices have been proposed and are currently being developed which employ integrated magnetic and semiconductor structures.

7.5. Resistivity in Metals

The electrical resistance of most metals at room temperature (~300 K) is caused by conduction electrons being scattered from lattice phonons, and at liquid helium temperature (~4 K) it is caused by scattering of the conduction electrons from impurity atoms and lattice imperfections.[30] A pure and perfect (ideal) crystal has infinite conductivity (zero resistivity), but when the atoms are displaced from their mean positions due to thermal motion, impurities and vacancies the electrons are scattered and the metal has a finite resistance. The resistance of a metal is proportional to the number of times per second an electron is scattered from one state to another state. In transition metals, resistivity is caused by electrons being scattered from the 4s to the 3d band, because the density of states in the d band is large and therefore electrons will jump more frequently from the s to the d band than from one s band state to another. Therefore all of the transition metals are relatively poor conductors compared to the elements that follow them in the periodic table. For example the conductivity of nickel and palladium (with unoccupied d bands) is much less than the conductivity of the neighbouring elements copper and silver which have occupied d bands. In transition metals the relaxation time is also shorter and the conductivity is smaller than for noble metals in which only s-s transitions can take place.[31]

7.6. Magnetoresistance

Magnetoresistance is defined as a fractional change in the electrical resistance of a material when subjected to a magnetic field,

$$MR = \frac{\Delta R}{R} = \frac{|R_o - R_s|}{R_s} \tag{6}$$

where R_o is the resistance at $H = 0$ and R_s is the resistance at saturation field H_s. At the saturation field, the magnetization of the sample will be saturated. Usually it is more convenient to measure the percentage change in the resistance instead of the resistance change.

$$MR(\%) = \frac{\Delta R}{R} \times 100 = \frac{|R_o - R_s|}{R_s} \times 100(\%) \tag{7}$$

If the applied field is not large enough to saturate the magnetization, the resistance at maximum applied field R_{Hmax}, is used instead of R_s.

7.6.1. *Ordinary magnetoresistance*

When an electric field (E) is applied to a metallic specimen, a current density (j) flows parallel to the electric field, from which the normal electrical resistivity (ρ) can be defined as (ρ = E/j). Now if a magnetic field H is applied in addition to the electric field, the conduction electrons are forced to have helical instead of linear paths between two collisions and the resistivity of the specimen will be increased. This phenomenon is called *ordinary magnetoresistance.* If the magnetic field is parallel to the current density, the fractional change in resistance that occurs is called *the longitudinal magnetoresistance*, and if the applied field is perpendicular to the current density, the fractional change in resistance is called *the transverse magnetoresistance.*

All non-magnetic metals exhibit an increase in their electric resistivities as a magnetic field is applied, and the transverse magnetoresistance is larger than the longitudinal one. In the first approximation the increase in the resistivity is proportional to H^2, but this can be more complicated when a high magnetic field is applied or at low temperature. The simple free electron model can not predict magnetoresistance quantitatively, and to obtain this it is necessary to consider more complicated models. The magnetoresistance itself is a second order effect compared to the normal resistance. The ordinary magnetoresistance has been treated theoretically by a number of authors.[32,33]

7.6.2. *Anisotropic magnetoresistance*

It was about 160 years ago that William Thomson discovered that the resistivity of nickel and iron changes when they are magnetised.[34] In ferromagnetic materials there is a decrease in the transverse magnetoresistance (current perpendicular to magnetic field), and an increase in the longitudinal magnetoresistance (current parallel to magnetic field), hence it is known as anisotropic magnetoresistance (AMR).

The magnitude of the change in resistance caused by magnetization is usually a few per cent and rarely reaches 2% at room temperature.

For a ferromagnetic specimen which is subjected to a magnetic field the resistivity $\rho(\xi)$ is given by[35]:

$$\rho(\theta) = \rho_{\perp} + \Delta\rho \cos^2(\theta) \quad (8)$$

where θ is the angle between the magnetization M and current density **j**, and $\rho_{\perp}$ and $\rho_{\parallel}$ are the resistivity of the specimen when M is perpendicular and parallel to j respectively. The initial difference $\Delta\rho$ is the magnitude of the anisotropic magnetoresistance;

$$\Delta\rho = \rho_{\parallel} - \rho_{\perp} \quad (9)$$

AMR has been observed in thin films as well as in bulk materials. The magnitude of the AMR in thin films depends on the thickness, grain size and film surface conditions.[36]

Large AMR values (up to 8.2% at 4.2 K in MBE-grown Co/Ni multilayers) have been reported in Co/Ni (ferromagnetic/ferromagnetic) multilayer thin films.[37] A review of AMR in bulk and thin film ferromagnetic 3d alloys is given by McGuire and Potter.[35]

7.7. Giant Magnetoresistance (GMR)

The phenomenon of giant magnetoresistance (GMR) has been observed in many multilayer structures of the form $n[(F(t_F \text{ Å})/NM(t_{NM} \text{ Å})]$ in which n is the number of bilayers, F is a ferromagnetic transition metal such as Fe, Co, Ni or one of their alloys, NM is a non-magnetic transition or noble metal such as Cr, Cu, Ag, Au, Pt and t_F and t_{NM} refer to the thickness of the magnetic and non-magnetic layer (in Å) respectively. The first observation of GMR was reported by Baibich *et al.* who found the resistivity of (001)Fe/(001)Cr superlattices prepared by molecular beam epitaxy (MBE) on (001)GaAs decreased as an external magnetic field was applied. The percentage change in resistivity increased with decreasing temperature, so that at 4.2 K the resistivity decreased by almost a factor of 2 in a magnetic field of 2T. Nearly at the same time, and independently, Binasch *et al.*[38] found that the resistivity of Fe-Cr-Fe sandwiches prepared by MBE on (110)GaAs decreased when a magnetic field was applied parallel to the film planes. The effect was much stronger than the usual anisotropic magnetoresistance, measured for a single Fe layer with the same thickness, (25 nm) which was prepared in similar growth conditions in order to enable comparison.

In contrast to AMR where the transverse MR decreases and the longitudinal MR increases as a magnetic field is applied, in GMR both the transverse and longitudinal MR decrease with increasing magnetic field. In most cases the magnitude of the transverse MR is larger than the longitudinal one, due to an anisotropic component in addition to the GMR.

It was noted earlier that the resistivity of metals is caused by scattering of conduction electrons. The amount of scattering and hence the resistivity is related to whether the scattered electrons are spin moment up (majority spin) or spin moment down (minority spin) i.e. whether their spin moments are parallel or antiparallel to the magnetisation. Electric current is carried by both spin moment up and spin moment down electrons which may be considered as moving in two different channels with associated resistivity of $\rho_\uparrow$ and $\rho_\downarrow$ respectively. The total current is therefore the sum of the currents carried by spin moment up and spin moment down electrons. In non-magnetic metals, the two currents and hence the resistivities of the two channels are equal, but in ferromagnetic metals and magnetic multilayers they are different.[39,40]

The effective resistivity ρ, can be determined by adding the spin moment up resistivity $\rho_\uparrow$ and the spin moment down resistivity $\rho_\downarrow$ in parallel

$$\rho = \frac{\rho_\uparrow \rho_\downarrow}{\rho_\uparrow + \rho_\downarrow}. \qquad (10)$$

It is generally accepted that GMR arises from spin dependent scattering with electrons of one spin moment scattered more strongly than those of the other. The fact that for the ferromagnetic transition metals the spin moment down electrons (minority spins) are usually scattered more strongly than the spin moment up electrons (majority spins), can be understood in term of the density of states at the Fermi level. In non-magnetic metals and in the absence of an external magnetic field, the densities of states for spin moment up and spin moment down electrons are equal. Application of a magnetic field splits the spin moment up and spin moment down bands. In ferromagnetic metals, however, the spin moment up and spin moment down bands are split due to the local magnetization, and the result is an asymmetry between the scattering of majority and minority spin conduction electrons. Figure 7.6 shows a schematic diagram of the density of states of the s and d bands in a ferromagnetic transition metal. The spin moment down electrons have more unoccupied d states available to be scattered into, than the spin moment up electrons, and hence they are scattered more strongly than the spin moment up electrons.

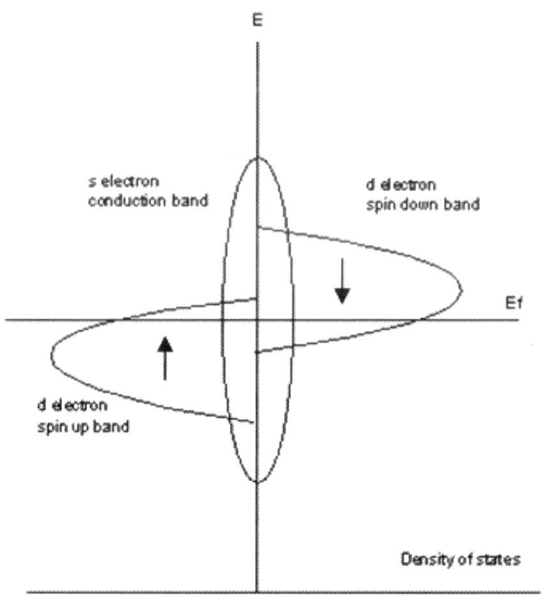

Fig. 7.6. A schematic diagram of the density of state for s and d bands in a transition metal (redrawn from Ref. 28).

The resistivity of a magnetic multilayer for each conduction channel (spin moment up or spin moment down) depends on the relative alignment of the magnetizations of neighbouring layers. If the magnetizations of successive magnetic layers are antiparallel Fig. 7.7(a), each spin moment direction is alternately strongly scattered when the spin moment direction and the magnetization are antiparallel, and weakly scattered when the spin moment direction and the magnetization are parallel. The effective resistivity of two successive ferromagnetic layers in an antiparallel configuration can be written as:

$$\frac{1}{\rho} = \frac{1}{\rho_\uparrow + \rho_\downarrow} + \frac{1}{\rho_\downarrow + \rho_\uparrow} \tag{11}$$

so $\rho \sim \frac{1}{2}\rho_\downarrow$ (assuming $\rho_\uparrow << \rho_\downarrow$) which is still large. If the magnetizations of successive layers in the multilayer specimen are parallel, (due to application of a magnetic field) spin moment up electrons are hardly scattered whereas spin moment down electrons are very strongly scattered Fig. 7.7(b). Thus the effective resistivity of successive layers in parallel alignment may be written as

$$\frac{1}{\rho} = \frac{1}{\rho_\uparrow + \rho_\uparrow} + \frac{1}{\rho_\downarrow + \rho_\downarrow} \tag{12}$$

so $\rho \sim 2\rho_\uparrow$ which is quite small.

Thus in magnetic multilayers, if the magnetizations of the magnetic layers are initially antiparallel to each other, application of an external magnetic field can rearrange them to be parallel to the field and decrease the resistivity of the sample.

This is the origin of GMR, as first proposed by Baibich *et al.* The discussion is only correct if the thicknesses of the individual layers are much less than the mean free path of the conduction electrons.

7.8. Oscillatory GMR in Superlattices

After the discovery of GMR and the attribution of the effect to the relative arrangement of the magnetization in the different magnetic layers, Parkin *et al.*[41,42] discovered that the magnetic coupling in the Co/Cu, Co/Ru Co/Cr and Fe/Cr systems oscillated between

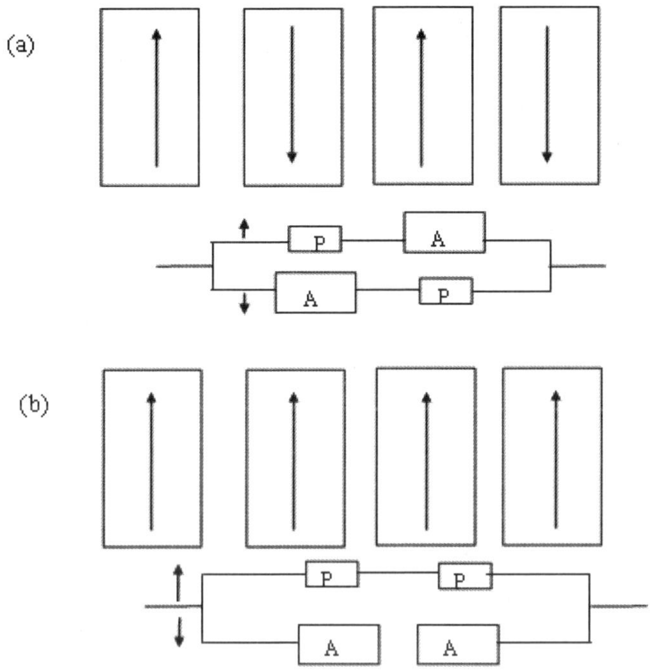

Fig. 7.7. Alignment of the magnetization in superlattices (a) in the antiparallel configuration and (b) in the parallel configuration. In each magnetic layer some of the electrons are strongly scattered (high resistance), and some are weakly scattered (low resistance).[28]

antiferromagnetic and ferromagnetic as the non-magnetic layer thickness was increased. GMR in Co/Cu superlattices also oscillates as a function of the Cu layer thickness. The GMR had a maximum value when the magnetic layers were antiferromagnetically coupled and was reduced (and even vanished) for ferromagnetic coupling.

Oscillations in the maximum percentage change in MR were observed later in many different multilayer and sandwich systems, grown by MBE or sputtering. For example, oscillatory GMR was then observed in different systems such as Co/Cu,[43] Fe/Mo,[44] Fe/Cr/Fe,[45]

Fe/Cu,[46] Co/Ag,[47,48] $Ni_{81}Fe_{19}$/Cu,[49] Fe/Cr,[50,51] Ni-Co/Cu,[52,53] and NiFe/Cu/CoFe/Cu.[54] However, while sputtered Co/Cu superlattices grown on (100) and (111)-orientation substrates exhibit oscillatory behaviour, the GMR in MBE grown superlattices on (111)-oriented substrates decreases continuously as the spacer layer thickness is increased.[49,55]

Oscillatory behaviour in electrodeposited superlattices was first reported by Lashmore *et al.*[56] who observed at least two peaks at Cu layer thicknesses of about 8 and 24Å. Hua *et al.*[57] reported the observation of an oscillation in the GMR for electrodeposited Co-Ni-Cu/Cu superlattices. The first peak that was centred at a copper thickness of ~10Å has been attributed to AFM coupling analogous to that observed in many Co/Cu and Co-Ni/Cu sputtered superlattices.[52,58,59]

Alper *et al.*,[60] who first reported the observation of GMR in electrodeposited superlattices, did not observe the oscillatory behaviour in their Co-Ni-Cu/Cu superlattices. They attributed the lack of oscillation in their samples to fluctuations in the space layer thickness which could lead to partial contact of successive ferromagnetic layers, and therefore lead to their being ferromagnetically coupled.

Electrodeposition of Co-Cu/Cu superlattices was first reported by Bird and Schlesinger,[61] who observed a remarkable 55% change in MR for a superlattice with a Cu layer thickness of ~8Å at a magnetic field of 0.45T. They also reported observation of oscillatory GMR in their Co-Cu/Cu superlattices, with three peaks observed in the graph of GMR as a function of Cu layer thickness. Bird and Schlesinger also reported a maximum percentage change in GMR as large as 7% for a series of electrodeposited Ni-Cu/Cu superlattices, with varying Cu layer thickness.

This result is surprising because it is nearly two times larger than that found by Lashmore *et al.*[56] Oscillatory GMR with three peaks in good agreement (in peak positions) with those reported by Lashmore *et al.*,[56] was observed in this series of superlattices. Nabiyouni and Schwarzacher[5] observed oscillatory behaviour for electrodeposited Co-Ni-Cu/Cu superlattices grown on Cu(100) single crystal substrate, but they did not observe any oscillatory behaviour for the superlattices grown

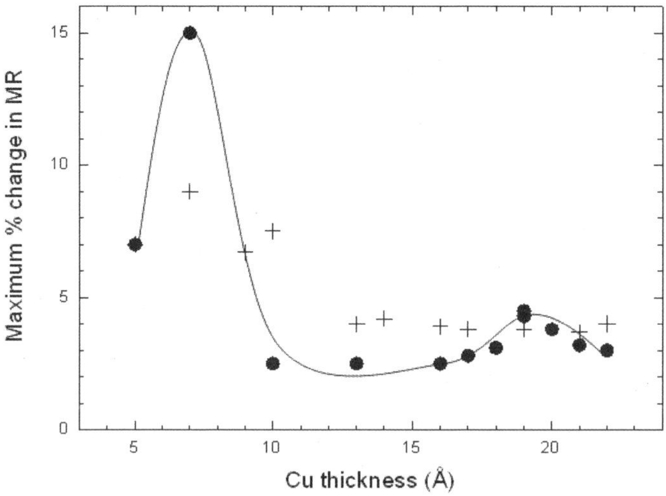

Fig. 7.8. Dependence of the maximum percentage change in 'transverse' (current predominantly perpendicular to the applied field) magnetoresistance on Cu layer thickness for electrodeposited Co-Ni-Cu/Cu superlattices grown on Cu (100) (●) and Cu (111) (+). The Co-Ni-Cu layers had a nominal thickness of 30Å. The solid line is a guide to the eye.[28]

on (111)-oriented single crystal and (100)-textured polycrystalline copper substrates. The result is illustrated in Fig. 7.8.

7.9. Research on GMR

Since 1988 when GMR was discovered, (Babich *et al.* and Binasch *et al.*) thousands of papers have been published on GMR in different multilayer, superlattice and sandwich systems. Araki and Shinjo[47] soon confirmed Baibich's results by growing Fe/Cr multilayers on MgO and glass substrates. The significant step of the observation of GMR in different multilayer systems was made by Parkin *et al.*,[41] who introduced two new metallic superlattice systems that exhibited the GMR effect, Co/Cr and Co/Ru. In these systems, as well as in Fe/Cr superlattices, the coupling between the ferromagnetic layers oscillates between antiferromagnetic and ferromagnetic as a function of the spacer layer thickness. The magnitude of the saturation MR is also found to oscillate with the spacer layer thickness. Parkin *et al.*[47] also demonstrated that the GMR amplitude in sputtered superlattices could be even larger than in MBE grown multilayers.

Two different multilayer and sandwich systems, the Fe/Cr[42,62,63] and Co/Cu[64–69] systems have been widely studied because they exhibit a large room temperature GMR as well as the oscillatory behaviour which caused much scientific interest in the origin of GMR. Co/Cu multilayers show the largest saturation MR value at room temperature. Parkin *et al.*[70] sputtered Co(10Å)/Cu(9Å) multilayers with a maximum room temperature GMR of 65% at 10 kOe. These multilayers exhibit a maximum of 115% GMR at 4.2 K. However the largest MR value at liquid helium temperature was reported by Fullerton *et al.*,[71] for sputtered Fe(14Å)/Cr(8Å) superlattices, with a maximum GMR of 150% at 20 kOe.

Although Co/Cu and Fe/Cr superlattices exhibit large GMR, a large magnetic field (usually greater than 1 kOe) is required. The antiferromagnetic coupling between the magnetic layers is so strong that only a large magnetic field can realign the moments from antiparallel to parallel. The coupling is stronger when the non-magnetic spacer layer is ultrathin. There is particular interest, however, in multilayer systems for which GMR is observed at low field, because of their potential applications in read heads for magnetic recording. The field range for a magnetoresistive head is typically from 0–20 Oe. The field sensitivity of a MR system is defined as the fractional change of resistivity per unit change in the applied magnetic field i.e. $(\Delta\rho/\rho\Delta H) \times 100\%$ where ΔH is the field required to change the resistivity by the amount of $\Delta\rho$.

The Co/Cu superlattices with the largest room temperature GMR exhibit a sensitivity as low as 0.01% Oe^{-1}, which is not high enough for this particular application. Using soft magnetic materials as the ferromagnetic layers and increasing the non-magnetic layer thicknesses, it is possible to fabricate multilayer structures which, although they do not exhibit a very large GMR value, are sensitive to small magnetic fields. $Ni_{80}Co_{20}$ and permalloy $Ni_{80}Fe_{20}$ are good candidates for soft magnetic multilayers. Jimbo *et al.*[72] observed a saturation field of as low as 20 or 30 Oe for $Ni_{66}Fe_{16}Co_{16}$(15Å)/Cu(22Å) multilayers with maximum 12% GMR. Relatively thick Cu layers in these multilayers prevent the magnetic layers from being strongly antiferromagnetically coupled and enhance the sensitivity. Further work has been done on soft magnetic multilayers. A sensitivity of 2.5% Oe^{-1} has been achieved by Jimbo *et al.*[73] in spin valve structures. The term 'spin valve' was introduced by Dieny *et al.*[74] In spin valve multilayer systems two uncoupled ferromagnetic layers with in-plane anisotropy are separated by a non-magnetic metal. The first layer is an antiferromagnetic material and there is an exchange coupling between this layer and the first magnetic layer. Dieny *et al.* showed that, the spin valve effect is related to the relative angle between the magnetizations of the two ferromagnetic layers. Electrodeposition of spin valve multilayers has been reported by a number of research groups.[75,76]

GMR was also observed in non-multilayer, but still magnetically heterogeneous media. It is well known that GMR can be observed in both antiferromagnetically coupled and uncoupled layer structures. In uncoupled layer structures, at the coercive field, the magnetizations are randomly oriented. Granular materials consist of ferromagnetic metallic particles dispersed in a non magnetic (metallic or non-metallic) matrix. Berkowitz *et al.*[77] prepared granular Cu-Co alloys in which the uncoupled layers were replaced by single domain magnetic particles in a non-magnetic matrix and thereby showed that GMR is not restricted only to multilayer systems. GMR then was observed in electrodeposited granular systems.[78,79] Fine grains play an important role in obtaining low coercivity for electrodeposited CoNiFe thin films and other soft magnetic materials.[2,80,81] Ross *et al.*[82] showed that the phosphorous content in electrodeposited Ni/NiP multilayers dictated the crystalline sizes. They also showed that in the case of Ni/Cu systems large grain size were observed when the copper layer was deposited well below the copper limiting current density, while nano-size grains were reported when the depositing copper current density was closed to its limiting value. GMR in granular magnetic systems has been invesigated theoretically by Sheng *et al.*,[83] Wiser[84] and Hichey *et al.*[85]

After GMR was discovered in MBE-grown superlattices,[23] a number of groups started to investigate whether it is possible to fabricate superlattices showing GMR using simpler and cheaper techniques. The advantages of electrodeposition over these vacuum deposition techniques encouraged a number of groups to grow superlattices by this method. Alper *et al.*[60] reported the first observation of GMR in electrodeposited superlattices. Their Co-Ni-Cu/Cu superlattices showed 15% GMR at room temperature in a magnetic field of up to 8 kOe. The largest GMR value in electrodeposited superlattices was reported by Bird and Schlesinger[61] who observed a maximum of 55% GMR in Co-Cu/Cu multilayers in a field of 0.45T.

GMR magnitude in the electrodeposited superlattices is highly related to electrochemical growth conditions such as electrolyte pH and temperature, cell geometry and electrodeposition mode (potentiostatic or galvanostatic control or a combination on both potentiostatic and galvanostatic controls). Alper et al.[86,87] studied the effect of electrolyte pH on structural and magnetoresistive properties of Co-Ni/Cu and Ni-Cu/Cu superlattices. They found that for superlattices with the same bilayer and total thicknesses, the GMR magnitude decreases as the electrolyte pH increases. They also found that superlattices grown at high pH, contain much more Cu compare to those deposited at low pH.

Two comprehensive review articles on GMR in electrodeposited films have been given by Schwarzacher *et al.*,[88] and, Bakonyi *et al.*[89]

7.10. Superparamagnetism Contribution to GMR in the Electrodeposited Superlattices

One the most important multilayer systems exhibiting GMR effect is Co/Cu superlattices. As it was explained before, it is not possible to electrodeposit pure cobalt/pure copper multilayers using an electrochemical single bath, but the grown material is Co-Cu/Cu instead. However, by keeping the Cu concentration in the electrolyte as low as possible one would expect to be able electrodeposite Co-rich/pure Cu superlattices. This system is appropriate for studying an imperfect (and in some cases discontinuous) multilayer with the rough magnetic/nonmagnetic interfaces. Under equilibrium condition Co and Cu atoms do not expect to mix and make a homogenous alloy, instead, some separation takes place when Cu is co-deposited during Co deposition, and when Co is dissoluted during Cu deposition. This is the case in all of the electrodeposition modes i.e. potentiostatic,[90] galvanostatic[91] and two pulse plating[92] modes. In spite of consisting magnetic atoms, in some of Co-Cu alloy the Co particle sizes are too small to exhibit ferromagnetic behaviour with a constant direction at the field applied. This region is called superparamagnetic. The superparamagnetism behaviour is generally related to small ferromagnetic entities for which the orientation of the magnetization rapidly fluctuates.[93] In such areas, anisotropy energy is less than thermal energy, so that in the absence of any external magnetic field the magnetization orientation can be randomly fluctuated. Since GMR is known to be related to spin dependent scattering, the scattering of electrons from (or to) the superparamagnetic region as well as scattering of electrons from (or to) the ferromagnetic layer has a contribution to GMR magnitude in the electrodeposited Co-Cu/Cu superlattices. According to the Wiser-Hickey theory,[84,85] for a given temperature there is a distribution of magnetic particle sizes with some particle being in the superparamagnetic regime and the rest in the ferromagnetic regime. With the simultaneous presence of both ferromagnetic (FM) and superparamagnetic (SPM) regions the GMR contain three contributions: (i) scattering of electrons from one superparamagnetic region to the next in the successive magnetic layer (SPM → SMP), (ii) scattering of electrons from one superparamagnetic region to the successive ferromagnetic layer and vice versa (SPM → FM = FM → SPM), and (iii) scattering of electrons from one ferromagnetic layer to the superparamagnetic region in the next successive ferromagnetic layer (FM → FM). The above contributions have different effects on the magnetoresistance. While FM to FM contribution leads to a magnetoresitance curve saturating in a magnetic field as small as 1 kOe, much stronger magnetic field (typically larger than 10 kOe) is required to saturate the contribution of SPM to SPM in the magnetoresistance curve. Attempt to describe the magnetoresistance of electrodeposited Co-Ni-Cu/Cu superlattices was made by Nabiyouni *et al.*[90] Based on the Wiser and Hickey theory on the effect of superparamagnetism on GMR in the granular metals, Bakonyi and Peter[94–98] showed that in electrodeposited Co-Cu/Cu superlattices the field dependence of the magnetoresistance at sufficiently high magnetic field can be successfully described by the Langvin function $L(\mu H/kT)$, with μ denoting to the average magnetic moment of superparamagnetic regions. Bakonyi and Peter described the field dependent of the measured magnetoresistance MR(H) for $H > H_s$ in the two separated terms:

$$MR = MR_{FM} + GMR_{SPM} \times L(\mu H/kT) \quad (13)$$

where;

$$MR_{FM} = GMR_{FM} + AMR \quad (14)$$

Liu *et al.*[95] found that, while the increase in the Cu content of the magnetic layer significantly increases contribution of the GMR_{SPM} in the magnetoresistance, MR_{FM} contribution is a function of Cu layer thickness. Regarding to the Eq. (13), while MR_{FM} is not a very sensitive function of temperature, both GMR_{SPM} and Langvin function $L(\mu H/kT)$, vary as a function of temperature. Peter *et al.*[99] made a comprehensive study on the effect of temperature on the GMR and magnetic properties of electrodeposited Co-Cu/Cu multilayers. They found that both GMR magnitude and coercivity increase as temperature decreases from room temperature down to liquid helium temperature. They attributed this result to the influence of superparamagnetism regime on the magnetic properties and in particular on the GMR value in electrodeposited superlattices.

7.11. General Remarks on Electrodeposited Superlattices

While very large GMR value has been reported in the multilayer and sandwich systems prepared by physical deposition techniques, the GMR magnitude in the electrodeposited superlattices is not high. One of the largest MR value at liquid helium temperature was reported by Fullerton *et al.*,[71] for sputtered Fe(14Å)/Cr(8Å) superlattices, with a maximum GMR of 150% at 20 kOe. On the best knowledge of author the largest GMR value in electrodeposited superlattices was reported by Bird and Schlesinger[61] who observed a maximum of 55% GMR in Co-Cu/Cu multilayers in a field of 0.45T. The fact that the magnitude of GMR in electrodeposited superlattices is almost by a factor of two smaller than which obtained for similar superlattice system prepared by physical deposition methods, can be attributed to the lack of complete antiferromagnetic coupling between two successive magnetic layers.

Very few research groups have observed oscillatory GMR as a function of spacer layer thickness. Nabiyouni and Schwarzacher[5] observed oscillatory behaviour for electrodeposited Co-Ni-Cu/Cu superlattices grown on Cu(100) single crystal substrate, but none was observed for the superlattices grown on (111)-oriented single crystal and (100)-textured polycrystalline copper substrates.

In most cases, the magnetoresistance curves of electrodeposited superlattices, do not saturate at low magnetic fields. They usually require magnetic fields of up to 10 kOe to saturate, while for superlattices grown by vacuum base techniques, only half of this magnetic field is enough to saturate.

Despite some limitations in using electrodeposition technique to fabricate superlattices with GMR properties, still electrodeposition promises a cheap, simple and more importantly a reliable method for preparation of thin films and superlattices.

Acknowledgments

I would like to thank those who to gave me permission to refer to their results. Special thanks to W. Schwarzacher and I. Bakonyi whose review articles provide me a good source of materials in electrodeposition of metallic multilayers. I would also like to thank my PhD student K. Hedayati who carefully and penitently typed, corrected and justified the manuscript to the right format of this book.

References

1. P. C. Andricacos and N. Robertson, *IBM J. Res. Dev.* **42**, 671 (1998).
2. T. Osaka, *Electrochimica Acta* **45**, 3311 (2000).
3. M. Alper, K. Attenborough, V. Baryshev, R. Hart, D. S. Lashmore, and W. Schwarzacher, *J. Appl. Phys.* **75**, 6543-6545 (1994).
4. K. Attenborough, R. Hart, S. J. Lane, M. Alper, and W. S. Schwarzacher, *J. Magn. Magn. Mater.* **148**, 335-336 (1995).
5. G. Nabiyouni and W. Schwarzacher, *J. Magn. Magn. Mater.* **156**, 355-356 (1996).
6. I. Bakonyi, J. Tóth, L. Goualou, T. Becsei, E. Tóth-Kádár, W. Schwarzacher, and G. Nabiyouni, *J. Electrochem. Soc.* **149**, C195-C201 (2002).
7. A. Yamada, T. Houga, and Y. Ueda, *J. Magn. Magn. Mater.* **239**, 272-275 (2002).

8. A. P. O'Keeffe, O. I. Kasyutich, W. Schwarzacher, L. S. de Oliveira, and A. A. Pasa, *Appl. Phys. Lett.* **73**, 1002-1004 (1998).
9. E. M. Kakuno, R. C. da Silva, N. Mattoso, W. H. Schreiner, D. H. Mosca, and S. R. Teixeira, *J. Phys. D: Appl. Phys.* **32**, 1209-1213 (1999).
10. H. Zaman, S. Ikeda, and Y. Ueda, *IEEE Trans. Magn.* **33**, 3517-3519 (1997).
11. I. Bakonyi, E. Tóth Kádár, Á. Cziráki, J. Tóth, L. F. Kiss, C. Ulhaq-Bouillet, V. Pierron-Bohnes, A. Dinia, B. Arnold, K. Wetzig, P. Santiago, and M.-J. Yacamán, *J. Electrochem. Soc.* **149**, C469-C473 (2002).
12. C. L. S. Rizal, A. Yamada, Y. Hori, S. Ishida, M. Matsuda, and Y. Ueda, *Phys. Stat. Sol. (c)* **1**, 1756-1759 (2004).
13. L. Péter, Z. Kupay, J. Pádár, Á. Cziráki, Zs. Kerner, and I. Bakonyi, *Electrochim. Acta* **49**, 3613-3621 (2004).
14. S. M. S. I. Dulal, E. A. Charles, and S. Roy, *Electrochim. Acta* **49**, 2041-2049 (2004).
15. A. Brenner, *Electrodeposition of Alloys: Principles and Practise.* Academic Press, New York (1963).
16. P. C. Andricacos, C. Uzoh, J. O. Horkans, and H. Deligianni, *IBM J. Res. Dev.* **42**, 567 (1998).
17. F. Spitzer, *Z. Elektrochem.* **11**, 345 (1905).
18. W. Blum, *Trans. Amer. Electrochem. Soc.* **40**, 307 (1921).
19. U. Cohen, F. B. Koch, and R. Sard, *J. Electrochem. Soc.* **130**, 1987 (1983).
20. D. Tench and J. White, *Metall. Trans. A.* **15A**, 2039 (1984).
21. J. Yahalom, D. F. Tessier, R. S. Timsit, A. M. Rosenfeld, D. F. Mitchell, and P. T. Robinson, *J. Mater. Res.* **4**, 755 (1989).
22. D. S. Lashmore and M. P. Dariel, *J. Electrochem. Soc.* **135**, 1288 (1988).
23. M. N. Baibich, J. M. Broto, A. Fert, F. Nguyen Van Dau, F. Petroff, P. Etienne, G. Creuzet, A. Friedrich, and J. Chazelas, *Phys. Rev. Lett.* **61**, 2472 (1988).
24. M. Alper, K. Attenborough, R. Hart, J. S. Lane, D. S. Lashmore, C. Younes, and W. Schwarzacher, *Appl. Phys. Lett.* **63**, 2144 (1993).
25. J. A. Switzer, R. P. Raffaelle, R. J. Phillips, C. J. Hung, and T. D. Golden, *Sci.* **252**, 1918 (1992).
26. I. Kazeminezhad, *PhD Thesis, University of Bristol* (2001).
27. G. Nabiyouni and W. Schwarzacher, *J. of Crystal Growth* **e275**, 1259 (2005).
28. G. Nabiyouni, *PhD Thesis, University of Bristol* (1997).
29. R. Hart, *PhD thesis, University of Bristol* (1996).
30. C. Kittel, *Introduction to Solid State Physics*, Wily Eastern, Eight Edition (2004).
31. N. F. Mott, *Proc. Phys. Soc.* **47**, 571 (1936).
32. J. P. Jan, *Solid states Physics, Vol 5*, F. Seitz and D. Turnlbull, Eds. New York, Academic Press (1957).
33. R. S. Allgaier, *Phys. Rev. B* **8**, 4470 (1973).
34. W. Thomson, *Proc. Roy. Soc.* **8**, 546 (1857).
35. T. R. Mc Guire and R. T. Potter, *IEEE. Trans. Magn.* **11 (4)**, 1081 (1975).

36. M. Tondra, D. K. Lottis, K. T. Riggs, Y. Chen, and E. D. Dahlberg, *J. Appl. Phys.* **73**, 6393 (1993).
37. J. M. Gallego, D. Lederman, T. J. Moran, and I. K. Schuller, *Appl. Phys. Lett.* **64**, 2590 (1994).
38. G. Binasch, P. Grünberg, F. Saurenbach, and W. Zinn, *Phys. Rev. B* **39**, 4828 (1989).
39. P. M. Levy, *J. Magn. Magn. Mater.* **140-144**, 485 (1995).
40. A. Barthélémy, A. Fert, R. Morel, and L. Steren, *Physics World* 34 (1994).
41. S. S. P. Parkin, S. Fan, and N. More, *J. Appl. Phys.* **67**, 5931 (1990).
42. S. S. P. Parkin, G. Z. Li, and D. J. Smith, *Appl. Phys. Lett.* **58**, 2710 (1991).
43. D. H. Mosca, F. Petroff, A. Fert, P. A. Schroeder, W. P. Pratt, and R. Laloee, *J. Magn. Magn. Mater.* **94**, L1 (1991).
44. M. E. Brubaker, J. E. Mattson, C. H. Sowers, and S. D. Bader, *Appl. Phys. Lett.* **58**, 2306 (1991).
45. J. Unguris, R. J. Celotta, and D. T. Piercfe, *Phys. Rev. Lett.* **67**, 140 (1991).
46. F. Petroff, A. Barthélemy, A. Hamzic, A. Fert, P. Etienne, S. Lequien, and G. Creuzet, *J. Magn. Magn. Mater.* **93**, 95 (1991).
47. S. Araki and T. Shinjo, *Jpn. J. Appl. Phys.* **29**, L621 (1990).
48. L. Wu, T. Shinjo, and N. Nakayama, *J. Magn. Magn. Mater.* **125**, L14 (1993).
49. S. S. P. Parkin, R. F. Marks, R. F. C. Farrow, G. R. Harp, Q. H. Lam, and R. J. Savoy, *Phys. Rev. B.* **46**, 9262 (1992). S. S. P. Parkin, *Appl. Phys. Lett.* **60**, 512 (1992).
50. P. Grunberg, S. Demokritov, A. Fuss, R. Schreiber, J. A. Wolf, and S. T. Purcell, *J. Magn. Magn. Mater.* **104-107**, 1734 (1992).
51. M. A. M. Gijs and M. Okada, *J. Magn. Magn. Mater.* **113**, 105 (1992).
52. X. Bian, J. Q. Ström-Olsen, Z. Altounian, Y. Huai, and R. W. Cochrane, *Appl. Phys. Lett.* **62**, 3525 (1993).
53. H. Kubota, S. Ishio, and T. Miyazaki, *J. Magn. Magn. Maert.* **126**, 463 (1993).
54. M. T. Kife, J. Bresowar, and Q. Leng, *J. Appl. Phys.* **79**, 4766 (1996).
55. M. J. Hall, B. J. Hickey, M. A. Howson, M. J. Walker, J. Xu, D. Greig, and N. Wiser, *Phys. Rev. B* **47**, 12785 (1993).
56. D. S. Lashmore, Y. Zhang, S. Hua, M. P. Dariel, L. Swartzendruber, and L. Salamanca-Riba, *Magnetic Materials, Processes and Devices*, Edited by L. T. Romankiw and D. A. Herman, *Electrochem. Soc. Symp. Proc.* **94-6**, 205 (1994).
57. S. Hou, C. Harell, L. Trofin, P. Kohli, and C. R. Martin, *J. Am. Chem. Soc.* **126**, 5674 (2004).
58. A. Fert, A. Barthélémy, P. Etienne, S. Lequien, R. Loloee, D. K. Lottis, D. H. Mosca, F. Petroff, W. P. Pratt, and P. A. Schroeder, *J. Magn. Magn. Mater.* **104-107**, 1712 (1992).
59. S. Honda, S. Ohmoto, R. Imada, and M. Nawate, *J. Magn. Magn. Mater.* **126**, 419 (1993).
60. M. Alper, P. S. Aplin, K. Attenborough, D. J. Dingley, R. Hart, S. J. Lane, D. S. Lashmore, and W. Schwarzacher, *J. Magn. Magn. Mater.* **126**, 8 (1993).

61. K. Bird and M. Schlesinger, *J. Electrochem. Soc.* **142(4)**, L65 (1995).
62. G. Oomi, Y. Uwatoko, Y. Obi, K. Takanashi, and K. Fujimori, *J. Magn. Magn. Mater.* **126**, 513 (1993).
63. R. Schad, C. D. Potter, P. Beliën, G. Verbanck, V. V. Moshchalkov, and Y. Bruyneraede, *Appl. Phys. Lett.* **64**, 3500 (1994).
64. M. Uhlemann, A. Gebert, M. Herrich, A. Krause, A. Cziráki, and L. Schultz, *Electrochim. Acta* **48**, 3005-3011 (2003).
65. H. Zhang and W. Wang, *Microsyst. Technol.* **9**, 436-440 (2003).
66. Y. Hayashi, C. G. Lee, B. H. Koo, T. Sato, M. Arita, and M. Masuda, *Phys. Stat. Sol. (a)* **201**, 1658-1661 (2004).
67. L. Péter, Q. X. Liu, Z. Kerner, and I. Bakonyi, *Electrochim. Acta* **49**, 1513-1526 (2004).
68. H. Sasaki, S. Kainuma, K. Takayanagi, K. Hisatake, and C. O. Kim, *J. Magn. Magn. Mater.* **281**, 53-57 (2004).
69. D. K. Pandya, P. Gupta, S. C. Kashyap, and S. Chaudhary, *J. Magn. Magn. Mater.* doi: 10.1016/j.jmmm. 2008.03.020 (2008).
70. S. S. P. Parkin, R. Bhadra, and K. P. Roche, *Phys. Rev. Lett.* **66**, 2152 (1991).
71. E. E. Fullerton, M. J. Conver, J. E. Mattson, C. H. Sowers, and S. D. Bader, *Appl. Phys. Lett.* **63**, 1699 (1993).
72. M. Jimbo, S. Tsunashima, T. Kanda, S. Goto, and S. Uchiyama, *J. Appl. Phys.* **74**, 3341 (1993).
73. M. Jimbo, K. Komiyama, and S. Tsunashima, *J. Appl. Phys.* **79(8)**, 6237 (1996).
74. B. Dieny, V. S. Speriosu, B. A. Gurney, S. S. P. Parkin, D. R. Wilhoit, K. P. Roche, S. Metin, D. T. Peterson, and S. Nadimi, *J. Magn. Magn. Mater.* **93**, 101 (1991).
75. Y. Jiang, S. Yao, and W. Zhang, *Thin Solid Films* **516**, 3210-3216 (2008).
76. K. Attenborough, H. Boeve, J. De Boeck, G. Borghs, and J. P. Celies, *Appl. Phys. Lett.* **74(15)**, 2206 (1999).
77. A. E. Berkowitz, J. R. Mitchell, M. J. Caarey, A. P. Young, S. Zhang, F. E. Spada, F. T. Parker, A. Hutten, and G. Thomas, *Phys. Rev. Lett.* **68**, 3745 (1992).
78. S. Kenane, J. Voiron, N. Benbrahim, E. Chainet, and F. Robaut, *J. Magn. Magn. Mater.* **297**, 99-106 (2006).
79. G. Nabiyouni and W. Schwarzacher, *J. Magn. Magn. Mater.* **198**, 116-118 (1999).
80. H. Iwasaki, R. Akashi, and Y. Ohsawa, *J. of Appl. Phys.* **73**, 8441 (1993).
81. H. Matsubara and A. Yamada, *J. Electrochem. Soc.* **141**, 2386 (1994).
82. C. Ross, L. M. Goldman, and F. Spaepen, *J. Electrochem. Soc.* **140**, 91 (1993).
83. L. Sheng, R. Y. Gu, D. Y. Xing, Z. D. Wang, and J.-X. Zhu, *J. Appl. Phys.* **79(8)**, 6255 (1996).
84. N. Wiser, *J. Magn. Magn. Matter.* **159**, 119 (1996).
85. B. J. Hckey, M. A. Howson, S. O. Musa, and N. Wiser, *Phys. Rev B* **51**, 667 (1995).
86. M. Alper, M. C. Baykul, L. Peter, J. Toth, and I. Bakonyi, *J. Appl. Electrochem.* **34**, 841-848 (2004).

87. M. Alper, W. Schwarzacher, and Lane S. L. Lane, *J. Electrochem. Soc.* **144**, 2346 (1997).
88. W. Schwarzacher and D. S. Lashmore, *IEEE Trans. Magn.* **32(4)**, 3133 (1996).
89. I. Bakonyi and L. Peter, Proc. 8th Int. Symp. *On Magnetic Materials, Processes and Divices (206th Electrochemical Society Meeting*, Honalulu, Hawaii, USA, 2004).
90. G. Nabiyouni, W. Schwarzacher, Z. Rolik, and I. Bakonyi, *J. Magn. Magn. Mater.* **253**, 77-85 (2002).
91. Á. Cziráki, J. G. Zheng, A. Michel, Z. S. Czigány, G. Nabiyouni, W. Schwarzacher, E. Tóth-Kádár, and I. Bakonyi, *Z Metallkde* **90**, 278-283 (1999).
92. Q. X. Liua, L. Peter, J. Toth, L. F. Kiss, A. Cziraki, and I. Bakonyi, *J. Magn. Magn. Mater.* **280**, 60-74 (2004).
93. B. D. Cultity, *Introduction to Magnetic materials*, Addison-Wesley, Reading (1972).
94. I. Bakonyi, L. Péter, Z. Rolik, K. Kiss-Szabó, Z. Kupay, J. Tóth, L. F. Kiss, and J. Pádár, *Phys. Rev. B* **70**, 054427 (2004).
95. Q. X. Liu, L. Péter, J. Pádár, and I. Bakonyi, *J. Electrochem. Soc.* **152**, C316-C323 (2005).
96. Q. X. Liu, L. Peter, J. Toth, L. F. Kiss, A. Cziraki, and I. Bakonyi, *J. Magn. Magn. Mater.* **280**, 60-74 (2004).
97. I. Bakonyi, J. Tóth, L. F. Kiss, E. Tóth-Kádár, L. Péter, and A. Dinia, *J. Magn. Magn. Mater.* **269**, 156-167 (2004).
98. L. Péter, V. Weihnacht, J. Tóth, J. Padar, L. Pogany, C. M. Schneider, and I. Bakonyi, *J. Magn. Magn. Mater.* **312**, 258-265 (2007).
99. L. Péter, Z. Rolik, L. F. Kiss, J. Tóth, V. Weihnacht, C. M. Schneider, and I. Bakonyi, *Phys. Rev. B* **73**, 174401/1-10 (2006).

Chapter 8

INTRODUCTION TO SPIN TRANSFER TORQUE

C. Baraduc, M. Chshiev* and U. Ebels†

SPINTEC, CEA/CNRS/UJF/INPG
CEA/INAC, 17 rue des martyrs
38054 Grenoble cedex 9, France
E-mail: claire.baraduc@cea.fr

Spin transfer torque corresponds to the interaction of a spin polarized electronic current with the local magnetization: magnetic moment is transferred from the conduction electrons to the magnetization, resulting in a change of the magnetization orientation.

This chapter introduces the physics of spin transfer torque in magnetic devices. It is meant as a tutorial and provides an elementary discussion of the basic concepts. It is not a review of the extensive theoretical and experimental work done on this subject in these last years.

8.1. Introduction

In 1996, it was theoretically demonstrated both by Berger[1] and Slonczewski[2] that a spin polarized electronic current can transfer magnetic moment to the local magnetization of a ferromagnet. In particular, this *spin transfer* can induce steady-state precession or reversal of the magnetization. In addition to its intrinsic scientific interest, the control of the magnetization state by an electrical current, instead of an applied field, may lead to several possible applications.

In order to appreciate this breakthrough, let us say a few words on spin electronics, also called "spintronics". Spintronic devices are usually composed of two ferromagnetic layers separated by a non magnetic spacer. Depending on the nature of the spacer, one consider two broad classes: spin valves when the spacer is a normal metal layer, and magnetic tunnel junctions when a spacer is an insulator (or semiconductor). In these

*mair.chshiev@cea.fr
†ursula.ebels@cea.fr

structures, one ferromagnetic layer is called the *fixed* or the *reference* layer, since its magnetization is pinned by magnetic coupling to an antiferromagnet, and the other is the *free* layer: the free layer magnetization can be switched to obtain either parallel or antiparallel alignment of the two magnetizations, which corresponds to two different resistance states of the device. This magnetoresistance has led to various applications such as field sensors (in particular read-heads for hard-disk) or two-state devices like magnetic memory (MRAM). Concerning MRAMs, significant progress can be achieved thanks to the spin transfer torque effect, since commutation between the two states of the memory can be directly driven by an electrical current.

Since it was theoretically predicted, the concept of spin transfer torque attracted a considerable interest and was tested on a large variety of devices, either metallic multilayer or magnetic tunnel junctions, with different geometries: mechanical or lithographically defined point contact, nanopillars, electrochemically grown nanowires etc. All these devices present the same essential characteristic: they have nanometric cross-sectional area with typical dimension about 100 nm. Such a small size is necessary since spin transfer torque can only be observed for very large current density of the order of 10^6 to 10^8 A.cm^{-2}. Soon after its theoretical prediction, a pioneer work showed experimentally the existence of spin transfer torque.[3] Then unambiguous demonstrations of current induced magnetization switching between the parallel and antiparallel states were performed on metallic systems during the years 2000-2001.[4,5] Similar results were later obtained on magnetic tunnel junctions.[6,7] Moreover, spectral measurements showed that spin transfer torque induces also steady-state precession of the magnetization at frequencies of a few GHz,[8,9] resulting in oscillations between the high and the low resistance level.

In the following, we will first introduce the concept of spin transfer torque. Then we will get more physical insight by considering this phenomenon at the microscopic level. Finally, we will address the influence of spin transfer torque on magnetization dynamics.

8.2. Spin Transfer Torque

In this section, we present a simple and intuitive approach of the physical underlying mechanism of spin transfer torque using a macrospin description. Let us consider two ferromagnetic layers F_1 and F_2, separated by a thin spacer (either a metallic layer or a tunnel barrier) with their magnetizations misaligned by an angle θ (see Fig. 8.1(a)). An electrical current passes through the structure, perpendicularly to the interfaces, with electrons flowing from F_1 to F_2. Electrons flowing in F_1 are spin polarized along the direction of the magnetization $\vec{M}_1$ of this ferromagnetic layer, whereas electrons flowing in F_2 are spin polarized along the direction of $\vec{M}_2$. So in the time span between crossing the imaginary surfaces A and B, the spin polarization of electrons changes. At this stage, we should distinguish between spin polarization and magnetic moment: the electron magnetic moment is opposite to its spin-angular momentum $\vec{\mu} = g_e \mu_B \vec{S}/\hbar$ with $g_e \approx -2$, where S is the spin, and likewise in transition-metal ferromagnets the magnetization is generally opposite to the spin density $\vec{M} = g\mu_B \vec{s}/\hbar$ where g is typically in the range -2.1 to -2.2.

Let us consider the fictitious box defined by the surfaces A and B: incoming electrons have a magnetic moment $\vec{\mu}_1$ parallel to $\vec{M}_1$ and outgoing electrons a magnetic moment $\vec{\mu}_2$ parallel to $\vec{M}_2$. Since the incoming and outgoing magnetic moment is not the same, some magnetic moment is transferred to the system per time unit. However magnetization vectors have a fixed magnitude, so any temporal variation of $\vec{M}$ has to be perpendicular to $\vec{M}$:

$$\frac{d|\vec{M}|^2}{dt} = 2\vec{M} \cdot \frac{d\vec{M}}{dt} = 0\,. \tag{8.1}$$

Consequently only the component perpendicular to $\vec{M}$ of the electrons magnetic moment can be transferred to the bulk magnetization. The total transferred moment $\vec{T}$ is then decomposed in two parts $\vec{T}_1$ and $\vec{T}_2$ (Fig. 8.1(b)). One part is transferred to $\vec{M}_1$ and the other to $\vec{M}_2$. The

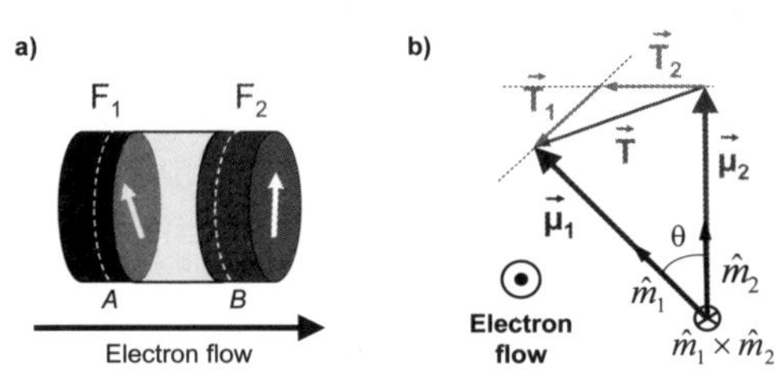

Fig. 8.1. a) The studied device composed of two ferromagnetic layers F_1 and F_2 separated by a spacer. The virtual planes A and B are far enough from the N/F interface so that the current is perfectly spin polarized along the ferromagnetic layer magnetization. b) Schematics of the ingoing ($\vec{\mu_1}$), outgoing ($\vec{\mu_2}$) and transferred ($\vec{T}$) magnetic moments, drawn in the plane that contains the magnetizations of the two layers. In the particular case studied here, this plane is parallel to the N/F interfaces.

magnetic moment transferred to $\vec{M}_1$ is then a vector perpendicular to $\vec{M}_1$ which lies in the plane defined by $(\vec{M}_1, \vec{M}_2)$. Since the unit vector normal to this plane is $\hat{m}_1 \times \hat{m}_2$,[a] $\vec{T}_1$ and $\vec{T}_2$ are necessarily written as: $\vec{T}_{1,2} \equiv T_{1,2}\hat{m}_{1,2} \times (\hat{m}_1 \times \hat{m}_2)$. By analogy to magnetization dynamics, where the time derivative of the magnetization is equal to the torque exerted by the magnetic field, this transferred moment per time unit was originally called "pseudo-torque" by Slonczewski.[10] It is now universally called *spin transfer torque*. Nevertheless we must keep in mind that this term is not exactly a torque but stems directly from the flow of spin momentum into the considered volume element.

In the case of a two-magnetic-layers device composed of a fixed and a free layer, the spin torque on the free-layer magnetization due to the misalignment with the fixed layer magnetization is:

$$\vec{T}_{\parallel} = \gamma_0 \frac{a_j}{M_s} \vec{M} \times (\vec{M} \times \hat{p}) \tag{8.2}$$

where M_s is the saturation magnetization and $\hat{p}$ the magnetic polarization of the fixed layer. The prefactor a_j was calculated in the case of a metallic multilayer.[11] The subscript $\parallel$ recalls that the torque is parallel to the plane that contains the two magnetization vectors.

8.3. A Microscopic Picture

We will try now to get more insight on the physics of spin transfer torque at the microscopic level. Again we consider the device composed of two ferromagnetic layers F_1 and F_2 separated by a normal metal spacer N, with their magnetizations misaligned by an angle θ. Let us consider a single electron, initially polarized by F_1 along $\hat{z}'$, arriving at the N/F_2 interface. The incoming electron is described as a plane wave with a wave-vector equal to k. With respect to the quantization axis of the second ferromagnet F_2 ($\hat{z}$), the electron wave-function is a superposition of spin-up and spin-down states:

$$\psi_{in} = \frac{e^{ikx}}{\sqrt{\Omega}} \left[\cos(\theta/2)\,|\uparrow\rangle + \sin(\theta/2)\,|\downarrow\rangle\right]. \tag{8.3}$$

This decomposition corresponds to a geometrical projection on an orthogonal basis, taking into account the fact that $|\uparrow\rangle$ and $|\downarrow\rangle$ define an orthogonal basis in spin-space, but with a π angle in geometrical space. Geometrical angles must then be divided by 2 when dealing with spin-space.

[a]Where $\hat{m}_i$ is a unit vector along $\vec{M}_i$ for $i = 1, 2$.

Following the Stoner model, spin-up and spin-down electrons are considered as free electrons but with a different energy potential: V_+ for spin-up electrons, and $V_- = V_+ + \Delta$ for spin-down electrons, where Δ is the exchange energy. So, depending on its spin, an incoming electron will experience a different potential step at the N/F_2 interface (Fig. 8.2). For simplicity, we assume that $V_+ = 0$. In that case, a spin-up electron is always transmitted whereas a spin-down electron may be either partially or fully reflected depending whether its energy is larger or smaller than Δ.

Now let us consider an incident electron polarized along $\hat{z}'$ with an energy larger than Δ. When entering into F_2, the spin-up and spin-down parts of its wave-function are submitted to a different potential, so their velocities will be different. We note $k_\uparrow$ (resp. $k_\downarrow$) the wave-vector of spin-up (resp. spin-down) electrons. By matching wave-functions and their derivatives at the interface, it is easy to calculate transmitted and reflected wave functions:

$$\begin{aligned}\psi_{trans} &= \frac{e^{ik_\uparrow x}}{\sqrt{\Omega}}\cos(\theta/2)\left|\uparrow\right\rangle + \frac{e^{ik_\downarrow x}}{\sqrt{\Omega}}\frac{2k}{k+k_\downarrow}\sin(\theta/2)\left|\downarrow\right\rangle \\ \psi_{refl} &= \frac{e^{-ikx}}{\sqrt{\Omega}}\frac{k-k_\downarrow}{k+k_\downarrow}\sin(\theta/2)\left|\downarrow\right\rangle . \end{aligned} \tag{8.4}$$

The spin flux $\Phi = (\Phi_x, \Phi_y, \Phi_z)$ is defined by $\Phi = \frac{\hbar^2}{2m} Im(\psi^* \sigma \otimes \nabla\psi)$[12] or equivalently by:[2]

$$\begin{aligned}\Phi_+ &= \Phi_x + i\Phi_y = i\frac{\hbar^2}{2m}\left(\psi_\downarrow \frac{d\psi_\uparrow^*}{dx} - \psi_\uparrow^* \frac{d\psi_\downarrow}{dx}\right) \\ \Phi_z &= \frac{\hbar^2}{2m} Im\left(\psi_\uparrow^* \frac{d\psi_\uparrow}{dx} - \psi_\downarrow^* \frac{d\psi_\downarrow}{dx}\right). \end{aligned} \tag{8.5}$$

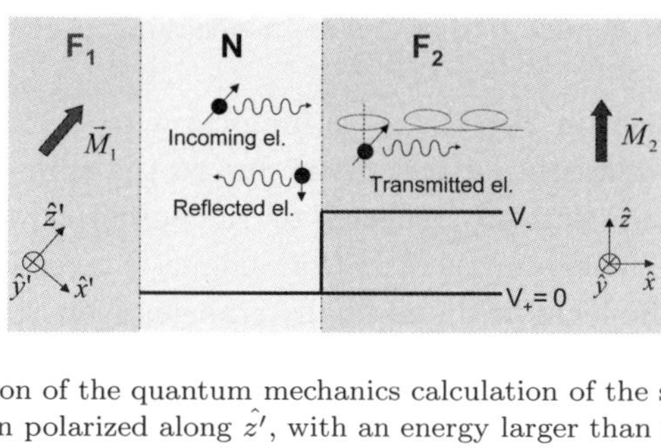

Fig. 8.2. Illustration of the quantum mechanics calculation of the spin transfer torque. An incident electron polarized along $\hat{z}'$, with an energy larger than the Stoner potential step is partially transmitted as a precessing spin and partially reflected as a spin-down electron.

So using the wave-function defined in Eq. (8.3) and (8.4), the spin flux is determined:

$$\Phi_{in} = \frac{\hbar^2}{2m\Omega}(k\sin\theta\hat{x} + k\cos\theta\hat{z})$$

$$\Phi_{trans} = \frac{\hbar^2}{2m\Omega}k\sin\theta\,(\cos[(k_\uparrow - k_\downarrow)x]\,\hat{x} - \sin[(k_\uparrow - k_\downarrow)x]\,\hat{y})$$

$$+\frac{\hbar^2}{2m\Omega}\left[k\cos^2(\theta/2) - k_\downarrow\left(\frac{2k}{k+k_\downarrow}\right)^2\sin^2(\theta/2)\right]\hat{z}$$

$$\Phi_{refl} = \frac{\hbar^2}{2m\Omega}k\left(\frac{k-k_\downarrow}{k+k_\downarrow}\right)^2\sin^2(\theta/2)\hat{z}\,. \tag{8.6}$$

Some points of physics can be illustrated by this example. First, when the energy of the incident electron is larger than V_-, spin flux is continuous at the interface: $\Phi_{in} + \Phi_{refl} = \Phi_{trans}$ at $x = 0$.

The second point is that when the incoming electron penetrates into the ferromagnet, its spin-up and spin-down parts do not travel at the same speed due to the difference between $k_\uparrow$ and $k_\downarrow$. This induces a phase shift between the spin-up and spin-down components of the wave-function. In other words the electron spin precesses around the local field with a very small spatial precession period (a few atomic spacings). So in metallic samples, when many electrons arrive at the interface from different directions, they follow different paths into the ferromagnet and the spin precessions of all these electrons rapidly average to zero due to classical dephasing.[13] Then within about 1 nm, the spin flux transverse to $\hat{z}$ has disappeared and has been transferred to the local magnetization. In magnetic tunnel junctions however, only the directions of incidence close to the perpendicular are selected, since the effective barrier thickness is much larger for grazing incidence. In that case, the classical dephasing between precessing electrons is less efficient. In a pure ballistic case, a decaying oscillating spin flux is expected.[14]

Finally, as shown in Fig. 8.2, it is important to notice that there is a reflected spin current. In our case, it is due to the spin-down part of the wave-function that is partially reflected on the potential step.[b] When these reflected electrons will enter the first ferromagnetic layer F_1, they will start to precess along the local field and finally transfer their transverse moment to the local magnetization. So reflected electrons are responsible for the torque on the first ferromagnetic layer F_1.

[b] Spin-down are traveling backwards, which is equivalent to spin-up traveling forwards. So the positive sign of the reflected spin flux is correct.

We may summarize the situation as the following: the incoming electrons arriving in F_2 tend to tilt the magnetization toward their own spin moment direction which is the direction of the magnetization of F_1. By contrast, the reflected electrons which are mostly polarized antiparallel to $\vec{M}_2$ tend to tilt the F_1 magnetization toward their own spin moment and therefore away from the F_2 magnetization. So the spin torque in both layers has the same direction.

8.4. Transverse Spin Transfer Torque

In a ferromagnetic structure $F_1/N/F_2$, there is a magnetic coupling between the two ferromagnets that is mediated by the conduction electrons, in absence of any bias current. This RKKY coupling is due to conduction electrons below the Fermi energy which travel through the structure along both directions.[15,16] The coupling energy is $E_{ex} = J\vec{M}_1 \cdot \vec{M}_2$. This energy corresponds to a field $\vec{h_i} = \partial E_{ex}/\partial \vec{M}_i \propto \vec{M}_j$ $(i \neq j)$ that exerts a torque $\vec{T}_{i_\perp}$ on the magnetization $\vec{M}_i$, in each ferromagnetic layer $i = 1, 2$:

$$\vec{T}_{i_\perp} = -\gamma_0 \vec{M}_i \times \vec{h_i} \propto -\vec{M}_i \times \vec{M}_j \qquad (i \neq j)\,. \tag{8.7}$$

Here the subscript $\perp$ emphasizes the fact that this torque is transverse to the plane that contains $\vec{M}_1$ and $\vec{M}_2$, contrary to the longitudinal spin transfer torque described in the previous sections. Let us also notice that the torques on both magnetizations are equal and opposite: $\vec{T}_{1_\perp} = -\vec{T}_{2_\perp}$. The same type of coupling also exists in magnetic tunnel junctions and is called *interlayer exchange coupling* or *conservative exchange coupling*.[17] This transverse torque, which exists in absence of bias current, can nevertheless be related to a spin current: electrons traveling from the right and from the left electrodes produce no net charge current but a non-zero spin current as soon as the magnetizations of the two ferromagnetic layers are misaligned. When the structure is biased, this transverse torque increases. The transverse torque exerted on the free layer magnetization is equal to:

$$\vec{T}_\perp = \gamma_0 b_j \vec{M} \times \hat{p} \tag{8.8}$$

where $\hat{p}$ is the polarization of the fixed layer.

In the following, we present a few hints to understand the origin of this transverse torque at the microscopic level. First, we reconsider the problem we studied in the previous section of a spin polarized electron impinging the N/F_2 interface. Now we choose a more realistic description of the ferromagnet, with two non-zero Stoner potential steps V_+ and V_-.

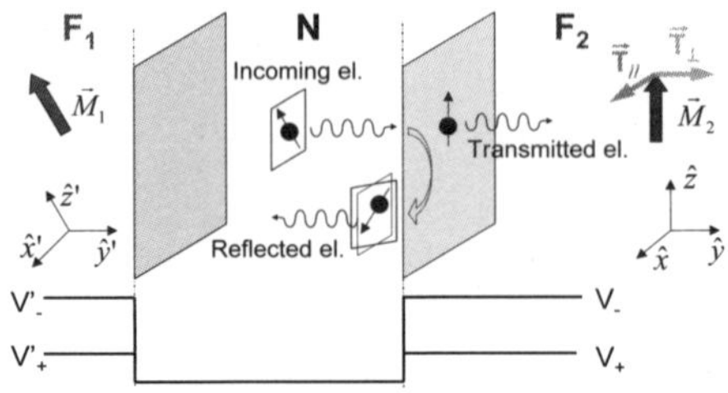

Fig. 8.3. Illustration of a spin-polarized incident electron with an energy between the two Stoner potential steps V_+ and V_-. The spin-down part of the wave function can only penetrate into the ferromagnet as an evanescent wave; it is fully reflected. This process induces a phase shift between the spin-up and spin-down part of the wave-function which is equivalent to a rotation of the spin moment. Note that V'_+ (resp. V'_-) is the Stoner potential of spin-up (resp. spin-down) electrons defined relatively to the quantization axis $\hat{z}'$.

Let us consider an incident electron with an energy comprised between V_+ and V_-. The spin-polarized electron wave function is a superposition of a spin-up and a spin-down states. The spin-up part experiences a small potential step so it is partially transmitted and partially reflected. The spin-down part is fully reflected at the interface since the electron energy cannot overcome the V_- step. Only an evanescent spin-down wave enters the ferromagnet. It can be demonstrated with simple quantum mechanics that this process induces a phase shift between the spin-up and the spin-down part of the reflected electron wave-function. This phase shift corresponds to a rotation of the spin around $\hat{z}$-axis. It may be viewed as a precession of the spin that enters the ferromagnet as a vanishing wave before being reflected. Since the reflected electron gained some magnetic moment along $\hat{y}$, it is responsible for a transverse component Φ_y of the reflected spin flux. Therefore, in a $F_1/N/F_2$ structure, electrons indefinitely reflected by the two N/F interfaces contribute to a transverse torque on both ferromagnets. In the case studied here, let us note that the spin flux may be considered as discontinuous when the evanescent contributions to the wave function are neglected.

From the previous section we already know that electrons with an energy larger than V_- enter the ferromagnet with their spins precessing around the local field. This induces an oscillatory spin flux on both $\hat{x}$ and $\hat{y}$ axis. Due to the various directions of the electrons motion, classical dephasing rapidly averages Φ_x and Φ_y close to the interface. A small amount of average transverse flux may remain and also contribute to the transverse torque. In metallic systems the transverse torque $T_\perp$ does not exceed a few percent of the longitudinal spin transfer torque $T_{\parallel}$. In tunnel junctions where wave vector close to the perpendicular incidence are selected, the transverse torque is much larger.

In the particular case of magnetic tunnel junctions, spin currents in the complete structure $F_1/barrier/F_2$ can be calculated, thus leading to a better understanding of spin transfer torque compared to the description of a single interface. In the limit of a thick barrier and with use of WKB approximation inside the latter, the spin torques can be written as follows:

$$T_{\parallel} \sim \sin\theta \frac{q_0 q_a (k_L^{\uparrow} - k_L^{\downarrow})(k_R^{\uparrow 2} - k_R^{\downarrow 2})(q_0^2 - k_L^{\uparrow} k_L^{\downarrow})}{\mid Den \mid^2} e^{-2\int_0^a q(y)dy}(f_L - f_R)$$
$$\times \left\{(q_a^2 + k_R^{\uparrow} k_R^{\downarrow}) \sin[(k_R^{\uparrow} - k_R^{\downarrow})y'] + q_a (k_R^{\uparrow} - k_R^{\downarrow}) \cos[(k_R^{\uparrow} - k_R^{\downarrow})y']\right\}$$

$$T_{\perp} = T_{\perp}^0 + T_{\perp}^1$$

with

$$T_{\perp}^0 \sim -\sin\theta \frac{q_0 q_a (k_L^{\uparrow} - k_L^{\downarrow})(k_R^{\uparrow 2} - k_R^{\downarrow 2})}{|Den|^2} e^{-2\int_0^a q(y)dy} 2 f_R$$
$$\times \left\{\left[q_0 q_a (k_L^{\uparrow} + k_L^{\downarrow})(k_R^{\uparrow} + k_R^{\downarrow}) - (q_0^2 - k_L^{\uparrow} k_L^{\downarrow})(q_a^2 - k_R^{\uparrow} k_R^{\downarrow})\right] \cos[(k_R^{\uparrow} + k_R^{\downarrow})y']\right.$$
$$\left. - \left[q_0 (k_L^{\uparrow} + k_L^{\downarrow})(q_a^2 - k_R^{\uparrow} k_R^{\downarrow}) + q_a (k_R^{\uparrow} + k_R^{\downarrow})(q_0^2 - k_L^{\uparrow} k_L^{\downarrow})\right] \sin[(k_R^{\uparrow} + k_R^{\downarrow})y']\right\}$$

$$T_{\perp}^1 \sim -\sin\theta \frac{q_0 q_a (k_L^{\uparrow} - k_L^{\downarrow})(k_R^{\uparrow 2} - k_R^{\downarrow 2})(q_0^2 - k_L^{\uparrow} k_L^{\downarrow})}{\mid Den \mid^2} e^{-2\int_0^a q(y)dy}(f_L - f_R)$$
$$\times \left\{\left(q_a^2 + k_R^{\uparrow} k_R^{\downarrow}\right) \cos[(k_R^{\uparrow} - k_R^{\downarrow})y'] - q_a (k_R^{\uparrow} - k_R^{\downarrow}) \sin[(k_R^{\uparrow} - k_R^{\downarrow})y']\right\}$$

where

$$\mid Den \mid^2 = (q_0^2 + k_L^{\uparrow 2})(q_0^2 + k_L^{\downarrow 2})(q_a^2 + k_R^{\uparrow 2})(q_a^2 + k_R^{\downarrow 2});$$

q_0 and q_a are the values of the evanescent wave vector inside the barrier $q(y)$ at the left ($y = 0$) and right interfaces ($y = a$) and $y' = y - a$; $k_R^{\uparrow}$ and $k_R^{\downarrow}$ (resp. $k_L^{\uparrow}$ and $k_L^{\downarrow}$) are the wave-vectors of spin-up and spin-down electrons in the right (resp. left) electrode. Again, these expressions are valid when the exponential factors are small. It is interesting to note that unlike $T_{\parallel}$ and $T_{\perp}^1$ which are proportional to a difference between the left and right Fermi-Dirac functions $(f_L - f_R)$, $T_{\perp}^0$ exists even in the absence of applied voltage and gives origin to the exchange coupling between the FM electrodes.[17] This term corresponds to the already mentioned interlayer exchange coupling. Some more physical insights can be gained from those equations. As already observed in the single interface model, both $T_{\|}$ and $T_{\perp}^{1}$ spatially oscillate with an inverse period $k_R^{\uparrow} - k_R^{\downarrow}$, which corresponds to the spin precession around the exchange field in the ferromagnetic layer. This oscillation is due to interferences between transmitted (or equivalently reflected) majority and minority wave functions. In contrast with that, $T_{\perp}^{0}$ oscillates with a smaller period inversely proportional to $(k_R^{\uparrow} + k_R^{\downarrow})$ which originates from the interference between original waves incident from the right and their reflected parts.[14]

Finally let us conclude on two remarkable properties of the transverse torque which are detailed in a review paper by Ralph and Stiles.[12] First, the transverse torque is always equal and opposite on the two ferromagnets. It can easily be understood by considering the incoming and outgoing magnetic moment in an imaginary box around the structure (see Fig. 8.4). Its borders A and B are far from the interfaces with the spacer, so that at these locations, the electrical current is perfectly spin polarized along the bulk magnetization. Thus, there is no transverse moment coming in or going out. Due to the conservation of angular momentum, the total transverse torque must be zero: $\vec{T}_{1_\perp} + \vec{T}_{2_\perp} = \vec{0}$. The second interesting property is that the transverse torque does not change sign when reversing the bias current in a symmetric structure. The picture represents the transverse torque when electrons flow from left to right (positive flow). The effect of a flow from right to left on F_2 can be imagined by rotating the system: it is the same as the effect of the positive flow on F_1; the torque still points outwards. So, the transverse torque does not depend on the sign of the bias in a symmetric structure. It is the reason why the transverse torque is expected to depend only on V^2 in symmetric magnetic tunnel junctions. However, deviations from such a behavior are expected when asymmetry

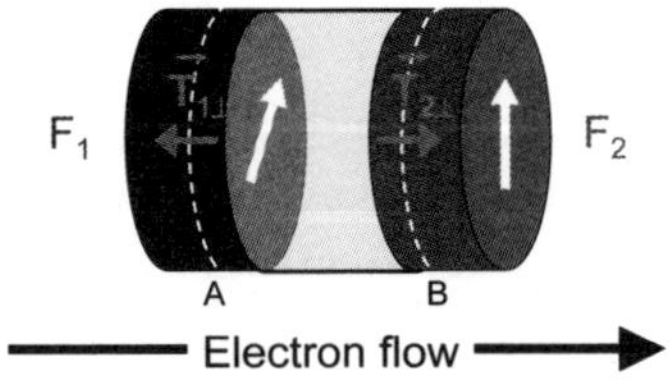

Fig. 8.4. The transverse torque is equal and opposite in the two ferromagnetic layers. It does not depend on the current direction in the case of a symmetric device. The torques here exist only on the surface of these thick ferromagnetic layers. Nevertheless the same considerations still apply for thin ferromagnetic layers in a $N/F_1/N/F_2/N$ structure (cf[12]).

is introduced which may be caused either by using different thicknesses and/or type of FM layers[14,18] or by disorder effects.[12,19]

8.5. Magnetization Dynamics

In this section we describe the effects of spin transfer torque on magnetization dynamics in the macrospin approximation. The magnetization dynamics is described by the Landau-Lifschitz Gilbert (LLG) equation that includes the precession and the damping torques (first two terms of Eq. (8.9)). In order to take the effect of spin transfer into account, the longitudinal spin torque $\vec{T}_{\|}$ is included to the LLG equation (third term of Eq. (8.9)), thus leading to the Landau-Lifschitz-Gilbert-Slonczewski (LLGS) equation:

$$\frac{d\vec{M}}{dt} = -\gamma_0 \vec{M} \times \vec{H}_{eff} + \frac{\alpha}{M_s}\vec{M} \times \frac{d\vec{M}}{dt} + \gamma_0 \frac{a_j}{M_s}\vec{M} \times (\vec{M} \times \vec{p}) \,. \quad (8.9)$$

Here γ_0 is the gyromagnetic ratio, α the natural damping constant, M_s the saturation magnetization and $\vec{H}_{eff}$ the effective field which is given by the variational derivative of the energy density with respect to $\vec{M}$, $\vec{H}_{eff} = -\partial E/\partial \vec{M}$. Including spin torque will change the static and dynamic states of a given system. However it does not change the fundamental properties of the magnetization dynamics. In order to see this, we will discuss the LLGS equation in three steps: 1) Conservative dynamics including the precession term only; 2) Damped dynamics including the precession and damping term; 3) Spin transfer induced dynamics for the full LLGS equation.

Here, we will consider a thin film platelet (see Fig. 8.5), characterized by a uniaxial shape anisotropy field $\vec{H}_u$ and subjected to an external bias field $\vec{H}_b$, applied parallel to the shape anisotropy easy axis. The magnetic energy of this system can be written as the sum of the anisotropy, exchange and demagnetizating energies:

$$E = \frac{M_s H_u}{2}\left(1 - m_x^2\right) - \vec{M}\vec{H}_b + 2\pi M_s^2 m_z^2 \quad (8.10)$$

$$\vec{m} = \frac{\vec{M}}{M_s} = (m_x, m_y, m_z) = (\cos\theta\cos\phi, \cos\theta\sin\phi, \sin\theta) \,.$$

In most cases the general solution of the LLG and LGGS equations cannot be given analytically, in particular if one considers micromagnetic configurations. Solution of LLG(S) then requires numerical approaches.[13,20] However, in order to bring out the essential features of the magnetization dynamics we will assume a homogenous magnetization configuration.

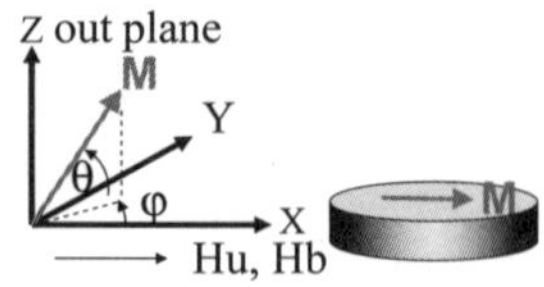

Fig. 8.5. Definition of the (x, y, z) coordinate system and the spherical angles θ, ϕ. Right: the free layer considered is an in-plane magnetized thin film platelet.

8.5.1. *Conservative dynamics*

The precession term is a conservative term that describes the precessional motion of the magnetization along constant energy trajectories. More generally, the static and periodic solutions of the precession term are directly related to the underlying energy surface which is plotted in Fig. 8.6 as a function of the spherical coordinates (θ, ϕ) for the energy given by Eq. (8.10). Obviously, there are three static states (defined by $\partial\vec{M}/\partial t = \vec{0}$) which are the in-plane energy minimum (along the in-plane easy axis), the out-of-plane energy maximum (representing the demagnetization energy cost when rotating M perpendicular to the film plane) and the saddle point (in-plane hard axis). Preparing the system initially in one of these three static states, the system does not change and remain in its initial state (constant energy). When perturbing the system, the magnetization will follow period orbits around either the energy minimum or the energy maximum as indicated by the colored lines in Fig. 8.6(a). These periodic orbits are better visualized in spherical coordinates, shown in Fig. 8.6(b). The precession around the energy minimum is generally called in-plane precession (IPP) since the energy minimum of a thin platelet is in-plane and the precession around the out-of-plane energy maximum is called out-of-plane precession (OPP).[8,13,20,21]

These trajectories require a few comments. 1) The IPP and OPP periodic orbits (or constant energy trajectories) are completely defined by their initial energy E_0 or equivalently by their initial state (θ_0, ϕ_0). 2) With increasing initial energy, the amplitude of the clamshell shaped IPP orbits increases. At the saddle point, the IPP orbit transforms into an OPP orbit whose amplitude, i.e. the out-of-plane magnetization component increases while at the same time the diameter decreases. 3) Due to the non-linearity of the equation of motion, there is a strong dependence of the precession frequency on the precession amplitude, which is different for the IPP and OPP orbits.[8,13] For IPP orbits, the frequency decreases with increasing amplitude since the length of the orbit increases faster than the local ve-

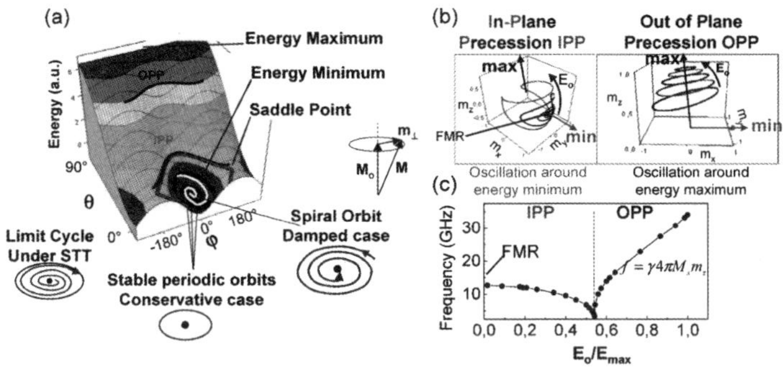

Fig. 8.6. (a) Energy surface as a function of (θ, ϕ) for a thin film platelet. In this example, the bias field amplitude H_b is larger than the uniaxial in-plane anisotropy field H_u, so that there is only one in-plane energy minimum at 0. The three stationary states (energy minimum, maximum and saddle point) are indicated. The IPP and OPP constant energy trajectories are indicated by the colored lines on top of the energy surface. The white spiral indicates the damped motion. Below the energy surface from left to right are schematically indicated a limit cycle, a constant energy trajectory (stable orbit) and a spiral orbit. (b) The IPP and OPP constant energy trajectories as a function of the magnetization components m_x, m_y, m_z. (c) The frequency vs. precession amplitude for the IPP and OPP orbits.

locity. Close to the saddle point, the local velocity goes to zero, explaining the zero frequency at the transition from IPP to OPP. For the OPP orbit the frequency is given by the out-of-plane demagnetization field which is proportional to the out-of-plane magnetization component m_z.

8.5.2. *Damped dynamics*

Adding the damping term to the precession term will change the stability of the three static solutions. Now, only the energy minimum is a stable solution, while the energy maximum and the saddle point correspond to unstable solutions. This means that they are not stable against small perturbations, which will relax the magnetization on a spiral orbit away from the energy maximum or the saddle point towards the stable energy minimum. On the other hand, perturbation of the system around the stable energy minimum leads to small amplitude damped oscillations of the form $m_\perp = m_0 e^{-\lambda t} e^{-i\omega t}$ where $m_\perp$ is a small transverse magnetization component, ω is the oscillation frequency and λ the damping rate ($\lambda > 0$). These small amplitude oscillations correspond to solutions of the linearized LLG equation around the static state $\vec{M}_0$ using $\vec{M} = \vec{M}_0 + \vec{m}_\perp$. These oscillations can be typically excited in ferromagnetic resonance experiments

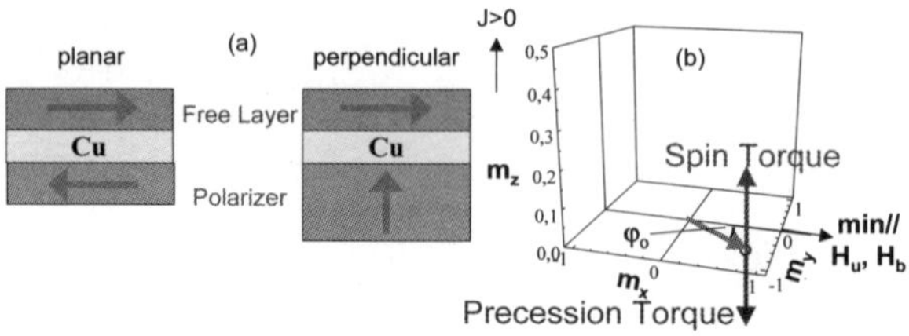

Fig. 8.7. (a) Schematics of a spin valve structure with a planar free layer and a planar (left) or perpendicular (right) polarizer. (b) In-plane rotation ϕ_0 of the magnetization for the perpendicular polarizer configuration.

(FMR) using a small radiofrequency pumping field.[22] Let us note that the same stability analysis around the energy maximum would give a negative damping rate $\lambda < 0$, indicating instability.

8.5.3. *Spin transfer torque induced dynamics*

The static and dynamic properties of the free layer magnetization including the spin torque term will be discussed here for two different configurations of the fixed layer, called planar and perpendicular polarizer, see Fig. 8.7(a).

8.5.3.1. *Static states*

As compared to the conservative and the damped case, the static (stable or unstable) states are not necessarily identical to the stationary states defined in Fig. 8.6. Following the LLGS equation, a static state is now obtained when the precession torque is equal and opposite to the spin torque, by contrast to previous cases where the precession torque is zero. Fortunately, in the planar case, the spin polarization $\hat{p}$ is collinear to the effective field $\vec{H}_{eff}$ of the free layer energy minimum. Thus the precession and spin torques are both zero when the magnetization is directed along its energy minimum ($\hat{p}//\vec{H}_{eff}//\vec{M}$). In the case of the perpendicular polarizer, $\hat{p}$ and $\vec{H}_{eff}$ are not collinear. Spin torque tends to push the magnetization $\vec{M}$ parallel to $\hat{p}$, i.e. out-of-plane. A static state is obtained when the precession torque counterbalances this out-of-plane spin torque: the magnetization then rotates in-plane away from the energy minimum, see Fig. 8.7(b). In this case, one can easily verify that $\vec{M} \times \vec{H}_{eff}$ is pointing out-of-plane since $\vec{H}_{eff} = \vec{H}_b + \vec{H}_u$ is oriented parallel to the x-axis. The corresponding rotation angle ϕ_0 increases with current and is negative for negative current and positive for positive current.[23]

8.5.3.2. *Stability*

We may now wonder whether the above mentioned static states are stable. To answer this question, the dynamics equation is linearized around the calculated static states. This procedure leads to solutions of the kind $m_\perp = m_0 e^{-\lambda t} e^{-i\omega t}$.[21,23–26] The damping rate depends now on the current density, $\lambda = \lambda(J)$ and can be either positive or negative depending on J. This means that the motion is either damped ($\lambda(J) > 0$) and the static state is stable,[c] or excited away from the static state ($\lambda(J) < 0$) and the static state is unstable. Thus upon increasing the current density from zero, an initially stable state can become unstable. The current value for which this occurs is called the critical current J_c and is given by $\lambda(J_c) = 0$. When the initial static state becomes unstable, the magnetization can transit either into another static state or into a dynamic state. For the planar and perpendicular polarizers the instability criteria describing the transition from the static state to a dynamic state are quite different. In the following we illustrate some of the general features of the dynamic state. For simplicity we will discuss in more detail only the planar configuration.

8.5.3.3. *Dynamic states*

At zero or low current, when $\lambda(J) > 0$, a damped motion of magnetization is observed under spin torque. In fact, the energy loss through the natural damping is only partially compensated by the energy gained from spin transfer. Once the current is strong enough, the energy gain through the spin transfer overcomes the energy loss. Magnetization then reaches a limit cycle where the energy gained from the spin torque during each cycle of precession is balanced by the energy lost to damping.[13,25] A schematic of such a limit cycle is shown at the bottom in Fig. 8.6(a). It is important to point out here, that the stabilization of the limit cycle at finite amplitude is a consequence of the non-linear dynamical system described by LLGS.[21,26,28] Non-linear means here that the spin torque and the damping torque depend on the precession amplitude. If this were not the case, then once the spin torque overcomes the damping torque at the static state, no condition for a stable periodic orbit could be realized.

Due to the compensation of the damping losses by spin torque, it seems at hand to suggest that the limit cycles are identical to the (zero current,

[c] In that case, the equilibrium state at zero temperature is unchanged compared to the zero current case. However, at finite temperature, a change of the thermal magnetization fluctuations is observed due to the evolution of the damping rate.[27]

zero damping) constant energy trajectories shown in Fig. 8.6(a). This suggestion is supported by the fact that both the spin torque and the damping torque are small compared to the precession torque and should thus be treated as perturbations.[13,21,25] It is fortunate that this suggestion holds well for the case of the planar and the perpendicular polarizer configuration, due to their special symmetry with respect to the energy minimum and maximum. However, in the general case for an arbitrary spin polarization direction, limit cycles can be stabilized that are quite different from constant energy trajectories. Furthermore, the amplitude of a constant energy trajectory is given by its initial condition (i.e. initial energy), while the amplitude of a limit cycle is defined by the balance between damping and spin torque (energy feedback) and thus by the current amplitude. Note also that the limit cycle is independent of its initial condition. As shown in the schematics of Fig. 8.6(a), preparing the system at any point inside (or outside) the limit cycle will produce a transient spiral motion of the magnetization until it reaches the stable limit cycle. To conclude, the planar polarizer can support limit cycles that are very close to IPP orbits and for larger current that are close to OPP orbits.[8,13,25] In contrast a perpendicular polarizer only supports limit cycles that are very close to OPP orbits[23] since the perpendicular polarizer does not have the symmetry of the energy minimum or of IPP trajectories and therefore cannot support limit cycles of this type. Finally, for a conservative trajectory there is no change of the energy with time, while for the spin current induced limit cycles the energy oscillates around a constant energy value.[20] To conclude, due to this correspondence between limit cycles and IPP and OPP orbits, the frequency of the limit cycles as a function of current follows the dependence given in Fig. 8.6(c), where the horizontal axis has to be replaced by the current. This property is another manifestation of the non-linearity of the LLGS equation.

8.6. State Diagram

In this part we will describe the state diagrams in the case of the planar and perpendicular polarizers. State diagrams define the different static and dynamic states in the current-field plane. As indicated above, the boundary between a static and a dynamic state can be obtained by finding the zeros of the damping rate $\lambda(J)$.

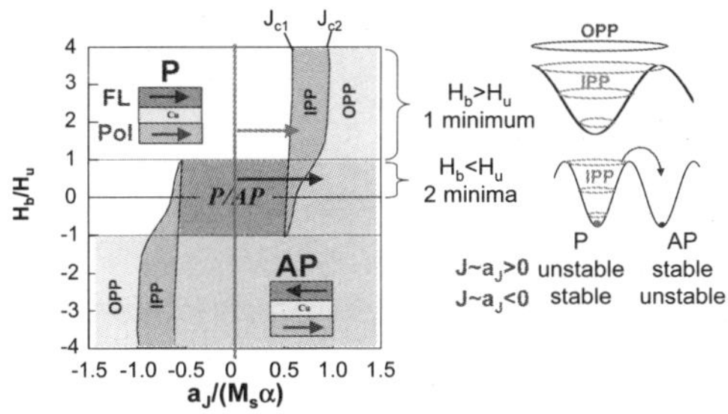

Fig. 8.8. Left: State diagram for the planar polarizer as a function of the normalized bias field H_b/H_u and the current density J. The static states correspond to the parallel (P) and antiparallel (AP) orientation of the free layer magnetization relatively to the polarizer. The critical current separating the static state and the IPP limit cycles is given by J_{c1} and the critical current separating IPP from OPP limit cycles is given by J_{c2}. Right: Schematic indication of the different states as a function of the energy vs.in-plane angle ϕ.

8.6.1. *Planar polarizer*

The full diagram for the planar polarizer configuration is shown in Fig. 8.8.[8,13] This diagram is asymmetric as a function of the sign of current and field. The spin torque acts such that it either enhances the damping rate $\lambda(J)$ for one sign of the current and thus stabilizes the initial state, or for the opposite current sign it counteracts natural damping and decreases $\lambda(J)$ thus destabilizing the initial state. In that case the damped oscillations become less and less damped until $\lambda(J)$ changes sign at the critical current J_{c1}. The static state then becomes unstable and gives way to IPP limit cycles. Upon further increasing the current one has to distinguish two regions for which the bias field H_b is either smaller (two in-plane energy minima) or larger than the anisotropy field H_u (only one energy minimum). When $H_b < H_u$, for increasing current density J, the IPP orbits pass the energy saddle point (in-plane hard axis) and the magnetization transits into the AP state. This essentially means that one can switch the magnetization orientation using a spin polarized current. In the case of $H_b > H_u$, there does not exist a second energy minimum and the magnetization therefore transits from IPP to OPP limit cycles at the saddle point. The corresponding critical current J_{c2} is shown in Fig. 8.8. It is calculated by considering that the total energy gain over one precession period is zero.[13,25] It should be noted that for $H_b > H_u$ (resp. $H_b < -H_u$) and very large positive (resp. negative) currents, the magnetization eventually stabilizes in a static state. However such high current densities cannot be reached experimentally.

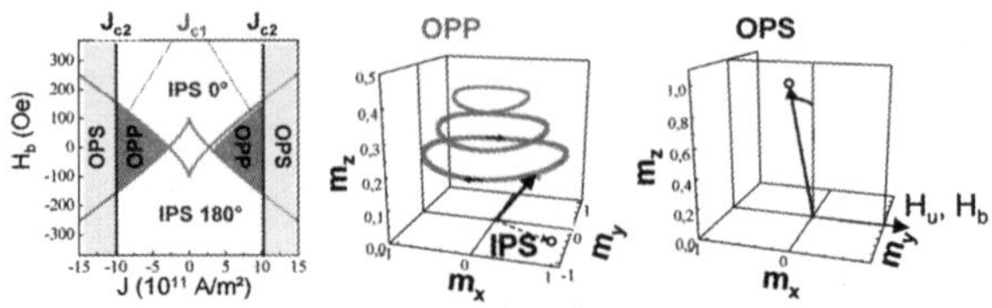

Fig. 8.9. State Diagram for the perpendicular polarizer, with three states: IPS (in-plane stable state), OPP (out-of-plane limit cycle) and OPS (out-of-plane stable state).

8.6.2. *Perpendicular polarizer*

For completeness, we briefly discuss the diagram for the perpendicular polarizer in Fig. 8.9.[23] In this case (i) the static states correspond to an in-plane rotation and (ii) only OPP limit cycles are supported. Here the critical current J_{c1} is reached when the spin torque becomes larger than the precession torque.[23] This transition from the in-plane static (IPS) state to the OPP limit cycles is accompanied by a relatively abrupt jump of the out-of-plane magnetization component m_z, see Fig. 8.9. By contrast to the planar case, the regions of OPP limit cycles are symmetric as a function of current and field. For very large currents, the free layer magnetization $\vec{M}$ is rotated either parallel ($J < 0$) or antiparallel ($J > 0$) to the polarizer direction $\hat{p}$ described as the out-of-plane stable (OPS) state in the diagram of Fig. 8.9. Finally, let us compare the effects of spin torque in the case of the planar and perpendicular polarizer. The planar polarizer has the symmetry of the energy minimum and can thus destabilize the in-plane static state when the spin torque is larger than the damping torque. For the perpendicular polarizer, the role of spin torque and damping torque are inverted, since the polarizer has the symmetry of the energy maximum. Thus the spin torque stabilizes the out-of-plane state, while it is the damping torque that destabilizes it. Thus the OPS state becomes unstable when the current is smaller than a certain critical current J_{c2}.[23,26]

8.7. Conclusions

At large enough current density, spin polarized electrons exert a significant torque on the local magnetization. This spin transfer torque corresponds to a direct transfer of magnetic moment from the conduction electrons to the local magnetization. This phenomenon strongly affects the magnetization dynamics. When the applied current it large enough, spin torque is responsible either for magnetization reversal or steady-state precession.

Acknowledgments

It is a pleasure to acknowledge our colleagues, S. Petit, N. de Mestier, D. Houssameddine, C. Thirion, I. Firastrau, D. Gusakova, L. Buda-Prejbeanu, B. Diény, A. Vedyayev. One of us (C.B.) thanks also M. Stiles for helpful discussions. This work has been supported by the French National Research Agency ANR (SpinChat, MagICO) and the by OSEO/Anvar.

References

1. L. Berger, Emission of spin waves by a magnetic multilayer traversed by a current, *Phys. Rev. B.* **54**(13), 9353–9358 (Oct., 1996).
2. J. C. Slonczewski, Current-driven excitation of magnetic multilayers, *J. Magn. Magn. Mater.* **159**, 1–7 (Jun., 1996).
3. M. Tsoi, A. G. M. Jansen, J. Bass, W. C. Chiang, M. Seck, V. Tsoi, and P. Wyder, Excitation of a magnetic multilayer by an electric current, *Phys. Rev. Lett.* **80**(19), 4281–4284 (May, 1998).
4. J. A. Katine, F. J. Albert, R. A. Buhrman, E. B. Myers, and D. C. Ralph, Current-driven magnetization reversal and spin-wave excitations in $co/cu/co$ pillars, *Phys. Rev. Lett.* **84**(14), 3149–3152 (Apr., 2000).
5. J. Grollier, V. Cros, A. Hamzic, J. M. George, H. Jaffres, A. Fert, G. Faini, J. B. Youssef, and H. Legall, Spin-polarized current induced switching in $co/cu/co$ pillars, *Appl. Phys. Lett.* **78**(23), 3663–3665 (Jun., 2001).
6. Y. Huai, F. Albert, P. Nguyen, M. Pakala, and T. Valet, Observation of spin-transfer switching in deep submicron-sized and low-resistance magnetic tunnel junctions, *Appl. Phys. Lett.* **84**(16), 3118–3120 (Apr., 2004).
7. G. D. Fuchs, N. C. Emley, I. N. Krivorotov, P. M. Braganca, E. M. Ryan, S. I. Kiselev, J. C. Sankey, D. C. Ralph, R. A. Buhrman, and J. A. Katine, Spin-transfer effects in nanoscale magnetic tunnel junctions, *Appl. Phys. Lett.* **85** (7), 1205–1207 (Aug., 2004).
8. S. I. Kiselev, J. C. Sankey, I. N. Krivorotov, N. C. Emley, R. J. Schoelkopf, R. A. Buhrman, and D. C. Ralph, Microwave oscillations of a nanomagnet driven by a spin-polarized current, *Nature.* **425**(6956), 380–383 (2003).
9. W. H. Rippard, M. R. Pufall, S. Kaka, S. E. Russek, and T. J. Silva, Direct-current induced dynamics in $co_{90}fe_{10}/ni_{80}fe_{20}$ point contacts, *Phys. Rev. Lett.* **92**(2), 027201 (Jan., 2004).
10. J. C. Slonczewski, Currents, torques, and polarization factors in magnetic tunnel junctions, *Phys. Rev. B.* **71**(2), 024411 (2005).
11. J. C. Slonczewski, Currents and torques in metallic magnetic multilayers, *J. Magn. Magn. Mater.* **247**(3), 324–338 (Jun., 2002).
12. D. Ralph and M. Stiles, Spin transfer torques, *J. Magn. Magn. Mater.* **320** (7), 1190–1216 (Apr., 2008).
13. M. Stiles and J. Miltat, *Spin transfer torque and dynamics.* vol. Spin Dynamics in Confined Magnetic Structures III (Springer Berlin/Heidelberg, 2006).

14. A. Manchon, N. Ryzhanova, A. Vedyayev, M. Chschiev, and B. Dieny, Description of current-driven torques in magnetic tunnel junctions, *J. Phys.: Condensed Matter.* **20**(14), 145208 (2008). ISSN 0953-8984.
15. M. D. Stiles, Exchange coupling in magnetic heterostructures, *Phys. Rev. B.* **48**(10), 7238 (Sept., 1993).
16. J. C. Slonczewski, Overview of interlayer exchange theory, *J. Magn. Magn. Mater.* **150**(1), 13–24 (Sept., 1995).
17. J. C. Slonczewski, Conductance and exchange coupling of two ferromagnets separated by a tunneling barrier, *Phys. Rev. B.* **39**(10), 6995–7002 (Apr., 1989).
18. M. Wilczynski, J. Barnas, and R. Swirkowicz, Free-electron model of current-induced spin-transfer torque in magnetic tunnel junctions, *Phys. Rev. B.* **77** (5), 054434–9 (Feb., 2008).
19. Z. Li, S. Zhang, Z. Diao, Y. Ding, X. Tang, D. M. Apalkov, Z. Yang, K. Kawabata, and Y. Huai, Perpendicular spin torques in magnetic tunnel junctions, *Phys. Rev. Lett.* **100**(24), 246602–4 (Jun., 2008).
20. Z. Li and S. Zhang, Magnetization dynamics with a spin-transfer torque, *Phys. Rev. B.* **68**(2), 024404 (Jul., 2003).
21. A. N. Slavin and V. S. Tiberkevich, Current-induced bistability and dynamic range of microwave generation in magnetic nanostructures, *Phys. Rev. B.* **72** (9), 094428–5 (Sept., 2005).
22. L. Baselgia, M. Warden, F. Waldner, S. L. Hutton, J. E. Drumheller, Y. Q. He, P. E. Wigen, and M. Maryko, Derivation of the resonance frequency from the free energy of ferromagnets, *Phys. Rev. B.* **38**(4), 2237 (Aug., 1988).
23. U. Ebels, D. Houssameddine, I. Firastrau, D. Gusakova, C. Thirion, B. Dieny, and L. D. Buda-Prejbeanu, Macrospin description of the perpendicular polarizer-planar free-layer spin-torque oscillator, *Phys. Rev. B.* **78**(2), 024436–16 (Jul., 2008).
24. J. Grollier, V. Cros, H. Jaffrès, A. Hamzic, J. M. George, G. Faini, J. Ben Youssef, H. Le Gall, and A. Fert, Field dependence of magnetization reversal by spin transfer, *Phys. Rev. B.* **67**(17), 174402 (May, 2003).
25. G. Bertotti, C. Serpico, I. D. Mayergoyz, A. Magni, M. d'Aquino, and R. Bonin, Magnetization switching and microwave oscillations in nanomagnets driven by spin-polarized currents, *Phys. Rev. Lett.* **94**(12), 127206–4 (Apr., 2005).
26. A. Slavin and V. Tiberkevich, Excitation of spin waves by spin-polarized current in magnetic nano-structures, *IEEE Trans. Magn.* **44**(7), 1916–1927 (2008). ISSN 0018-9464.
27. S. Petit, C. Baraduc, C. Thirion, U. Ebels, Y. Liu, M. Li, P. Wang, and B. Dieny, Spin-torque influence on the high-frequency magnetization fluctuations in magnetic tunnel junctions, *Phys. Rev. Lett.* **98**(7), 077203 (2007).
28. S. M. Rezende, F. M. de Aguiar, and A. Azevedo, Spin-wave theory for the dynamics induced by direct currents in magnetic multilayers, *Phys. Rev. Lett.* **94**(3), 037202–4 (Jan., 2005).

Chapter 9

SPINTRONICS POTENTIAL OF RARE-EARTH NITRIDES

Ben J. Ruck
School of Chemical and Physical Sciences
Victoria University of Wellington
P.O. Box 600
Wellington, New Zealand
*E-mail: Ben.Ruck@vuw.ac.nz**

The rare-earth nitrides are an interesting class of materials that have recently received attention due to their strong coupling between spin and charge degrees of freedom. This Chapter presents a review of our current understanding of these materials. Various means have been devised to produce rare-earth nitrides in both bulk and thin film form, but at present the materials still suffer from rather large defect concentrations that inhibit the study of their fundamental properties. Their electronic and magnetic properties have been studied both theoretically and experimentally. It is now clear that some members of the series are intrinsic ferromagnetic semiconductors, while others may be half-metals in their ground state. Their magnetic interactions remain less well understood and present a fascinating avenue for further study. The present state of knowledge is used to formulate a discussion of the rare-earth nitrides potential for use in spintronics devices.

9.1. Introduction

The rapidly developing field of spintronics, in which the spin degree of freedom is utilised as well as the electric charge, has generated substantial interest in materials that show strong coupling between their electronic and magnetic properties.[1–3] The most striking example is the magnetic read head found in all modern hard disk drives, which is now based on the tunneling magnetoresistance phenomenon,[4] and which has lead to enormous

*Also at MacDiarmid Institute for Advanced Materials and Nanotechnology, Victoria University of Wellington.

increases in magnetic storage capacity. But there also exists the possibility of even more advanced spintronics devices. Numerous proposals have been put forward for creating magnetic random access memories with rapid read/write times and low power consumption,[2,3] in which the information is retained even after the device is powered down. The first such devices are now appearing on the market.[5] Other possibilities include devices in which the electron spin plays either an active or a passive role, such as spin field effect transistors, spin-logic devices, novel polarised photon sources, or even quantum computers.[3,6]

The giant magnetoresistance and tunneling magnetoresistance devices in hard disks are based around conventional materials such as iron and magnesium oxide. Rapid uptake of spintronics technology will be aided if more novel materials can be developed that show the strong spin-charge coupling required to effectively manipulate the electron spins. The most intensely studied class of novel materials for spintronics are dilute magnetic semiconductors (DMS) in which magnetic elements are added to conventional semiconductors to induce ferromagnetism.[7] GaAs doped with a few percent manganese is known to become ferromagnetic, but only at temperatures below[8] about 170 K. There are numerous claims of room temperature ferromagnetism in other DMS systems, such as $Ga_xMn_{1-x}N$,[9] $Ga_xGd_{1-x}N$[10] or transition metal doped ZnO,[11] but these claims are complicated by the lack of a strong theoretical underpinning and issues associated with nano-sized impurity phases. Furthermore, the application of these materials is somewhat limited because the magnetic elements heavily dope the semiconductor preventing independent control of the carrier concentration, and their main proposed use is as spin injection layers coupled to conventional semiconductor devices.

Against this background it is clearly interesting to consider materials that are both intrinsically semiconducting and magnetic. In principle it is then possible to control the carrier concentration through doping independent of the magnetism, opening new possibilities for spintronics applications and for fundamental spin-transport research. At present only a few intrinsic ferromagnetic semiconductors are known. They include the first ferromagnetic semiconductor discovered, $CrBr_3$, the europium chalcogenides EuO and EuS, which are perhaps the most studied, a few chalcogenide spinels some of which have Curie temperatures near room temperature (e.g., $CdCr_2S_4$ with T_c of 97 K or $CuCr_2Te_3I$ with T_c of 294 K), halides such as DyF_3 or $GdCl_3$, and perovskite oxides such as $EuLiH_3$.[12,13] The parent compounds of the manganite perovskites that show colossal magnetoresistance (CMR) are semiconducting, but they are antiferromagnetic, only displaying ferromagnetism and the CMR effect upon doping to degeneracy.[13]

A particularly interesting class of materials that contain intrinsic ferromagnetic semiconductors are the rare-earth nitrides (RE-Ns), binary alloys of a rare-earth element and nitrogen with formula RN (R = rare-earth) which crystallise in the simple rock-salt structure. Early investigations of these compounds were performed in the 1960s and 1970s,[14,15] but without significant focus on their spin-charge coupling. The advent of modern band-structure calculation techniques has seen a renewal of interest in the RE-Ns, driven by predictions that the series includes both magnetically ordered half-metals and semiconductors, with potential application in spintronics.[16–19] This has been coupled with renewed experimental attention aimed at addressing contradictions amongst the theoretical predictions. Many outstanding questions remain regarding the electronic structure and the nature of the magnetic interactions, but the combination of theory and experiment is beginning to present a clear picture of some of the RE-Ns more fundamental properties.

This review will focus on recent developments in the study of rare-earth nitrides, especially the experimental aspects. An excellent review covering more of the theoretical aspects can be found in Ref. 19. The preparation of rare-earth nitrides will be discussed in Section 9.2, followed by a review of the theoretical and experimental investigations of their band structures in Sections 9.3.1 and 9.3.2. Section 9.4 describes their observed magnetic ordering and models of the exchange mechanisms responsible for the ordering. The chapter is concluded in Section 9.5 with a discussion of the potential for these materials to be used in spintronics devices.

9.2. Rare-Earth Nitride Preparation

The difficulty of preparing rare-earth nitride samples with the correct stoichiometry, low defect density, and low impurity concentration is one of the most significant challenges that must be overcome if a full understanding of their properties is to be achieved. In particular, their rapid reaction with atmosphere means that any *ex-situ* studies must be performed with the samples protected from exposure to air. A range of techniques have been applied to rare-earth nitride preparation, resulting in varying levels of sample quality.

A significant amount of work was performed in the 1960s and 1970s aimed at determining the bulk magnetic properties of the RE-Ns, for which samples were prepared in both thin film and bulk form. Bulk samples can be prepared by high temperature reaction of the lanthanide metal with either nitrogen gas or ammonia, or through chemical pathways such as the decomposition of lanthanide amides (e.g., $RE(NH_2)_3$). Early attempts were also made at thin film growth, with reported epitaxial layers prepared on heated substrates by decomposition of a volatile lanthanide chloride with ammonia. Films were also prepared by sublimation of fine powders. A summary of these early preparation techniques can be found in Ref. 14.

Perhaps the highest quality samples studied to date have been a series of YbN single crystals grown by Degiorgi *et al.*[20] in 1990. Considerable care was taken to avoid exposing the crystals to atmosphere, with all experimental characterisations being carried out after transferring the samples only via an argon-filled glove box. This allowed them to undertake the most comprehensive study of well-characterised RE-N samples yet reported. Various stoichiometries were achieved, ranging from $YbN_{0.96}$ to fully stoichiometric YbN (within $\pm 0.5\%$) as determined by a micro Kjeldahl method. The stoichiometric sample had a lattice constant of 4.781 Å, which is the lowest value reported for this material. The $YbN_{0.96}$ sample had a slightly expanded lattice constant of 4.784 Å, with the increase being comparable to the 0.005 Å increase for 5% off-stoichiometry reported as typical for the RE-Ns.[21] The stoichiometric material has a lower conductivity than off-stoichiometric material which is consistent with the expectation that nitrogen vacanices add free electrons to the system (YbN was found here to be semi-metallic).

More recent attempts to prepare rare-earth nitride samples for study have returned focus to thin film growth, which is the likely form that would be required for technological applications of these materials. This has been encouraged by the rapid improvements in nitride growth technology that have accompanied the rise of gallium nitride as a wide-band-gap electronic material, including the development of nitrogen plasma sources for molecular beam epitaxy growth.

Leuenberger *et al.* have successfully prepared GdN films on silicon [100], sapphire [$11\overline{2}0$], kapton, or mylar substrates using reactive ion-beam sputtering in a partial pressure of $2 - 3 \times 10^{-5}$ mbar N_2, with the substrate chosen to suit the subsequent characterisation method.[22] Reactive nitrogen was supplied by an ECR plasma source, and the growth temperature was varied from room temperature to 700°C. For *ex situ* studies capping layers of 50 Å Al or W were deposited, and a thin buffer layer of Cr or W was also deposited onto the silicon substrates prior to film growth to prevent reaction of the film with the substrate. Films grown above 400°C are single phase polycrystalline GdN with crystallite diameter of approximately 10 nm and a lattice constant of 5.05 Å, slightly larger than the bulk value. They are stoichiometric and free of oxygen to within the uncertainty of the resonant nuclear reaction analysis and X-ray photoelectron spectroscopy techniques used to characterise them.

One of the simplest preparation techniques involves evaporation (either thermal or electron beam) of the rare-earth element in a background partial pressure of approximately 10^{-4} mbar pure N_2 gas onto substrates held at room temperature.[23] Surprisingly, even in the absence of any excitation source for the nitrogen, the depositing rare-earth film reacts with the N_2 to form a nitride. After growth the films were capped with a layer of GaN grown by ion-assisted deposition. Thus far this has been demonstrated for GdN,[23] SmN, DyN,[24] ErN, and LuN,[25] using substrates of sapphire [100], silicon [100] or [111], or YSZ [100]. The stoichiometry of films thus prepared was investigated using Rutherford backscattering spectroscopy (RBS) and X-ray diffraction. RBS results indicated the films consist of equal atomic concentrations of rare-earth and nitrogen to within the uncertainty of a few percent, while the lattice constants measured by XRD are in agreement with previously reported values. A Scherrer analysis of the xrd peak widths indicates a small crystallite size of $\approx$ 10 nm, with evidence of weak [111] texturing in all materials except for SmN which appears to be fully randomly oriented. Further evidence for the stoichiometric nature of the films is obtained from the value of 70 K measured for the Curie temperature of GdN, which is in agreement with other reported values for stoichiometric GdN. Growth of GdN under similar conditions but using nitrogen excited by an ion-source produced films with smaller crystallites ($\approx$ 3 nm) and with a lattice constant expanded by 2.4%.[26]

Clearly some minimum nitrogen pressure must be supplied to grow RE-Ns using this technique. The pressure dependence of the films' stoichiometry has been investigated by measuring the resistance of films during growth on substrates onto which electrical contacts had first been deposited.[23] For growth at 10^{-4} mbar the resistance falls in inverse proportion to the film thickness, as expected for a uniform layer, with a resistivity of 0.25 $\mu\Omega$cm characteristic of a semiconducting material (see Section 9.3 below for further discussion of the electronic structure). However, when the nitrogen pressure was steadily lowered during growth the conductivity steadily increased. The overall conductance S of the film is related to the conductivity

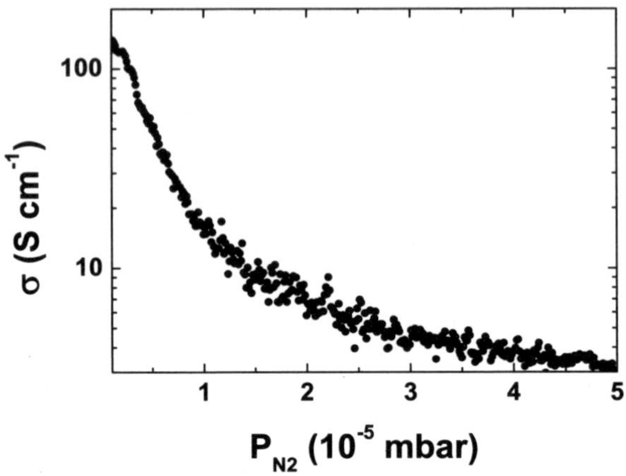

Fig. 9.1. The conductivity of a GdN film measured as a function of N_2 partial pressure during growth. Below about 10^{-5} mbar the GdN becomes metallic, while at higher pressures the film acts as a heavily doped semiconductor, with nitrogen vacancies the likely dopants (from Ref. 23).

$\sigma(z)$ at depth z by

$$S = \frac{W}{L}\int \sigma(z)dz,$$

$$\sigma(z) = \frac{L}{W}\frac{dS}{dz},$$

where W is the width of the film and L is the length between the contact pads. The conductivity thus extracted is plotted as a function of growth pressure in Fig. 9.1. The clear implication is that with decreasing nitrogen pressure the number of nitrogen vacancies increases, with evidence for a transition to a metallic state for growth pressures below $\approx 10^{-5}$ mbar.

The simplicity of this technique has been exploited to perform *in situ* X-ray absorption and emission studies of SmN and DyN as described below, where the films were grown in a preparation chamber attached to the synchrotron beamline and then transferred to the measurement chamber under vacuum without exposing them to atmosphere.[24]

More recently Gerlach *et al.*[27] have demonstrated epitaxial growth of GdN by molecular beam epitaxy onto [100] MgO substrates at a temperature of 750°C. A hollow anode plasma-beam source was used to provide the nitrogen. Reflection high-energy electron diffraction proved the epitaxial nature of the films, which remained smooth for thicknesses up to 17 nm, but which became three-dimensional at greater thicknesses. The crystalline quality of these films, as determined by XRD, was found to be higher than for any other reported films, although further optimisation of the growth parameters is still required to reduce the degree of twinning and mosaic spread. Epitaxial films of GdN, SmN, and LuN have also been produced recently using pulsed laser deposition, although these tend to suffer from the presence of metallic nanoparticulates produced by the laser ablation.[28] An optimised growth temperature window around 500°C was found for SmN, while for GdN and LuN no clear temperature dependence of the film quality was found between 500 and 850°C. This is presumably related to the much higher vapour pressure of Sm metal at these temperatures.

Given the dramatic enhancements in film quality that have been achieved with more conventional nitrides in recent years the prospects for improving rare-earth nitride film growth appear very strong. A more detailed exploration of the parameter space, including choice and preparation of substrate material, growth temperature, the nature of the nitrogen source, and the rare-earth to nitrogen flux ratio, should enable significant improvements in crystal quality and a reduction in defect density. Alternative growth techniques, such as chemical vapour deposition, may also prove fruitful in the search for better quality films. Of particular importance will be an improved understanding of the dynamics of the nitrogen incorporation, as nitrogen vacancies are an important source of free carriers that influence the material properties. At present no deliberate attempts to dope RE-N films have been made, but as we show below the band structure of these materials allows the possibility of doping both n- and p-type if the residual doping levels can be reduced sufficiently.

9.3. Electronic Structure

9.3.1. *Band structure calculations*

Until recently there was little reliable experimental information regarding the electronic structure of the rare-earth nitrides, and fundamental questions such as whether they are metals or insulators in their ground state remained unanswered. In a simple ionic model one would expect the rare-earth ion to be in the +3 charge state (with the exception of tetravalent

Table 9.1. Physical and magnetic properties of the rare-earth nitrides. Both antiferromagnetic and ferromagnetic ground states have been attributed to SmN, TbN, and DyN, as indicated by the inclusion of both measured Curie (T_c) and measured Néel (T_N) temperatures. The quoted conductivities are for the ground state, where SM = semimetallic, SC = semiconducting, HM = half metallic, M = metallic. Radioactive promethium (Pm) is omitted.

RE-N	r (Å)	Magn. order (expt.)	T_c/T_N (K)	Conduct. (theory)	Conduct. (expt.)
LaN	5.30	none[15]	-	SC[17,29]	SC[14,30]
CeN	4.87	mixed val.[15]	-	SM[16,29] SC[17]	-
PrN	5.17	unknown	-	HM[16,17,29]	M[14,30]
NdN	5.15	FM[15]	$T_c = 27.6 - 35$	HM[16,29] SC[17]	M[14]
SmN	5.04	AFM[15]/FM[31]	$T_N = 18$ $T_c = 26 - 30$	HM[16,17,29]	SC[24]
EuN	5.00	unknown	-	HM[16,17,29,32]	M[14,30]
GdN	4.98	FM[15,22,23]	$T_c = 58 - 69$	HM[16,18,33–35] SC[17,36,37]	SC[23]
TbN	4.92	FM/AFM	$T_c = 34 - 42$ $T_N = 31$	SC[16,17]	-
DyN	4.90	FM[31]/AFM[15]	$T_c = 17 - 26$ $T_N = 15$	SC[16,17]	SC[24]
HoN	4.87	FM[15]	$T_c = 11 - 18$	SC[16,17]	-
ErN	4.83	FM[15,38]	$T_c = 3.4 - 7$	SM[16] SC[17]	SC[38]
TmN	4.80	none	-	SM[16] SC[17]	-
YbN	4.78	AFM[15,20]	$T_c = 0.75$	SM[16] SC[17]	SM[20]
LuN	4.76	none	-	SC[17]	M[14,30] SC[25]

Ce), having donated the two $6s$ electrons and either a $5d$ or a $4f$ electron to leave nitrogen in a -3 state. The resulting electronic structure would have a valence band derived from the occupied nitrogen $2p$ states and a conduction band made from the empty rare-earth states. In this picture the $4f$ levels remain partially occupied, and thus one may naively expect them to lie in a narrow band at the Fermi level resulting in a metallic state. Indeed basic calculations within the local density approximation (LDA) to density functional theory (DFT) do result in such a band structure.[19] However, the picture is made considerably more complex when the strong correlations amongst the $4f$ electrons are taken into account, something which LDA fails to do. Furthermore, for materials possessing a band gap there is the well-known failure of the LDA to adequately account for the screening of the long-ranged Coulomb repulsion between the band electrons, which results in underestimation of the band gap in semiconductors.[39] Finally, the strong exchange interaction between the $4f$ and the band electrons requires the two different spin orientations to be treated separately. Thus more advanced treatments are required, and the rare-earth nitrides have become an ideal testing ground for techniques such as the the local spin density approximation with a self-interaction correction (LSDA-SIC),[40,41] LSDA with a Hubbard-U correction (LSDA+U),[29,34,35,42–45] advances on the GW correction to DFT,[36,46] or other means of accounting for many-body effects.[37,47]

The relatively simple half-filled $4f$ shell of Gd in GdN has led band structure calculations to focus the most attention on this RE-N. One of the first band structure calculations was performed by Hasegawa and Yanase,[48] who predicted a semiconducting state with a band gap of ~ 1 eV, but who noted the possibility that exchange splitting of the bands could conceivably close the gap and make GdN metallic. In this calculation they treated the $4f$ levels as core states. Determining the best way to treat the $4f$s and answering the question of whether the ground state is metallic, half-metallic, semi-metallic, or semiconducting, has dominated subsequent calculation attempts. A series of papers by Lambrecht *et al.* has done a great deal to elucidate the physics of the GdN band structure. An LSDA calculation published in 1996 gave the first prediction that GdN is half-metallic,[49] but a follow-up paper in 2000 which used LSDA with an approximate GW correction found instead a semiconducting ground state.[50] This was confirmed by an LSDA+U calculation that included a Hubbard-like U_f to shift the $4f$ levels away from the Fermi level, as well as a quasiparticle correction to the d levels that accounts for the usual underestimation of band gaps by LSDA.[51] The U_f values were determined by extrapolating photoemission and inverse photoemission data from GdP, GdAs, GdSb, and GdBi.[52] The quasiparticle correction was included using a U_d value in a similar fashion to the Hubbard U_f, but it was stressed that it actually represents quite different physics, being a correction for long-ranged Coulomb effects rather than the localised $4f$ correlations. The importance of including the U_d correction is highlighted by the fact that without it the band structure is half-metallic.

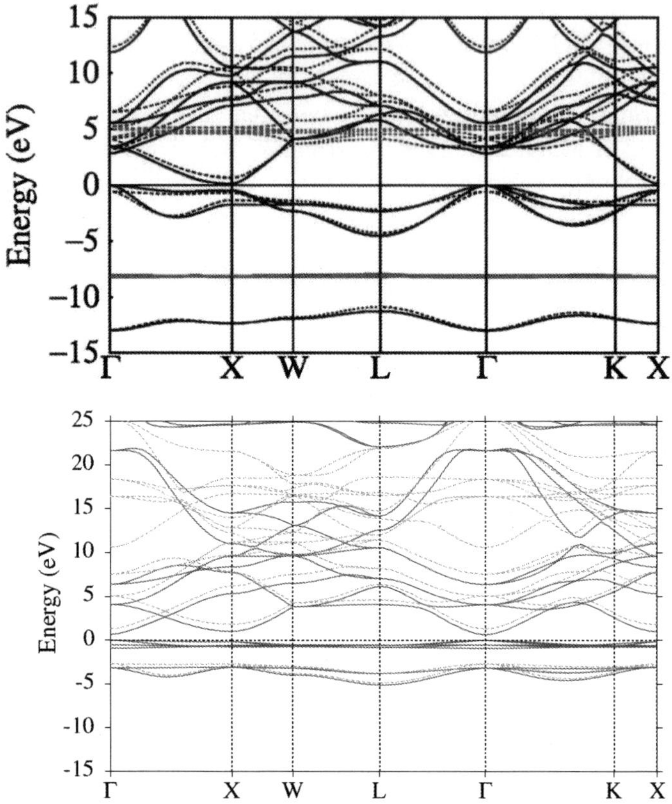

Fig. 9.2. (a) Spin-resolved band structures of (a) GdN (from Ref. 17) and (b) EuO (from Ref. 53) calculated by LSDA+U. The majority spin states are solid lines and the minority states in dashed lines.

The resulting calculated band structure of GdN is reproduced in Fig. 9.2(a). The occupied majority spin $4f$ levels, whose location is determined by U_f, are seen as the very flat bands about 8 eV below the Fermi level (defined to be the top of the valence band). The minority spin $4f$ levels, located 5 eV above the Fermi level, show slightly more dispersion due to greater hybridisation with the conduction band states. Both the conduction and valence bands show strong exchange splitting of approximately 0.5 eV, but it is important to note that the splitting is of opposite sign, so that the upper valence band and lower conduction band states are both of majority spin character. The resulting band gap is indirect. The smallest indirect gaps from Γ to X for majority and minority spins are 0.09 and

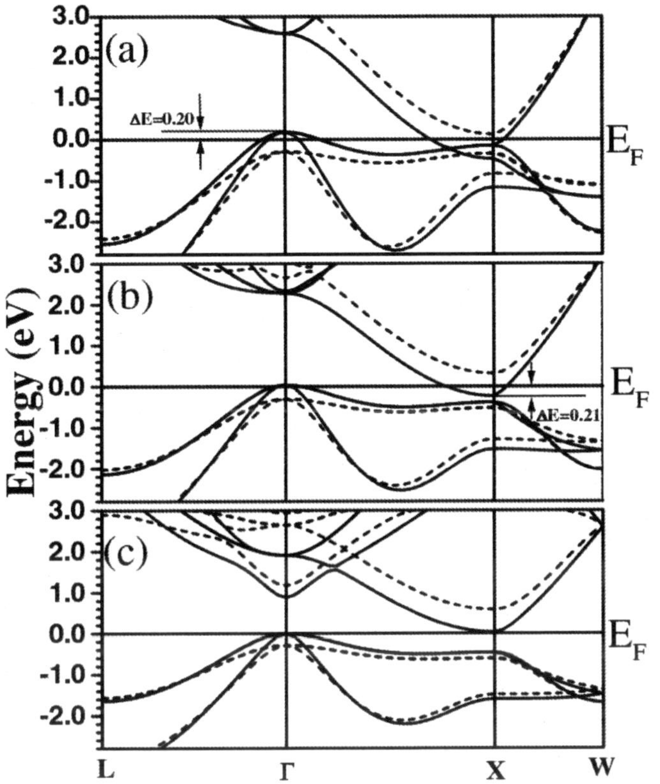

Fig. 9.3. The dependence of the band structure of GdN on lattice parameter. As the lattice constant increases from (a) 4.92 Å through (b) 5.16 Å to (c) 5.63 Å the band structure evolves from half-metallic to semiconducting (from Ref. 18).

0.82 eV, respectively, while the corresponding smallest direct gaps at X are 0.50 and 1.41 eV (note that these values have since been refined slightly in light of new experimental data, as will be described in Section 9.3.2). For comparison to experimental data it is important to note that these values are appropriate for the ferromagnetic ground state. In the room temperature paramagnetic state the band structure will be approximately an average of the majority and minority spin bands.

The subtleties involved in obtaining the correct band structure near the Fermi level are illustrated by contrasting the above results with an LSDA-SIC calculation by Aerts *et al.* which found a half-metallic ground state.[16] A very interesting transition from a half-metallic to a semiconducting state with increasing lattice constant was predicted by Duan *et al.*[18] using LSDA+U, with the main results reproduced in Fig. 9.3. At the equilibrium lattice constant the orbital overlap is large enough that the Gd 5d majority spin conduction bands cross the majority spin N 2p valence bands, giving a half-metal. As the lattice expands the orbital overlap decreases and a band gap opens up. The exchange parameters were also found to depend on the lattice constant, implying that both the conduction state and the Curie temperature could be tuned by applying stress.

Possibly the most advanced band structure calculation has been provided by Chantis *et al.*[36] using a recently developed quasiparticle self-consistent *GW* correction to LSDA. This approach is able to cope with both the localised nature of the 4f levels and the band-gap underestimation issues. The band structure of GdN was found to be semiconducting, and in good agreement with the LSDA+U_f+U_d results,[51] but the proximity to the critical point of a (half-) metal to insulator transition was emphasised.

It is interesting to compare the band structure of GdN in Fig. 9.2(a) to that of EuO shown in Fig. 9.2(b), which was calculated within the LSDA+U framework, with U shifts applied to both the 4f and 5d electrons.[53] Like GdN, EuO is calculated to be semiconducting with a (spin-averaged) band gap of 0.97 eV, and there is a strong exchange splitting of the Eu 5d-derived conduction band of 0.66 eV. However, EuO and GdN differ significantly in the location of the majority 4f levels. As described above, for GdN these are well below the N 2p-derived valence band, but for EuO the occupied 4f levels form a rather flat band just below the Fermi level with the O 2p-derived valence band states lying approximately 2 eV further down. Furthermore, the sense of the exchange splitting of the p levels is opposite in EuO to that in GdN, with the minority O p states lying above the majority states.

There have also been attempts to calculate the electronic structures of other members of the rare-earth nitride series, including SmN,[54] CeN,[55] ErN,[56] and EuN[32,57] using similar theoretical frameworks to those applied to GdN. The first systematic attempt to calculate the band structure across the entire RE-N series was provided in 2004 by Aerts *et al.*,[16] using LSDA-SIC. This work predicted ground states ranging from metallic to insulating, and including half-metallic for the RE-Ns from PrN to GdN, which heightened interest in the RE-Ns and their alloys as potential spintronic materials. Following the focus on refining the calculation techniques that was applied to GdN, Larson *et al.*[17] used their LSDA+U framework to calculate the band structures of all of the RE-Ns. This work emphasised the importance of allowing the $4f$ orbitals to break the crystal field symmetry, and they are generally found instead to obey Hund's rules (the exceptions being EuN and YbN). PrN, PmN, EuN, and SmN are found to be half-metallic, while the other RE-Ns possess small indirect band gaps that tend to increase as the rare-earth atomic number increases. EuN proved the most difficult to study from a theoretical perspective, partly because divalent and trivalent solutions have similar energy. A review article by Duan *et al.* has provided an excellent summary of the theoretical progress, including a more detailed discussion of the calculation methods and their relative strengths and weaknesses.[19]

9.3.2. *Experiment*

The rare-earth nitrides are presently in the unusual position that there have been more attempts to calculate their band structure than there have been experimental investigations to guide the theory. There were early attempts to measure their room-temperature optical transmission, with the majority of the series exhibiting an absorption minimum near 1 eV that was interpreted as a band-gap.[14,58] In some cases this was supported by resistivity data that appeared to show semiconducting behaviour, albeit with very large carrier concentrations. However, it is far from clear from these measurements that the rare-earth nitrides are in fact semiconductors. First, it should be remembered that the optical absorption edge generally measures the direct band gap, but in RE-Ns elementary symmetry considerations show that the indirect gap is smaller and may in fact be negative[20] (see Section 9.3.1 above). The measured transmission minimum is rather weak, so it was not completely clear whether the low-frequency absorption implied that the materials were heavily doped semiconductors or in fact semi-metals. Furthermore, as described above, a room temperature band gap does not imply the existence of a band gap in the ground state of ferromagnetic RE-Ns, as the exchange splitting tends to narrow the gap. Finally, the quality of the early samples was compromised by the difficulties achieving stoichiometry and in protecting the samples from oxidation, so it was clear that a definitive determination of the conduction states required improved samples.

The most comprehensive experimental investigation of any member of the RE-N series was carried out using single crystals of YbN,[20] which at the time was motivated by suggestions that YbN might be a heavy fermion system. This was found not to be the case, but the main result was that

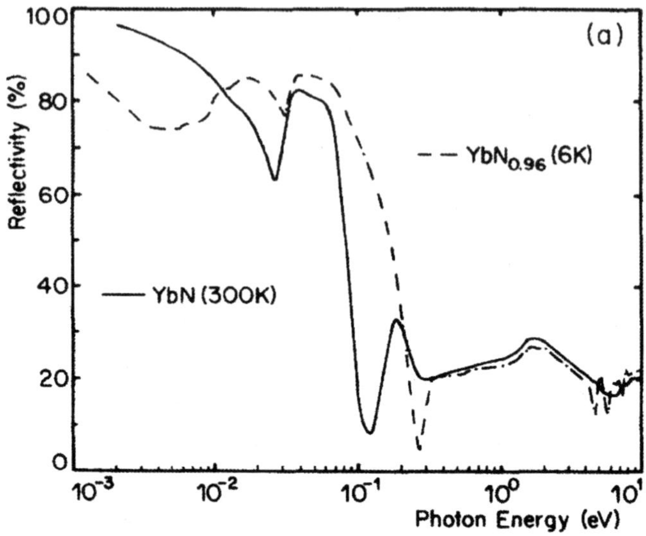

Fig. 9.4. Reflectivity of a YbN single crystal across a broad energy range. Also shown is the reflectivity from a non-stoichiometric sample where the resulting increased carrier concentration shifts the plasma edge to higher frequencies (from Ref. 20).

YbN is a self-compensated semimetal in which the ytterbium d states cross below the nitrogen p states giving rise to a moderate carrier concentration of about 10^{20} cm^{-3}. This conclusion was drawn from a study of the resistivity and Hall effect, which show metallic behaviour, and measurements of the optical constants covering a very broad energy range from 1 meV to 12 eV (see Fig. 9.4). The reflectivity of a stoichiometric sample shows a clear Drude edge at about 0.1 eV, and this moves out to about 0.2 eV in a nitrogen deficient sample due to an increased carrier concentration. Direct interband transitions dominate at higher energies, and they were able to identify the feature at 0.2 eV as arising from transitions between Yb d sates at the Fermi level and the unoccupied $4f^{14}$ state. The close proximity of the the empty $4f^{14}$ level to E_f is in agreement with the band structure calculation in Ref. 17, although there the calculation predicts a semiconducting ground state.

This comprehensive study of YbN is extremely useful as a guide to theory, but YbN becomes antiferromagnetic only below 0.75 K. Thus, some of the interesting aspects of the RE-N band structures, such as the exchange splitting, are not accessible. As with theoretical work, the more recent experimental investigations have been largely focussed on GdN. Figure 9.5

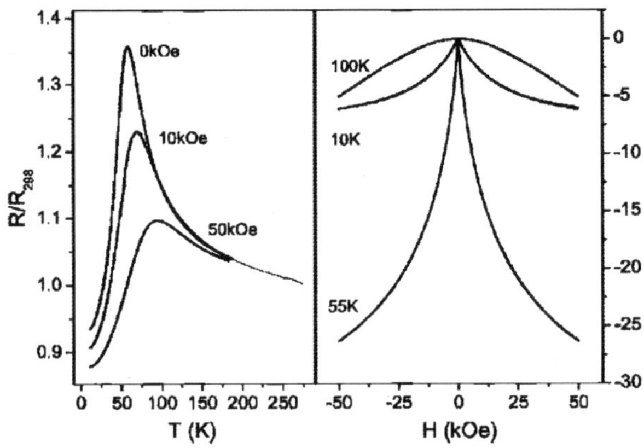

Fig. 9.5. Temperature dependent resistance and magnetoresistance of a GdN film (from Ref. 22).

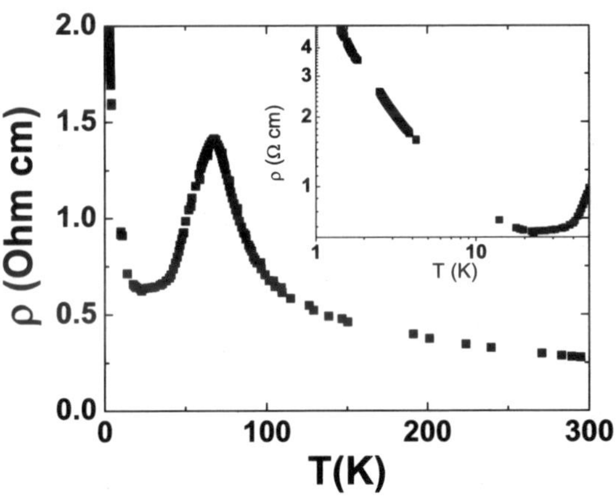

Fig. 9.6. Temperature dependent resistivity of a GdN film demonstrating a semiconducting ground state in this material (from Ref. 23). This film had fewer nitrogen vacancies than the sample described in Fig. 9.5.

shows the temperature dependent resistivity and magnetoresistance of GdN thin films grown by reactive-ion sputtering as measured by Leuenberger *et al.*[22] The resistivity goes through a peak at the Curie temperature (T_c) of about 59 K, and a rather strong negative magnetoresistance is observed in the vicinity of T_c. They attribute the behaviour to a transition from a degenerately doped semiconductor above T_c to a metallic state below T_c. The nature of the metallic state is similar to that observed in non-stoichiometric EuO,[59] namely the exchange splitting does not fully close the band gap but instead it drives the bottom of the conduction band down below a rather large density of nitrogen vacancy induced donor states, which lie in the gap for temperatures above T_c. Further work by Granville *et al.*[23] on samples with higher resistivity, and presumably fewer nitrogen vacancies, found a stronger peak at T_c (which was 69 K in their films) and a semiconducting ground state at low temperature (see Fig. 9.6). These findings are in agreement with the band structure calculations in Refs. 36, 51 and highlight the importance of adequately treating the long-range Coulomb interactions.

More direct evidence for the presence of a band gap in the ground state of GdN was obtained from the temperature dependent optical transmittance data of Trodahl *et al.*[60] shown in Fig. 9.7. At room temperature there is a relatively clear absorption edge at around 1.3 eV, corresponding to direct transitions at X in the band structure, and there is little change as the temperature is lowered to 75 K. However, at 6 K there is a clear red shift of the absorption edge to 0.9 eV, which is exactly as expected when exchange splitting narrows the majority spin band gap in the ferromagnetic

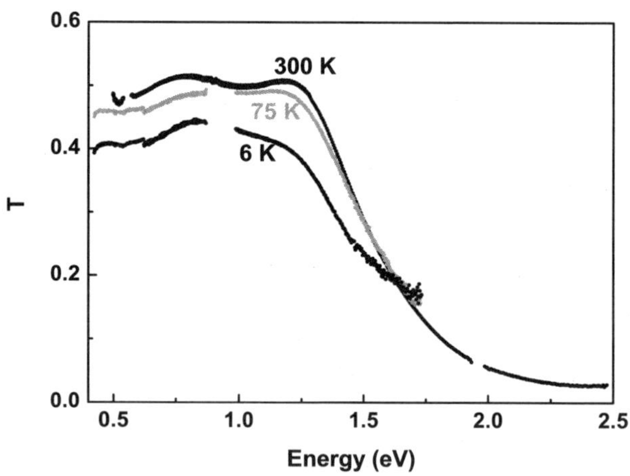

Fig. 9.7. Optical transmittance of a GdN thin film showing a red shift of the absorption edge at low temperature due to narrowing of the majority spin band gap in the ferromagnetic phase (from Ref. 60).

phase. These data were used to retune the U_d parameter in the calculated LSDA+U band structure and thereby obtain excellent agreement between theory and experiment. Recently, calculations of the full optical conductivity of GdN have been presented,[35,61] and it will be interesting in future to compare quality experimental measurements to these calculations.

Transport and optical studies address the band structure near to the Fermi level. This can be complemented by synchrotron-based X-ray absorption and emission spectroscopy[62] (XAS and XES), which measure the nitrogen p-projected partial density of empty and occupied states, respectively. XAS and XES studies have been performed for GdN by Leuenberger *et al.*[22,63] and for SmN and DyN by Preston *et al.*[24] Results from the latter study are shown in Fig. 9.8 where they are compared to the appropriate partial densities of states (PDOS) obtained from LSDA+U band structure calculations performed according to the method described in Ref. 17. The XES data show a relatively narrow nitrogen-derived valence band which is in excellent agreement with the calculated PDOS. For the XAS data the presence of the N 1s core hole in the final state must be considered, which is achieved using the so-called final state rule.[64,65] Thus the XAS data are in excellent agreement across a broad energy range with a calculation that includes a core hole on every fourth nitrogen site. The data for SmN and DyN are similar, with only subtle differences between relative peak locations and intensities, highlighting the similarity between the materials electronic structures (more striking differences would be observed using a technique that directly probed the 4f PDOS). For both materials there is a gap of around 1 eV between the filled and empty states, but due to the effects of the core hole and the absence of strong N p character at the bottom of the conduction band this cannot be taken as direct evidence for the presence of a room temperature band gap. However, resistivity data from these two films both showed a semiconductor-like temperature dependence with a peak at the magnetic ordering temperature similar to that shown in Fig. 9.6 for GdN, providing evidence that these two materials are indeed semiconductors both at room temperature and at low temperature.

There is clearly a reasonable agreement between the available experimental data and the most recent band structure calculations for the RE-N series. However, some issues still require addressing, such as the disagreement between the calculated and measured conduction states of SmN and YbN, and the absence of any systematic experimental data that provides information about the location of the 4f orbitals. In the case of the former it will be interesting to see whether further tuning of the U_d parameter

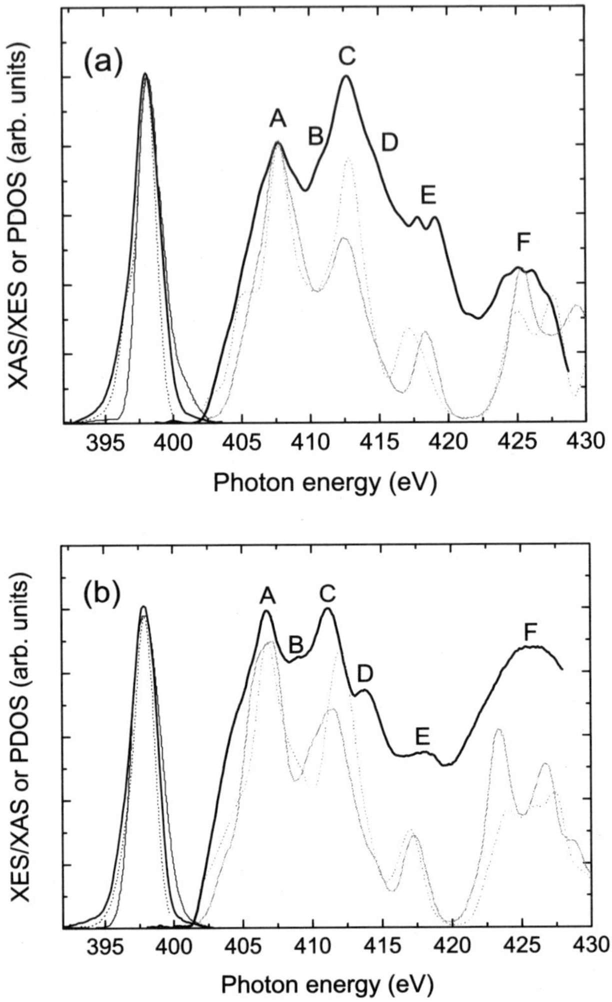

Fig. 9.8. Measured X-ray absorption (thick black line) and emission (thin black line) for (a) DyN and (b) SmN, compared to the calculated nitrogen PDOS. The calculations are performed both with (solid lines) and without (dotted lines) including the core hole (from Ref. 24).

in LSDA+U will be sufficient to describe the band structures of all of the RE-Ns near the Fermi level, or whether perhaps further new physics must be considered.

9.4. Magnetic Properties

The magnetic properties of rare-earth ions have remained at the forefront of condensed matter physics for nearly 80 years, for reasons of both fundamental and technological interest. The highly localised nature of the $4f$ electrons is in contrast to the more itinerant nature of the d electrons in transition metal-based magnetic materials, and thus provides an interesting contrast to those systems. The atomic picture generally provides an excellent starting point for a description of their magnetic moments, but the lack of direct orbital overlap between the $4f$ electrons on neighbouring ions means that the exchange interactions that lead to magnetic ordering can be quite complex.

Much of the focus of the early work on the RE-Ns was on determining their magnetic ground states and transition temperatures, with the results summarised in Table 9.1. As with the electronic properties, issues with sample stoichiometry and oxygen contamination have contributed to considerable uncertainty in the magnetic transition temperatures, and even in the exact nature of the magnetic ordering. For example, it has been noted that quantities of oxygen greater than about 5% can destroy ferromagnetism and give rise to antiferromagnetic correlations in GdN,[66,67] and it has even been reported that pure GdN with no nitrogen vacancies is metamagnetic.[68] Inclusion of carbon has been reported to raise T_c to as much as 190 K.[67] More recent experimental work has found ferromagnetism in GdN samples with low levels of oxygen impurities and nitrogen vacancies, although there is still some variation in the measured Curie temperatures.[22,23,26,63]

The strong coupling between the electronic and magnetic properties of the RE-Ns has stimulated renewed interest in their magnetism. Numerous attempts have been made to model the exchange mechanism in GdN[69–72] starting from band-structure calculations (there are no recent predictions of the transition temperatures of the other RE-Ns). The focus on GdN is in part prompted by the simplifying effect of having a half-filled $4f$ shell and hence no orbital component to the $4f$ magnetism. It is generally accepted that competing exchange interactions exist,[72] with an antiferromagnetic Gd-Gd superexchange contribution mediated through the nitrogen ions and a ferromagnetic Gd-Gd interaction mediated by the Gd d orbitals. The first interaction is reasonably well understood, and calculations are able to reproduce the measured Néel temperatures of the related pnictides GdP, GdAs, GdSb, and GdBi for which the same antiferromagnetic contribution is the dominant exchange mechanism.[69] The ferromagnetic contribution has proven more difficult to interpret. It has been assumed that a Ruderman-Kittel-Kasuya-Yosida-type carrier mediated exchange is important in Gd-monopnictides,[19] but this is unlikely to be relevant for semiconducting GdN at the experimentally observed carrier concentrations, and higher order mechanisms have also been explored.[71]

Some insight into the magnetic interactions in GdN has been provided by recent experiments. Leuenberger *et al.* found a surprisingly strong x-ray magnetic circular dichroism (XMCD) signal at the nitrogen K-edge in GdN, which could only be understood based on the presence of $p-d$ exchange between the Gd and N sites which induces a magnetic moment on the nitrogen. Two groups have modelled the XMCD with some success, although their calculations are both based on LSDA+U band structure calculations that predict half-metallic behaviour in GdN.[33,34] The influence on T_c of an expanded lattice constant has also been investigated using XMCD[63] and using ferromagnetic resonance.[26] In both cases a reduced T_c was observed, in agreement with expectations of a reduction in the ferromagnetic exchange interaction as the lattice constant increases.[18]

A very detailed theoretical analysis has recently been provided by Mitra and Lambrecht.[69] Their crucial finding is that the Gd $5d$ and N $2p$ states carry equal and opposite moments, so the material is essentially an antiferromagnet. This coupling can be directly interpreted in terms of the band structure, where the Gd $5d$ – N $2p$ hybridisation is greater for the majority spin states as these lie closer together (see Fig. 9.2(a)), with the result being a transfer of majority spin to the d levels leaving a net minority spin on the nitrogen site. However, the p and d moments are small, and the ferromagnetic $f-d$ coupling on the Gd site ensures that the strong $4f$ moments are all aligned with the $5d$ moments giving overall ferromagnetism amongst the $4f$ spins.

Although the intuitive nature of this model is highly appealing, when the extracted exchange parameters are used to calculate the Curie temperature the resulting value of 11 K is significantly lower than the experimental values (see Table 9.1). It may be that this discrepancy is related to the changes to the band structure induced by defects such as nitrogen vacancies, which could alter the rather subtle antiferromagnetic $p-d$ exchange. This again is evidence of the sensitivity of the RE-N electronic structure to perturbations, as was also shown in Fig. 9.3. Clarifying the exact nature of the exchange interaction may have to wait for further advances in sample preparation to enable systematic studies of the role of defects.

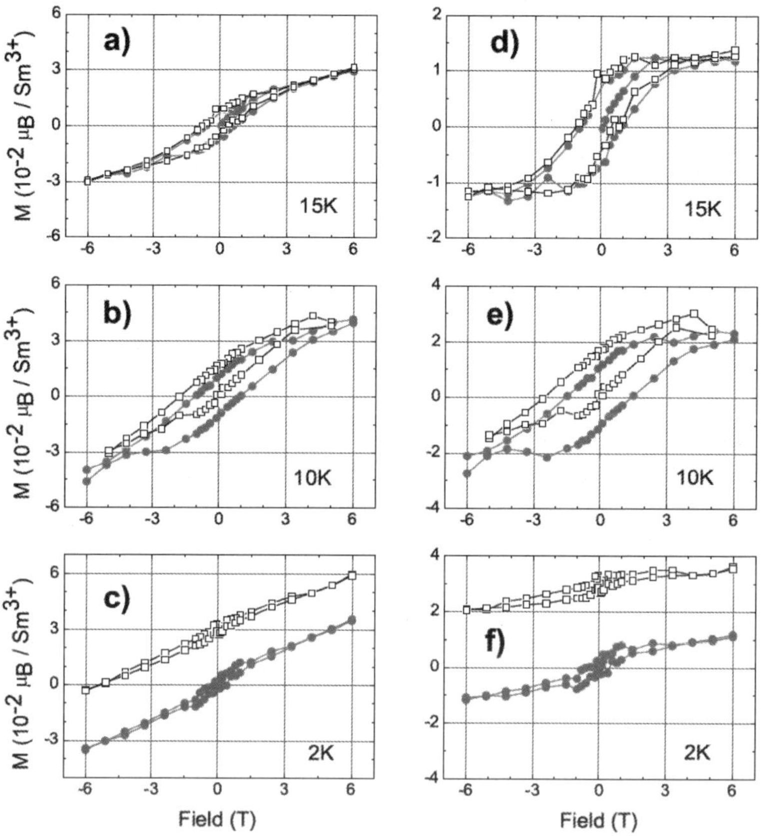

Fig. 9.9. Magnetic hysteresis loops from a SmN thin film taken after cooling in zero field (filled circles) or after cooling in 6 T (open squares). At 2 K the coercive field is larger than the maximum applied field of 6 T, so no hysteresis is evident. The curves on the right have been corrected for a paramagnetic contribution from the GaN capping layer and Si substrate (from Ref. 31).

The two rare-earth ions with one or two electrons less than a half-filled $4f$ shell, Sm and Eu, are also particularly interesting. According to Hund's rules Eu^{3+} has a ground state with $J = 0$ and hence it should be nonmagnetic (i.e., the magnetic moment should have zero projection along an applied field). However, Johannes and Pickett[32] have pointed out that this does not mean that EuN lacks long-ranged spin or orbital correlations. For EuN they have raised the possibility of a "hidden ferromagnetic order", which provides a fascinating experimental challenge to detect.

Sm^{3+} has a $4f^5$ configuration and according to Hund's rules a total orbital angular momentum $L = 5$, a spin angular momentum $S = 5/2$, and in the ground state these oppose each other giving $J = |L - S| = 5/2$. The magnetic moment per ion is given by $|\vec{M}| = \mu_B|(\vec{L} + 2\vec{S})|$ so for SmN the orbital and spin contributions to the magnetic moment are then expected to almost cancel. This has been highlighted in the band structure calculations of Ref. 17, and recently confirmed experimentally by Meyer *et al.*[31] Magnetisation data measured on a thin film of SmN are shown in Fig. 9.9. Clear hysteresis is observed at temperatures below about 30 K indicative of ferromagnetism. However, below about 10 K the hysteresis loops close, and loops taken after cooling in a 6 T applied field are shifted relative to those taken after zero-field cooling. This is understood based on the very small moment of the Sm^{3+} ion, which couples weakly to the applied field such that at low temperatures the magnetic domains cannot be rotated even in the highest available field of 6 T. Near-vanishing magnetic moments are not unknown in Sm compounds,[73] but the observation of this in a semiconducting material appears thus far to be unique to SmN.

It should be noted that early experiments concluded that SmN is antiferromagnetic[15] below about 20 K, although these were not confirmed by neutron scattering.[74] This is not surprising considering the weak moment in SmN. A peak in the resistance is observed at the magnetic transition temperature similar to that shown for GdN in Fig. 9.6, giving further support to the conclusion that SmN is indeed ferromagnetic.

The magnetic properties of rare-earth nitrides offer rich prospects for future research. It is also interesting to consider the relationship between RE-Ns and rare-earth doped conventional nitride semiconductors, such as $Ga_xGd_{1-x}N$, for which remarkably high magnetic moments have been observed.[10] Although the local structure around a Gd ion in GdN is different to that in GaN, an understanding of the exchange mechanism and electronic states in GdN may provide insight into the magnetism in $Ga_xGd_{1-x}N$, shedding light on aspects such as whether the Gd introduces significant defect states.

9.5. Device Prospects and Future Challenges

The discussion presented above makes it clear that, although some highly interesting questions remain, a consensus is emerging about many of the RE-Ns electronic and magnetic properties which make them of considerable interest for use in future spintronics devices. The series contains several ferromagnetic semiconductors which can in principle be doped to show very high spin polarisation, and other members of the series may be half-metallic in their intrinsic state. This range of electronic structure, combined with their chemical similarity, makes it possible to imagine tuning their properties by alloying, or through growth of heterostructures consisting of layers of different rare-earth nitrides. The possible applications that can be considered range from simple field sensors to more exotic devices that rely on the RE-Ns novel band structures.

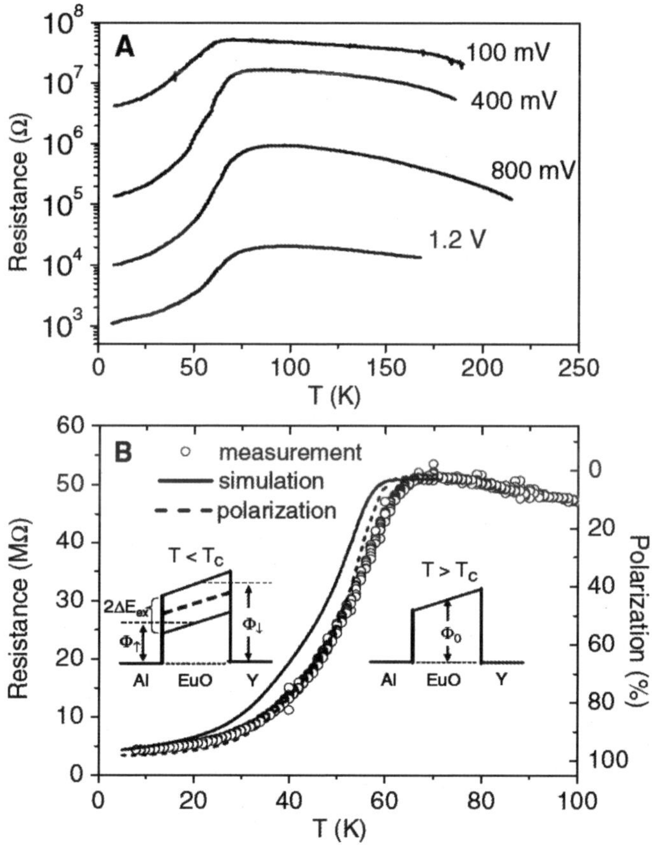

Fig. 9.10. Demonstration of spin filtering using a thin EuO tunnel barrier between two metallic electrodes (Al/EuO/Y). (a) The tunneling current decreases significantly below T_c and is dependent on the applied bias. (b) The tunnel barrier is determined by the exchange splitting and is thus spin dependent. The measured resistance is best fitted using a model in which spin polarisation reaches 98% at low temperature (from Ref. 75).

The results of Section 9.3.2 show that GdN has a large magnetoresistance, especially at temperatures close to the Curie temperature, and it is quite likely that this can be further enhanced by controlling the defect states that dominate the electrical conductivity. This material thus becomes of interest for use in magnetic field sensors. The major requirements for a resistive field sensor are a large magnetoresistance over a broad field range, small coercive field, and high transition temperature. GdN satisfies the first, and as its magnetic moment is almost entirely spin-derived it also has a small coercive field which can be minimised by improving crystalline quality. The transition temperature is low, so a GdN-based field sensor would be restricted to cryogenic operation, performing such tasks as sensing the field in a superconducting magnet. However, it is possible that steps can be taken to increase the transition temperature through alloying or strained growth. As noted above carbon doping of GdN has been reported[67] to raise the T_c to 190 K, and the T_c of the ferromagnetic semiconductor EuO can be raised from 69 K to 118 K by La doping,[76] to 170 K by Gd doping,[77] or to over 200 K under pressure.[78]

A far more novel proof-of-concept spintronic device relying on the strong exchange splitting in EuO to produce a highly efficient spin filter has recently been demonstrated by Santos *et al.*[75] Figure 9.10 demonstrates the device architecture and performance. A thin layer (1–6 nm) of ferromagnetic EuO is placed between two metallic contacts (Al and Y in the present example), with the exchange splitting of the EuO conduction band used to produce a spin-dependent tunneling barrier for electrons passing from the Al to the Y electrode. The very strong dependence of the tunneling current on the barrier height produces a tunneling current with 98% spin polarisation. The rare-earth nitrides exhibit comparable exchange splitting, so similar devices could be made for example by replacing the EuO with GdN. Furthermore, GdN also has a spin-split valence band (see Fig. 9.2), so a GdN-based spin filter could also be used as a source or detector of spin-polarised holes.

This polarisation of both the valence and conduction bands also raises the possibility of utilising semiconducting rare-earth nitrides to modify conventional semiconductor devices. If controlled doping can be achieved then it should be possible to fabricate diodes or transistors in which the currents are fully spin polarised. While the concept of a so-called spin bipolar junction transistor has been explored,[3] that proposal only includes a spin-polarised base and does not include polarisation of both carrier types. An investigation of such a device and an analysis of its efficacy would be both novel and interesting.

It was shown in Section 9.4 that SmN is a ferromagnetic semiconductor with a near-vanishing magnetic moment. As such it can potentially be used as a ferromagnetic element in spintronics devices in which stray fields are undesirable. One such specialist application would be as a magnetic coating on a scanning tunneling microscope tip, and more conventional spintronics applications could be imagined.

The above discussion lists only a few potential applications of rare-earth nitrides in spintronics devices, and it is certain that more examples could be imagined. For uptake of these devices several materials constraints must be overcome. Room temperature device operation is clearly desirable in most applications, and here the relatively low T_cs of the RE-Ns are an obvious limitation. However, as already mentioned there is scope for increasing the T_c, and there are niche markets for spintronics devices operating at low temperature. Furthermore, even cryogenic demonstration of highly novel proof-of-concept devices exploiting the rare-earth nitrides band structures would clearly represent an exciting development in the field of spintronics, perhaps paving the way to commercial uptake of similar devices based on materials with related properties.

Improving the material quality is another immediate area to address. At present even the highest quality samples have significant densities of defects such as nitrogen vacancies and grain boundaries, and these are capable of having a significant impact on the observed properties of the RE-Ns. Further investigation of epitaxial growth is required to determine the growth conditions that produce highly crystalline and stoichiometric material. This will require systematic studies of the influence of parameters such as substrate temperature, active nitrogen source, and rare-earth to nitrogen flux ratios. Of particular use would be schemes to grow RE-N films on existing semiconductors, such as silicon or gallium arsenide, which could allow integration into existing device architectures. Following optimisation of the growth it will be possible to move towards deliberate doping, both n- and p-type, to control the carrier concentration, from which point multilayered devices could be produced. The situation in this respect very much mirrors that of GaN in the early 1990s, where a few key breakthroughs paved the way for commercial applications.[79] Further development of robust encapsulation techniques will also be required to prevent degradation of the RE-Ns in air. The growth studies should be combined with theoretical investigations of the electronic structure of defects and potential dopants. These will be extremely useful in guiding the growth optimisation, but are likely to test the limits of the calculation techniques.

9.6. Conclusions

This Chapter has provided a summary of the present state of knowledge concerning the properties of the rare-earth nitrides. They exhibit some highly interesting physics, with strongly correlated electronic structures that lie near the boundary between what can and cannot be treated adequately with modern band structure calculation tools. Their number includes several intrinsically ferromagnetic semiconductors, which can in principle be doped either n- or p-type, marking a key difference between the rare-earth nitrides and other commonly studied intrinsic ferromagnetic semiconductors or dilute magnetic semiconductors. If sample preparation techniques can be optimised to improve crystal quality and minimise defect density then their novel properties may one day be utilised in spintronics devices.

Acknowledgments

The author gratefully acknowledges useful discussions with Prof. H. Joe Trodahl, Prof. Claire Meyer, Prof. Walter R. L. Lambrecht, Prof. Kevin Smith, Dr. James Downes, Dr. Jan Richter, Andrew Preston, and Bart Ludbrook.

References

1. I. Žutič, J. Fabian, and S. Das Sarma, *Rev. Mod. Phys.* **76**, 323, (2003).
2. D. D. Awschalom and M. E. Flatte, *Nat. Phys.* **3**, 153, (2007).
3. S. Bandyopadhyay and M. Cahay, *Introduction to Spintronics.* (CRC Press, 2008).
4. S. S. P. Parkin, C. Kaiser, A. Panchula, P. M. Rice, B. Hughes, M. Samant, and S.-H. Yang, *Nat. Mat.* **3**, 862, (1994).
5. http://www.freescale.com; http://www.everspin.com.
6. I. Žutič, J. Fabian, and S. C. Erwin, *J. Phys.: Condens. Matter.* **19**, 165219, (2007).
7. T. Dietl, H. Ohno, F. Matsukura, J. Cibert, and D. Ferrand, *Sci.* **287**, 1019, (2000).
8. T. Jungwirth, K. Y. Wang, J. Masek, K. W. Edmonds, J. Konig, J. Sinova, M. Polini, N. A. Goncharuk, A. H. MacDonald, M. Sawicki, A. W. Rushforth, R. P. Campion, L. X. Zhao, C. T. Foxon, and B. L. Gallagher, *Phys. Rev. B.* **72**, 165204, (2005).
9. G. T. Thaler, M. E. Overberg, B. Gila, R. Frazier, C. R. Abernathy, S. J. Paerton, J. S. Lee, S. Y. Lee, Y. D. Park, Z. G. Khim, J. kim, and F. Ren, *Appl. Phys. Lett.* **80**(3964), (2002).

10. S. Dhar, O. Brandt, M. Ramsteiner, V. F. Sapega, and K. H. Ploog, *Phys. Rev. Lett.* **94**(037205), (2005).
11. M. Venkatesan, C. B. Fitzgerald, J. G. Lunney, and J. M. D. Coey, *Phys. Rev. Lett.* **93**(177206), (2004).
12. E. L. Nagaev and M. Samokhvalov (Translator), *Physics of Magnetic Semiconductors.* (Mir Publishers, 1983).
13. E. L. Nagaev, *Colossal Magnetoresistance Materials and Other Magnetic Semiconductors.* (World Scientific Publishing Company, 2002).
14. F. Hulliger, *Handbook on the Physics and Chemistry of Rare Earths*, vol. 4, chapter edited by Karl A. Gschneider Jr., and LeRoy Eyring, pp. 153–236. (North-Holland Physics Publishing, New York, 1979).
15. O. Vogt and K. Mattenberger, *Handbook on the Physics and Chemistry of Rare Earths*, vol. 17, chapter edited by K. A. Gschneider Jr., L. Eyring, G. H. Lander and G. R. Choppin, pp. 301–407. (Elsevier Science Publishers, New York, 1993).
16. C. M. Aerts, P. Strange, M. Horne, W. M. Temmerman, Z. Szotek, and A. Svane, *Phys. Rev. B.* **69**, 045115, (2004).
17. P. Larson, W. R. L. Lambrecht, A. Chantis, and M. van Schilfgaarde, *Phys. Rev. B.* **75**, 045114, (2007).
18. C.-G. Duan, R. F. Sabiryanov, J. Liu, W. N. Mei, P. A. Dowben, and J. R. Hardy, *Phys. Rev. Lett.* **94**, 237201, (2005).
19. C.-G. Duan, R. F. Sabiryanov, W. N. Mei, P. A. Dowben, S. S. Jaswal, and E. Y. Tsymbal, *J. Phys.: Condens. Matter.* **19**, 315220, (2007).
20. L. Degiorgi, W. Bacsa, and P. Wachter, *Phys. Rev. B.* **42**, 530, (1990).
21. R. Hauger, E. Kaldis, G. von Schulthess, P. Wachter, and C. Zürcher, *J. Magn. Magn. Mater.* **103**, 3, (1976).
22. F. Leuenberger, A. Parge, W. Felsch, K. Fauth, and M. Hessler, *Phys. Rev. B.* **72**, 014427, (2005).
23. S. Granville, B. J. Ruck, F. Budde, A. Koo, D. Pringle, F. Kuchler, A. Bittar, G. V. M. Williams, and H. J. Trodahl, *Phys. Rev. B.* **73**, 235335, (2006).
24. A. R. H. Preston, S. Granville, D. H. Housden, B. Ludbrook, B. J. Ruck, H. J. Trodahl, A. Bittar, G. V. M. Williams, J. E. Downes, A. DeMasi, Y. Zhang, K. E. Smith, and W. R. L. Lambrecht, *Phys. Rev. B.* **76**, 245120, (2007).
25. A. R. H. Preston, *unpublished.* (2008).
26. K. Khazen, H. J. von Bardeleben, J. L. Cantin, A. Bittar, S. Granville, H. J. Trodahl, and B. J. Ruck, *Phys. Rev. B.* **74**, 245330, (2006).
27. J. W. Gerlach, J. Mennig, and B. Rauschenbach, *Appl. Phys. Lett.* **90**, 061919, (2007).
28. B. Ludbrook, *unpublished.* (2008).
29. E. Burzo, N. Bucur, H. Allmaier, and L. Chioncel, *J. Optoelectron. Adv. Mater.* **10**, 389, (2008).
30. W. Stutius, *Phys. Kondens. Mater.* **10**, 152, (1969).
31. C. Meyer, B. J. Ruck, J. Zhong, S. Granville, A. R. H. Preston, G. V. M. Williams, and H. J. Trodahl, *Phys. Rev. B.* **78**, 174406, (2008).
32. M. D. Johannes and W. E. Pickett, *Phys. Rev. B.* **72**, 195116, (2005).

33. V. N. Antonov, B. N. Harmon, A. N. Yaresko, and A. P. Shpak, *Phys. Rev. B.* **75**, 184422, (2007).
34. S. Abdelouahed and M. Alouani, *Phys. Rev. B.* **76**, 214409, (2007).
35. D. B. Ghosh, M. De, and S. K. De, *Phys. Rev. B.* **72**, 045140, (2005).
36. A. N. Chantis, M. van Schilfgaarde, and T. Kotani, *Phys. Rev. B.* **76**, 165126, (2007).
37. A. Sharma and W. Nolting, *J. Phys.: Cond. Matt.* **18**, 7337, (2006).
38. C. Meyer, *unpublished.* (2008).
39. R. P. Martin, *Electronic Structure: Basic Theory and Practical Methods.* (Cambridge University Press, 2004).
40. J. P. Perdew and A. Zunger, *Phys. Rev. B.* **23**, 5048, (1981).
41. W. M. Temmerman, A. Svane, Z. Szotec, and H. Winter, *Electronic Density Functional Theory: Recent Progress and New Directions.* (Plenum, New York, 1997).
42. V. I. Anisimov, J. Zaanen, and O. K. Anderson, *Phys. Rev. B.* **44**, 943, (1991).
43. V. I. Anisimov, F. Aryasetiawan, and A. I. Lichtenstein, *J. Phys.: Condens. Matter.* **9**, 767, (1997).
44. A. I. Lichtenstein, V. I. Anisimov, and J. Zaanen, *Phys. Rev. B.* **52**, R5476, (1995).
45. G. Kotliar, S. Y. Savrasov, K. Haule, V. S. Oudovenko, O. Parcollet, and C. A. Marianetti, *Rev. Mod. Phys.* **78**, 865, (2006).
46. M. van Schilfgaarde, T. Kotani, and S. Faleev, *Phys. Rev. Lett.* **96**, 226402, (2006).
47. S. Bhattacharjee and S. M. Jaya, *Eur. Phys. J. B.* **49**, 305, (2006).
48. A. Hasegawa and A. Yanase, *J. Phys. Soc. Jpn.* **42**, 492, (1977).
49. A. G. Petukhov, W. R. L. Lambrecht, and B. Segall, *Phys. Rev. B.* **53**, 4324, (1996).
50. W. R. L. Lambrecht, *Phys. Rev. B.* **62**, 13538, (2000).
51. P. Larson and W. R. L. Lambrecht, *Phys. Rev. B.* **74**, 085108, (2006).
52. H. Yamada, T. Fukawa, T. Muro, Y. Tanaka, S. Imada, S. Suga, D. X. Li, and T. Suzuki, *J. Phys. Soc. Jpn.* **65**, 1000, (1996).
53. P. Larson and W. R. L. Lambrecht, *J. Phys.: Condens. Matt.* **18**, 11333, (2006).
54. A. Svane, V. Kanchana, G. Vaitheeswaran, G. Santi, W. M. Temmerman, Z. Szotek, P. Strange, and L. Petit, *Phys. Rev. B.* **71**, 045119, (2005).
55. M. S. S. Brooks, *J. Magn. Magn. Mater.* **47&48**, 260, (1985).
56. T. Komesu, H.-K. Jeong, J. Choi, C. N. Borca, P. A. Dowben, A. G. Petukhov, B. D. Schultz, and C. J. Palmström, *Phys. Rev. B.* **67**, 035104, (2003).
57. M. Horne, P. Strange, W. M. Temmerman, Z. Szotek, A. Svane, and H. Winter, *J. Phys.: Condens. Matter.* **16**, 5061, (2004).
58. G. Busch, G. E. Kaldis, E. Schaufelberger-Teker, and P. Wachter. In *Coll. Int. CNRS No. 180*, vol. 1, p. 359, Paris-Grenoble, (1970).
59. Y. Shapira, S. Foner, and T. B. Reed, *Phys. Rev. B.* **8**, 2299, (1973).

60. H. J. Trodahl, A. R. H. Preston, J. Zhong, B. J. Ruck, N. Strickland, C. Mitra, and W. R. L. Lambrecht, *Phys. Rev. B.* **76**, 085211, (2007).
61. C. Mitra and W. R. L. Lambrecht, *Phys. Rev. B.* **78**, 195203, (2008).
62. J. Stöhr, *NEXAFS Spectroscopy.* (Springer, 2003).
63. F. Leuenberger, A. Parge, W. Felsch, F. Baudelet, C. Giorgetti, E. Dartyge, and F. Wilhelm, *Phys. Rev. B.* **73**, 214430, (2006).
64. U. von Barth and G. Grossmann, *Phys. Rev. B.* **25**, 5150, (1982).
65. G. D. Mahan, *Phys. Rev. B.* **21**, 1421, (1980).
66. R. J. Gambino, T. R. McGuire, H. A. Alperin, and S. J. Pickart, *J. Appl. Phys.* **41**, 933, (1970).
67. M. Kuznietz, *J. Appl. Phys.* **42**, 1470, (1971).
68. R. A. Cutler and A. W. Lawson, *J. Appl. Phys.* **46**, 6, (1976).
69. C. Mitra and W. R. L. Lambrecht, *Phys. Rev. B.* **78**, 134421, (2008).
70. K. Doll, *J. Phys.: Condens. Matter.* **20**, 075214, (2008).
71. T. Kasuya and D. X. Li, *J. Magn. Magn. Mater.* **167**, L1, (1997).
72. C.-G. Duan, R. F. Sabiryanov, W. N. Mei, P. A. Dowben, S. S. Jaswal, and E. Y. Tsymbal, *Appl. Phys. Lett.* **88**, 182505, (2006).
73. H. Adachi and H. Ino, *Nature.* **401**, 148, (1999).
74. R. M. Moon and W. C. Koehler, *J. Magn. Magn. Mater.* **14**, 265, (1979).
75. T. S. Santos, J. S. Moodera, K. V. Raman, E. Negusse, J. Holroyd, J. Dvorak, M. Liberati, Y. U. Idzerda, and E. Arenholz, *Phys. Rev. Lett.* **101**, 147201, (2008).
76. A. Schmehl, V. Vaithyanathan, A. Herrnberger, S. Thiel, C. Richter, M. Liberati, T. Heeg, M. Rockerath, L. F. Kourkoutis, S. Muhlbauer, P. Boni, D. Muller, Y. Barash, J. Schubert, Y. Idzerda, J. Mannhart, and D. Schlom, *Nat. Mater.* **6**, 882, (2007).
77. H. Ott, S. J. Heise, R. Sutarto, Z. Hu, C. F. Chang, H. H. Hsieh, H. J. Lin, C. T. Chen, and L. H. Tjeng, *Phys. Rev. B.* **73**, 094407, (2006).
78. D. DiMarzio, M. Croft, N. Sakai, and M. W. Schafer, *Phys. Rev. B.* **35**, 8891, (1987).
79. S. Nakamura, G. Fasol, and S. J. Pearton, *The Blue Laser Diode: The Complete Story.* 2nd edition, (Springer, 2000).

Chapter 10

DILUTE MAGNETIC OXIDES: CURRENT STATUS AND PROSPECTS

Karen Yates

The Blackett Laboratory, Physics Department
Imperial College London, SW7 2AZ, UK
E-mail: k.yates@imperial.ac.uk

Over the past ten years, a new field of dilute magnetic semiconductors and dilute magnetic oxides has emerged. These materials, in which a host semiconductor such as ZnO is doped with a transtion metal ion in order to produce ferromagnetic behaviour, have been extremely controversial. Numerous reports have claimed intrinsic ferromagnetism above room temperature while a large number of others attribute these room temperature effects to impurities. Certainly, the observation of room temperature ferromagnetism in insulating samples is a challenge both theoretically and experimentally. Recent developments have shown that some of these dilute magnetic oxides carry a spin polarised transport current. These materials hold significant potential for spintronics devices if the ferromagnetism can be understood and controlled.

10.1. Introduction

Dilute magnetic semiconductors (DMS) are materials such as GaAs, AlN, InSb that become ferromagnetic (FM) when doped with transition metal (TM) ions such as Mn, Fe, Co etc. In recent years, the field of dilute magnetic semiconductors has developed a sub-field, "dilute magnetic oxides" (DMO) in which the TM ions are doped into semiconducting oxides (eg. ZnO, SnO_2, TiO_2) with, or without, additional doping in order to introduce carriers. Many of these oxides have been reported to show FM behaviour above room temperature and a significant portion of these FM oxides are highly resistive. This has caused some authors to introduce a further classification "Dilute magnetic dielectrics" (DMD) when FM is observed in insulating samples. Research into DMS, DMO and DMD materials has, in recent years, been extraordinarily active. The volume of papers published in the field (more than 290 on the magnetic properties of ZnO in the past year alone) has meant that there are, by necessity, frequent reviews of the subject. In the past few years there have been excellent reviews on growth conditions, theoretical models and device applications and the reader is referred to these for in depth discussion of these points.[1–5] A room temperature (or above) ferromagnetic spin injector that could easily be incorporated into existing electronic architecture, opens up many device applications for spintronics and has been a motivating factor in this research drive.[1] Furthermore, the prospect of transparent DMOs opens the door to a whole new field of magneto-optical devices.[6] Early research on the carrier mediated dilute magnetic semiconductor (Ga,Mn)As quickly led to devices being made from this DMS material.[7,8] However, despite intensive effort to raise the Curie temperature, T_c, the low T_C, of (Ga,Mn)As (172 K) limits this material's applications potential. When calculations by Dietl *et al.*, suggested that p-type Mn doped ZnO would be ferromagnetic at temperatures up to or above room temperature[9] a frenzy of activity followed, where numerous semiconductors were investigated for their room temperature DMS potential.

DMS materials typically show very small magnetic moments relative to their sample size, of the order of $\sim 10^{-6}$ emu. The magnitude of these magnetic signals has inevitably led to conflicting results. One difficulty is that such magnetic signals can easily arise from impurity sources, either extrinsic (eg. stainless steel tweezers[10]) or intrinsic (eg. clusters[11–13]). Additionally, even in thoroughly characterised and carefully handled samples, results are highly variable and contradictory.[2,14–62] It is becoming understood that defects are necessary for the observation of room temperature ferromagnetism, particularly in the insulating DMD materials.[4,54] The implicit irreproducibility of defects may then go some way to explaining the amount of disagreement in the literature. Recently, it has even been suggested that under certain conditions, some undoped materials become ferromagnetic[10,63–67] and that adding the magnetic dopant, Co, into ZnO actually reduces the natural magnetic moment of the host material.[66] These results are often explained within the developing theoretical context of d^0 ferromagnetism.[64,68,69] However, before assigning the inconsistencies in the literature to intrinsic defects or to carrier concentration, it is important to rule out the potential impurities that can contribute to the magnetic signal observed. It should be anticipated that some of the papers which report intrinsic ferromagnetism may contain impurities yet limitations in detection technique available prevent their observation.[70] In this field of DMS, Carl Sagan's much used quote holds more than ever: "absence of evidence is not evidence of absence".

10.2. Impurities

A major problem of DMS research is that the moments typically measured in thin film samples are at the detection limit of the magnetometers employed. This becomes particularly important when these small FM signals are superimposed onto a large diamagnetic signal which originates from the substrate. Small amounts of extrinsic magnetic impurity can easily show magnetic moments ~5 μemu.[31] Thorough characterisation of the samples for potential impurities is thus key to producing intrinsically FM samples. The low concentration of TM ions and therefore impurity phases means that traditional detection techniques such as X-ray diffraction may not be adequate to detect impurities that cause measureable magnetic properties. In this section, expected impurities and ways to detect them are reviewed.

10.2.1. *Types of impurity in DMS systems*

10.2.1.1. *Extrinsic impurities*

Different types of impurities can contribute to the magnetic properties of DMS candidate systems. At the simplest level, impurities have been found to originate from the type of tweezers used to handle the thin films as they were grown.[10] Abraham *et al.*,[10] undertook a systematic study of the growth of HfO_2 thin films. It has recently been suggested that this material is intrinsically magnetic if oxygen deficient.[64,67,68] Abraham *et al.*, found no ferromagnetic signal in any of the films that they grew when Teflon tweezers were used to handle the films during and after growth. However, stainless steel tweezers used during film handling led to the observation of a magnetic signal of the same order of magnitude as that found in the material itself by Venkatesan *et al.*[67] Similarly, by subjecting a series of blank substrates to the same thermal treatment as used in thin film growth of DMS, Golmar *et al.*,[71] and Belghazi *et al.*,[72] showed that magnetic signals were present which were due to extrinsic impurties originating from the substrate holders. These facts emphasise the caution with which the DMS materials must be handled during growth and characterisation when the magnetic moments observed are close to the detection limit. Nonetheless, even while avoiding extrinsic impurities, the DMS/DMO materials are host to a minefield of potential magnetic impurities which need to be considered and eliminated as potential sources of ferromagnetism in any claimed DMS/DMO material.[3]

10.2.1.2. *Clusters*

The most common method of obtaining a DMS or DMO material has been the incorporation of TM ions into a host lattice.[2,11,13,15,16,18,23,27,28,31,33–37,39,41–43,46–48,52,54,58,62,66,73–128] Nanoclusters of the TM ion, which is frequently a magnetic element (eg. Co, Fe) therefore form the most obvious candidates for impurity sources of magnetism. If these clusters are small enough, the clustering results in superparamagnetic behaviour which can be easily detected by temperature dependent magnetisation measurements[70,129] although if formed from Co, the clusters only need to be between 4–8 nm for the blocking temperature to be higher than room temperature.[11] It may be suggested that an easier dopant to work with would be one that could not form ferromagnetic impurities such as Mn. However, even this dopant has been suggested to cluster and be a source of spurious magnetic moments in thin films of DMS and DMO material.[12,130–132]

Ferromagnetic coupling has been observed in very small clusters of Mn below 68 K,[131] although it has been argued as to whether Mn clustering would result in an increased, or decreased T_c in DMS.[130,132] The low level of cluster impurity expected limits the ability of techniques such as XRD to detect them. Nonetheless, owing to the extreme likelihood of TM cluster formation, it is worth taking long XRD scans to determine whether such clustering exists.[70] In a recent study Zhou *et al.*, showed that clusters of certain TM ions were easily detected with conventional XRD owing to their tendency to grow orientated with the host lattice,[70,133] this is shown in Fig. 10.1. For example, Co clusters can form both hcp and fcc structures. Zhou *et al.*, found that in ZnO, Co clusters orientated with the ZnO host matrix such that hcp-Co(0001)[1$\underline{1}$00]//ZnO(0001)[1$\underline{1}$00] and as a consequence, Co cluster formation within ZnO was detectable by conventional XRD with the hcp-Co(0002) diffraction peak being present at ~44° (Fig. 10.2). A similar relation was found for Ni implanted ZnO[133] indicating that both Co and Ni can be detected by conventional XRD if the scan time is long enough.

The situation however is quite different for Co implanted TiO_2 where the Co clusters are orientated at 49.82° compared with the crystal axis, corresponding to (b) in Fig. 10.1. In this case conventional XRD was not sufficient to detect the Co clusters and Zhou *et al.* needed synchrotron XRD to detect the presence of the clusters. For Mn into Si (forming $MnSi_{1.7}$) and Fe into ZnO (bcc structure), the magnetic clusters had random orientation with respect to the host lattice ((c) in Fig. 10.1).

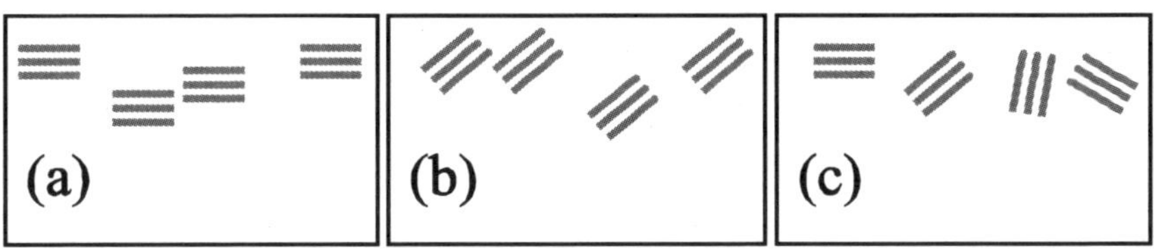

Fig. 10.1. Schematic of the orientation of clusters in DMS systems with respect to the substrate. (a) orientated, typical of Co and Ni in ZnO, (b) orientated but not with substrate, typical of Co in TiO_2, (c) random orientation, typical of Fe in ZnO, Reprinted with permission from Ref. 70 (Copyright 2008, American Institute of Physics).

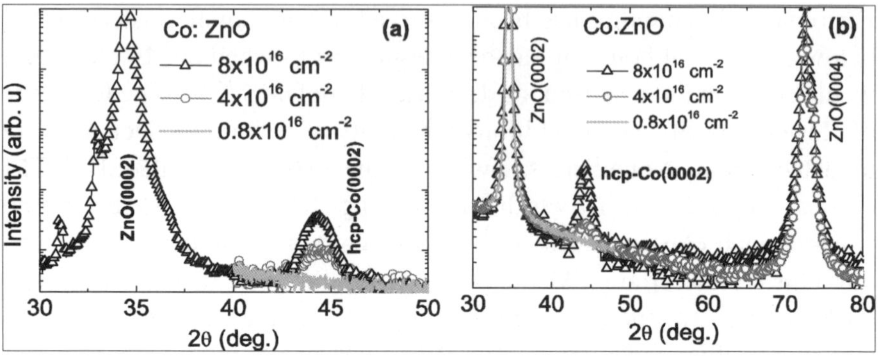

Fig. 10.2. (a) XRD diffraction of Co implanted ZnO as a function of ion implantation dosage. (b) Syncrotron XRD of Co doped ZnO. Reprinted with permission from Ref. 70 (Copyright 2008, American Institute of Physics).

Again, conventional XRD failed to detect the clusters. However for all of these samples in which cluster formation was observed, the temperature dependent magnetic properties indicated superparamagnetic behaviour. Therefore, even if the clusters were not indicated by conventional XRD, the temperature dependent magnetisation curves should signify that a degree of caution is necessary.

Transmission electron microscope (TEM) results are invaluable for cluster identification. Figure 10.3 shows a high resolution TEM image of a Mn_5Ge_3 cluster within a GeMn thin film. This system is of significant interest because, although room temperature ferromagnetism has been observed in nanowires of Mn doped Ge, FM is not observed in thin films of similar composition.[134] Kazakova *et al.*,[134] showed that in thin films, nanoclusters of Mn_5Ge_3 were formed (<10 nm diameter) which were absent in the corresponding nanowires. In Co doped ZnO, Ivill *et al.*, used TEM to show that Co cluster formation tended to occur near the film-substrate interface as shown in Fig. 10.3(b).

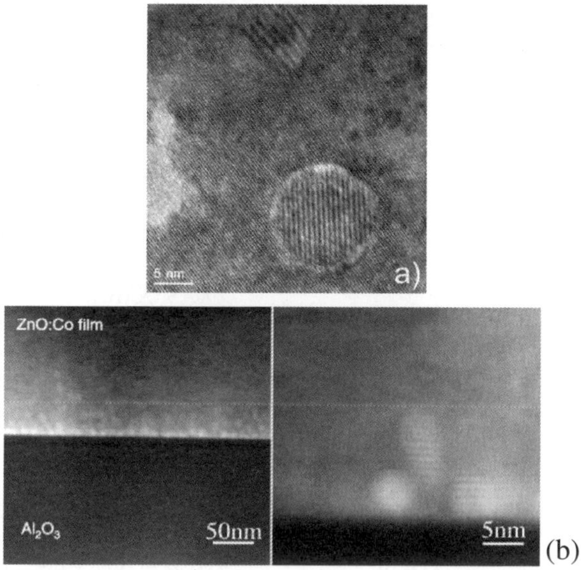

Fig. 10.3. High resolution TEM of a Ge:Mn film showing Mn_5Ge_3 clusters. Reprinted with permission from Ref. 134 (Copyright (2008) American Physical Society) (b) Co clusters forming near the substrate-film boundary in Co doped ZnO, reproduced with permission from Ref. 92, (copyright (2008) Institute of Physics Publishing).

However, observation of clusters by TEM requires the cross section made by the TEM to be coincident with a region of nanoclusters. The dilution of TM ions used to dope these DMS materials means that it is quite conceivable that the cross section does not contain any such impurities. Therefore, again, absence of evidence from TEM should not be taken as evidence of absence.[113,135]

10.2.1.3. *Solubility*

Although the concentration of TM ions doped into host materials for DMS/DMO research is very low, the solubility limit of the ions in the semiconductor host can be even lower.[136] These materials readily phase segregate into a variety of possible magnetic phases that are challenging to detect by techniques such as XRD. A summary of the more common types of magnetic impurity and the detection methods used to isolate them is given in Table 10.1.

Table 10.1. Dopants and possible impurities in DMO systems.

Dopant	Impurity	Magnetic properties	Detection method	Ref.
Co	Co clusters	FM, ~1373 K	XRD, RHEED, Auger, TEM, XAS, XMCD	13, 98, 99, 137
	Co metal at interfaces	FM	TEM	116
	$ZnCo_2O_4$	AFM if n-type FM if p-type (low Temp.)	Raman, XRD	138, 139
	Co_3O_4	AFM, T_N 30–40 K[140]	Raman, HRTEM	53
	CoO	AFM	XRD at high conc	34
	$CoAl_2O_4$	AFM		92
Mn	Mn clusters	Variable		131, 132, 141
	$Mn_{2-x}Zn_xO_{3-\delta}$	FM at rt	Measurements as a function of O content	142
	Mn_xN	Predicted FM		12, 143, 144
	Mn_4N	Ferrimagnetic, $T_c = 745$ K		145
	MnSi (Si from substrate)	Complex	Synchrotron radiation XRD	70
	$ZnMnO_3$	Spin glass		100
Ni	Ni clusters	FM	XRD, Synchrotron radiation XRD	70
Fe	Fe clusters	FM	Synchrotron radiation XRD, Mössbauer	40, 70
	Fe_3O_4	Ferrimagnetic	Mössbauer	146
	Fe_xO	Variable	XRD at Fe concentrations > 0.1	34
	$(ZnFe)_3O_4$	FM, $T_c = 440$ K	XRD at Fe concentrations > 0.1	34

Therefore, even if magnetic cluster formation is avoided, there is still an intrinsic difficulty that many of the transition metal ions used to dope these materials magnetically have a low solubility threshold.[30,34,41,47]

Preparation methods and off-equilibrium growth techniques such as pulsed laser deposition (PLD) may result in a greater incorporation of magnetic dopant ions than equilibrium growth methods. Nonetheless, the solubility of TM ions into the host materials is still relatively low and close to that used in many studies. For bulk samples, solubility levels of <10% have been reported.[47,100] Riyadi *et al.*, monitored the lattice parameter shift with increasing Mn content to show that the solubility of Mn into ZnO could even be as low as 6%.[47] In bulk materials it has also been shown that the ferrimagnetic phase (Mn,Zn)Mn_2O_4 is easily formed.[28] In contrast in thin films, Jin *et al.*, reported that although the equilibrium solubility of Mn in ZnO was ~20%[30] it was up to 36% in laser assisted processes far off equilibrium.[23,30] For Co doping the situation is fairly similar with very low solubility limits reported.[53,147,148] Raman spectroscopy was used to show that the solubility of Co into ZnO was only ~5% in films deposited by metal-organic decomposition,[53] 10% in bulk material[139] and up to ~15% in nanoalloys prepared by the solvothermal technique[147] or PLD,[149] Prellier *et al.*, showed that the c-lattice parameter in Co doped ZnO obeyed Vegard's law until Co doping of 10% while Kim *et al.*, measured a solubility of Co in ZnO of 40%.

10.2.1.4. *Spinels as secondary phases*

A key phase that is formed beyond the solubility phase are the spinels such as $ZnCo_2O_4$.[150] This phase is known to be magnetic under certain oxygenation conditions[138] and so is a significant problem phase for DMS materials doped with Co. The magnetic properties of $ZnCo_2O_4$ as a function of oxygenation is shown in Fig. 10.4.

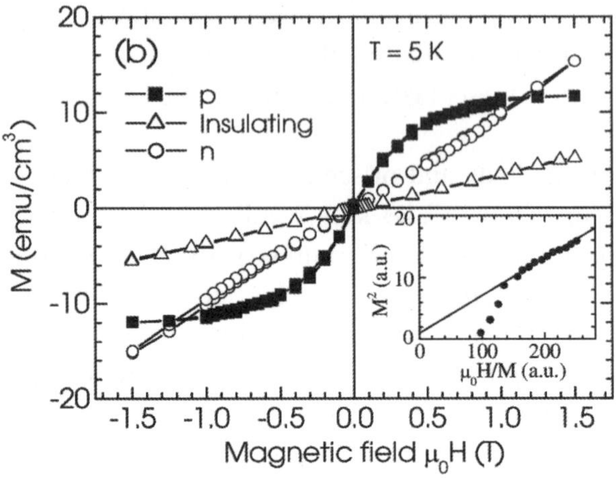

Fig. 10.4. As the oxygen conditions in the deposition chamber are varied $ZnCo_2O_4$ goes from n type to p type and an appreciable magnetic moment is observed in the films at low temperatures. Inset shows the Arrott plot for the p-type sample, taken with permission from Ref. 138 (Copyright (2004) Wiley-VCH Verlag GmbH & Co. KGaA).

Again, the low concentration of the spinel limits the ability of many detection methods to reliably confirm or eliminate its existence in a film.[92,151,152] It has been shown that for the spinel phase, Raman spectroscopy is a very sensitive probe for small quantities of impurity elements and detects this particular impurity form at concentrations much less than that detectable by XRD.[139,149,151] For example, Wang *et al.*, found that in Co doping into ZnO, the impurity spinel phase $Zn_yCo_{3-y}O_4$ was found by XRD at a dopant concentration of Co ≥ 0.391. In contrast, from Raman spectroscopy, it was clear that this phase was already present at Co concentrations of $x \geq 0.155$.[147] Figure 10.5(a) shows the effect on the Raman spectra of increasing the Co dopant concentration into ZnO in PLD grown thin films. Above a critical level of doping (5% in Ref. 149, 10% in Ref. 139), Raman modes owing to impurity phases can be observed. Figure 10.5(b) compares the Raman spectrum of a PLD target of Co doped ZnO and the spinel $ZnCo_2O_4$ showing clear indications that the spinel is present in the target material.[139,151] This makes the Raman technique very attractive for

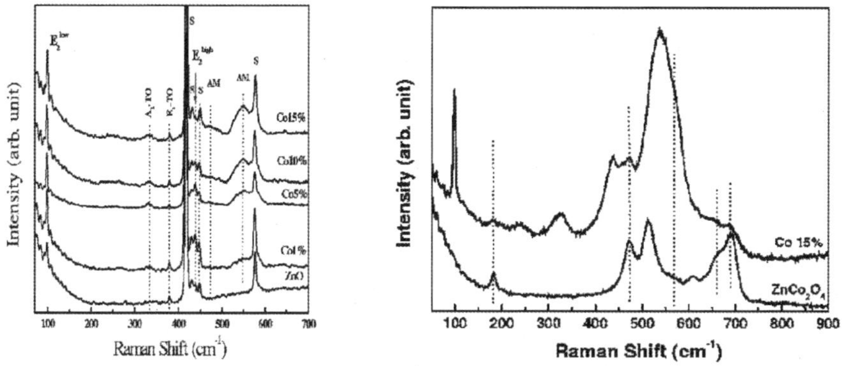

Fig. 10.5. (a) Raman spectra of Co doped ZnO as a function of Co concentration showing development of impurity modes (labeled AM) (b) Raman spectrum of a Co doped ZnO PLD target and the spinel $ZnCo_2O_4$ showing clear evidence of this impurity phase in the target. Taken with permission from Ref. 139 (Copyright (2006) American Physical Society).

detecting-possible spinel accumulation. For the n-type materials where this spinel phase is FM, this may prove particularly useful.[150] Raman spectroscopy has also been used to search for the existence of the magnetic impurity Co_3O_4.[53,147,152]

Although the similarity in Raman spectra between the spinel Co_3O_4 and the spinel $Zn_yCo_{3-y}O_4$ has been noted.[147,152] In all of these studies, XRD had shown the samples to be single phase at low dopant concentrations.

10.2.1.5. *Other secondary phases*

Even in the absence of specific precipitates such as Mn clusters, Co clusters or the spinel phase, it has been reported that metastable phases exist and are responsible for the observed magnetic phases. Such a metastable phase is oxygen vacancy stabilised $Mn_{2-x}Zn_xO_{3-d}$ which can form at the interface of thin films.[142] The ferromagnetism observed is then not due to an intrinsic carrier mechanism but because of this metastable phase.[142] Similarly, uncompensated spins at the surfaces and interfaces of thin films can result in the observation of a ferromagnetic moment.[110] Additional use of common experimental tools can indicate the presence of metastable or impurity phases. Temperature dependent magnetisation measurements indicate the presence (or absence) of superparamagnetic metallic clusters while time dependent magnetisation measurements can reveal information regarding the metastable phases.[110,153] Furthermore, growing films of the same composition but varying the thickness (or Ar-ion milling a film and making measurements as a function of film thickness) can indicate whether interface effects and defects are important to the ferromagnetic coupling mechanism.[36,154–157] Other characterisation tools such as transport measurements can reveal details regarding defects in the grains and grain boundaries,[158] however, the Anomalous Hall Effect (AHE) is not necessarily a good indicator of intrinsic ferromagnetic mechanisms in these oxides,[4] because AHE has been observed in non-magnetic semiconductors that contain magnetic secondary phases.[114,159]

Many other techniques such as magnetic circular dichromism (MCD), X-ray absorption (emission) spectroscopy (XA(E)S), X-ray magnetic circular dichromism (XMCD) and XANES give crucial information regarding the valency and environment of the TM dopant ions and the spin splitting of the band structure. XAS has recently been used to show that in magnetic Co or Mn doped ZnO, the TM ion is in the 2+ valency[16] while XMCD was used in TiO_2, to detect the presence of Co metal present in clusters.[4,99] Also by XMCD, Tietze *et al.*, were able to pinpoint oxygen vacancies as the source of ferromagnetic behaviour in their films.[54]

These results indicate that a combination of detection methods, are required before a new material can be considered an intrinsic dilute magnetic semiconductor and some crucial experimental checks (such as cycling the substrate without depositing the film) need to be undertaken routinely.[160]

10.3. Intrinsic Mechanisms for Magnetic Behaviour

Once all impurities have been accounted for, there remains a large number of conflicting reports either of ferromagnetism[1,4] or the absence of ferromagnetism.[11,34,47,73,100,122,153] Many studies have indicated that oxide DMS were only magnetic in the absence of carriers, that is, when the films were insulating.[83,161] Theoretical predictions however had indicated that carriers were necessary for FM behaviour and that p-type ZnO in particular would be ferromagnetic at room temperature when further doped with Mn.[150] In the absence of p-type carriers, neither Co or Mn doped ZnO were expected to be ferromagnetic.[6] Nonetheless, attempts to introduce carriers into the material were often unsuccessful, partly due to the intrinsic difficulty of doping p-type carriers into ZnO[6] and often resulted in non-magnetic films.[31] Attempts at systematic doping of ZnO with both p and n-type carriers did produce results that agreed with theory.[32] Nonetheless reports were contradictory, and ferromagnetism at room temperature has now been observed in hole doped and electron doped systems[18,37,54,59,85,90,92,102,104,162–165] as well as being absent in similarly doped and prepared materials.[15,31] Results therefore seem contradictory. In DMO materials, oxygen stoichiometry has been shown to be of enormous importance in producing n-type carriers.[125] A partial explanation for the wide range of behaviours observed on doping with carriers, in addition to TM ions, was suggested by the results of Behan *et al.*[15] By deliberately introducing Al as an electron dopant with oxygen defects, Behan *et al.*, showed that there exist three distinct regions of magnetic behaviour as a function of carrier concentration, Fig. 10.6.[15] At low carrier concentrations ($<2 \times 10^{18}$ cm^{-3}), appreciable magnetic moments were observed in films doped with Co. As the concentration of carriers increased, the magnetic moment decreased until no magnetism was observed at room temperature at intermediate carrier concentrations. As the carrier concentration increased to beyond the Mott limit ($>1 \times 10^{20}$ cm^{-3}), ferromagnetism at room temperature was regained. Surprisingly, the results were similar for films codoped with Al and Mn, as well as Al and Co, despite the fact that n-type Mn doped ZnO is not expected to be ferromagnetic at room temperature.[6,9,150] However, as seen by the scatter on the data in

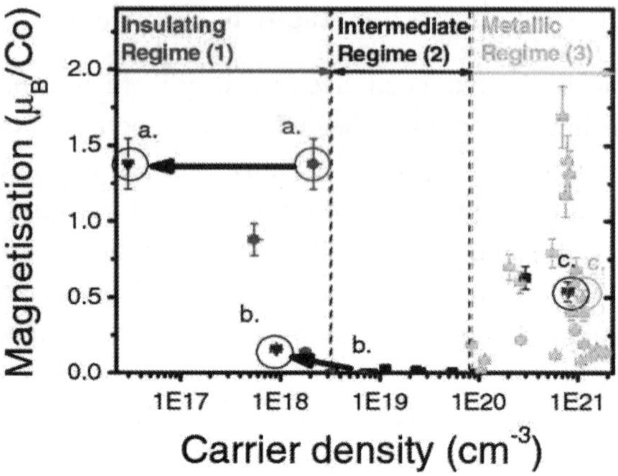

Fig. 10.6. For ZnO doped with Al, there are three regimes of magnetic behaviour as a function of carrier concentration. Arrows indicate the magnetisation at low temperature of two films. Taken with permission from Ref. 15 (Copyright (2008) American Physical Society).

Fig. 10.6 and the contradictory reports of magnetism (or not) in similarly prepared films, the situation is more complex.

10.3.1. *Insulating regime*

Both insulating DMS and DMOs have been reported. The absence of free carriers, in these materials means that conventional carrier mediated models for ferromagnetic exchange such as the Ruderman-Kittel-Kasuya-Yosida (RKKY) interaction cannot be adopted to explain the results.[96] It was proposed that the bound magnetic polaron (BMP) model could be used to explain the magnetic behaviour of the materials.[160] Nonetheless, the T_c calculated with this model was far below that observed. Recent experimental[4,152,158] and theoretical[6,83,161,166] works have emphasised the role that defects play in mediating the FM exchange in these DMD materials. The role of defects is further suggested by the fact that, well characterised samples, or films which were better structurally, showed either AFM or weaker FM coupling compared to more disordered films.[97,167] The magnetic properties are highly sensitive to the growth conditions.

10.3.1.1. *Theoretical treatments*

The Bound Magnetic Polaron (BMP) model was first proposed to explain the magnetic behaviour of the insulating material EuO[168] and has since been extended to deal with the specific situation of the dilute magnetic semiconductors.[160,169,170] In the theory, a shallow donor defect (such as an oxygen vacancy) provides an electron which is confined to a hydrogenic orbital of radius $r_m = \varepsilon_r(m/m^*)a_0$ where ε_r is the high frequency dielectric constant, m is the electron mass and m^* the effective mass of the donor electron. a_0 is the Bohr radius.[160] The exact ε_r will be dependent on growth parameters but assuming reasonable values of ε_r, the polaron size is estimated to be 7.6Å in ZnO, 4.8Å in TiO_2 and 8.6Å in SnO_2.[160] As the polaron density increases, bulk FM behaviour is observed when the polarons are large enough and plentiful enough to overlap. Although qualitatively there is experimental support for the BMP model,[53,120] quantitatively, the T_c predicted by the BMP model is below that experimentally observed in these materials.

Theories have been developed that incorporate the intrinsic irreproducibility of these materials to defects. The associated sensitivity of the magnetism to materials preparation has even been suggested as being integral to the ferromagnetic state of the DMS/DMO/DMD materials.[83,161,171,172] In these models, an oxygen vacancy-TM ion complex allows a vacancy mediated indirect exchange mechanism to develop that is long range and strong enough to provide room temperature ferromagnetism.[83,161] With only a 1% vacancy-TM ion complex, the extended defect states lead to a band of donor levels. As in the BMP model, the effective radius of the magnetic interaction is governed by the extended defect states. However, in this case, as the impurity band becomes full, the vacancy mediated exchange weakens and so a maximum should be observed in both the T_c and the magnetic saturation with increasing vacancy band occupation.[83,161] Although the theory was developed for TiO_2, experimental support for the theory was claimed in ZnO.[125]

10.3.1.2. *Experimental results*

In the insulating regime, where host semiconductors are doped only with a single TM ion, the maximum saturation magnetisation is observed at different doping levels for different TM ions.[4] Specifically, the most robust saturation magnetic moments in ZnO occur at 3–5% Co; 2 at% Mn; 1 at% Cr, Ni or Cu doping.[4] The differences between the ions were associated with differences in the solubility of the TM ion. For Mn doping into ZnO, low concentration led to above room temperature ferromagnetism[4] whereas higher concentrations led to FM ordering below room temperature (around 45 K),[4,173] presumably due to the formation of magnetic impurities.

Qualitative agreement with the BMP model comes from a number of studies[16,17,40,52,66,95,167,174] and the low carrier concentration region of the results of Ref. 15. Various studies have used carrier introduction as a means to test the BMP theory.[18] Chakraborti *et al.*, showed that as the carrier concentration increased, the resistivity decreased, as did the room temperature FM.[18] Similarly Ji *et al.*, doped ZnO with Mn and the p-type dopant Sb.[95] All of these films were insulating at all temperatures and so it was suggested that in accordance with theoretical expectations, p-type carriers localised on the TM site form BMPs that result in the observed room temperature ferromagnetism. Support for defect based models comes from the fact that films of good crystallinity, with narrow XRD rocking curves and in which there is evidence of TM ion substitution into the lattice are frequently non-magnetic, antiferromagnetic, or at least less magnetic than their disordered counterparts.[4,39,97,158,175,176] Although defects have been considered to be essential to the observation of ferromagnetism in TM doped thin films[4] the nature of these defects is debated. Many studies have shown that room temperature magnetic moments can be induced in previously non-magnetic films when the film is subjected to a reducing anneal leading to the conclusion that oxygen vacancies are a prime candidate defect for the development of a ferromagnetic exchange.[36,44,53,54,83,87,117,152,161] Evidence for oxygen vacancies being of critical importance for FM in these DMDs also comes from XMCD measurements. Tietze *et al.*, used XMCD to show that in Co doped ZnO, there was no FM signal on Co or O elements and therefore that oxygen vacancies were the origin for the room temperature ferromagnetism observed in their films.[54] Similar results were observed by Sudakar *et al.*, using Raman spectroscopy.[152] Sudakar *et al.*, studied the Raman spectra and magnetic behaviour of a 1.8% Co doped ZnO film as a function of a series of vacuum anneals. The spectra are shown in Fig. 10.7. A mode at ~690 cm^{-1}, which is attributed to disorder activated Raman scattering is alternatively switched on (off) when annealed in air (vacuum). The activation of this mode with air annealing suggests that it may be linked with oxygen vacancies. The films that were air annealed (few oxygen vacancies) did not show room temperature ferromagnetism, whereas the vacuum annealed films (many oxygen vacancies) showed room temperature ferromagnetism with a maximum value of 0.3 μ_B/Co for $Zn_{0.988}Co_{0.012}O$ at 300 K.[152] Hydrogen interstitials have been proposed as an alternative to oxygen defects,[48,66] although other studies have indicated that oxygen vacancies of themselves are not sufficient to activate the FM in Cr doped TiO_2.[97] Defects such as Zn interstitials have also been proposed to be activated on annealing the samples.[93,177,178] Defects that are produced not on annealing the samples, but in the growth process have also been shown to be of importance in determining which films show FM behaviour. Song *et al.*, studied the influence of varying the substrate on the magnetic behaviour of Co doped ZnO.[155] They found that the orientation of the substrate could have considerable influence on the room temperature FM properties indicating that the Co-O bond length was important in the manifestation of room temperature FM. Similarly strain was shown to be important by Liu *et al.*, who showed that the films subjected to the greatest tensile strain showed the largest saturation magnetizations.[179] These results are consistent with theories suggesting oxygen vacancy-TM ion complexes as a mediator in FM exchange.[83,161]

The role of defects in general to FM behaviour has been indicated by a variety of other measurement techniques. XRD was used to monitor structural defects and correlate these with magnetic behaviour and growth conditions,[115] while Impedance spectroscopy was used by Hsu *et al.*, to show that when films were annealed in Ar, defects occurred mainly at the grain boundaries, whereas films annealed in Ar-H introduced defects to both grains and grain boundaries.[158] It was only films with defects in both grains and grain boundaries that were strongly magnetic at room temperature.[158] Raman spectroscopy has also been used to correlate defect modes in the Raman spectra with magnetic behaviour. Duan *et al.*, showed that in ZnO nanoparticles co-doped with Co and Mn the room temperature ferromagnetism observed in the film was associated with a high ratio of two particular defect modes to the main ZnO wurtzite mode at 436 cm^{-1},[180] suggesting a possible detection method for defect mediated FM.

The results of Griffin *et al.*, suggest an alternative reason for the absence of FM in samples of good crystallinity.[176] Using high resolution TEM, Griffin *et al.*, showed that Co doped TiO_2 films of high crystallinity had a tendency towards Co enrichment near surfaces (grain boundaries, interfaces etc). In contrast, lower crystalline quality films had a more homogenous Co distribution.[176] These results indicate that while DMD materials do show a marked dependency of magnetic behaviour on precise details of the growth conditions, the defects produced, and the effect that these have on the films magnetic character, are far from understood.

10.3.1.3. *"d^0" ferromagnetism*

The observation of ferromagnetism in undoped, insulating oxide thin films such as HfO_2, TiO_2 and In_2O_3 suggests the possibility of a new form of magnetism, so called d^0 magnetism.[64] It is so called because of the absence of d or f electrons which are traditionally thought to be essential for ferromagnetism.

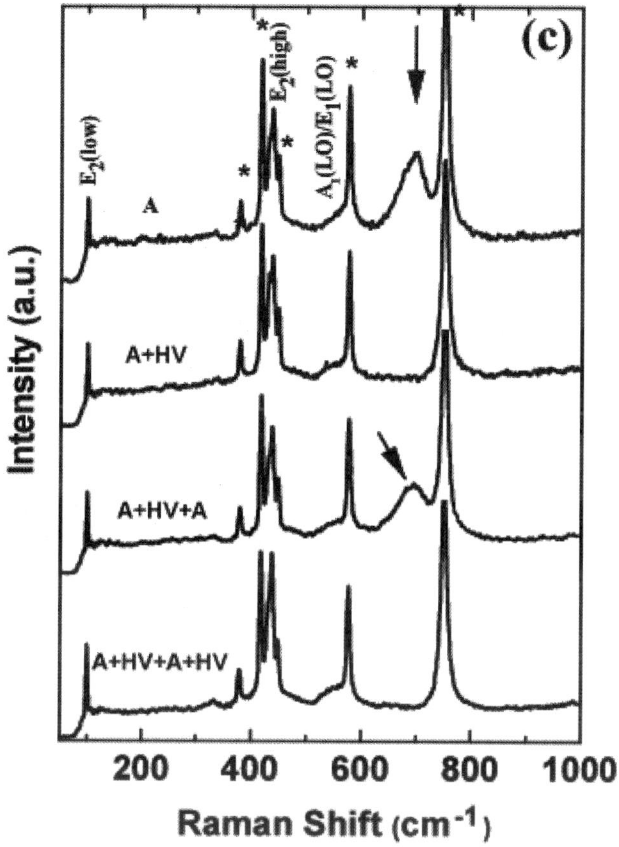

Fig. 10.7. The Raman spectra of $Zn_{0.982}Co_{0.018}O$ films subjected to repeated annealing in air (A) or vacuum (HV). * indicates substrate peaks while the arrow indicates the presence of a mode attributed to disorder induced Raman scattering, reproduced with permission from Ref. 152 (Copyright (2007) Institute of Physics Publishing).

Ferromagnetic behaviour at room temperature has now also been observed in nanocrystalline MgO,[181] GaN,[182,183] CeO_2[65] CdS, $TiO_{2\text{-d}}$, SnO_2 and ZnO[63,76,182] and thin films and nanocrystals of HfO_2[10,64,67,68] and a number theoretical models have been proposed.[184–187] The low magnetic moment observed in these materials also means that caution should be applied.[10,71,72] Nonetheless, the study of Hong *et al.*,[68] where FM was observed at room temperature in In_2O_3 grown on MgO but not on Al_2O_3 does suggest that some form of intrinsic, but defect mediated ferromagnetism can occur in such wide gap semiconducting materials. Again, there is debate about the role of oxygen deficiency in these materials. Hong *et al.*, and Coey *et al.*, showed that oxygen deficiency was crucial in HfO_2 thin films and nanoparticles respectively and that after annealing in oxygen the magnetic moment at room temperature was reduced. Other, contradictory results have been observed by Liu *et al.*, where it was size effects that were proposed to be imperative to observing FM behaviour in nanoparticles of CeO_2.[65] For thin films of these undoped materials, the defects were found to originate at the film-substrate interface as the magnetic moment increased as a function of decreasing film thickness in In_2O_3.[68] Although of course such apparent enhancement of the magnetic moment could result from extrinsic sources of magnetic impurity as discussed earlier.[3,71,72]

10.3.2. *Magnetism at high carrier concentrations*

In order for DMO materials to be useful in spintronics devices it is desirable that they can be used as spin injectors or sources of spin polarised transport carriers,[1] the absence of spin polarised carriers could limit the spintronics applications of these materials.[188] A carrier mediated Double Exchange or RKKY type exchange was originally proposed to explain the ferromagnetism in DMS and DMOs.[37,150,157,160,189,190] As with data in the insulating regime, there is a large amount of irreproducibility and variability in the data for films that are grown with high carrier concentrations and lower resistivity. N-type oxides are relatively easy to grow and electron dopants have included oxygen vacancies[32,125,176,191–194] and/or co-doping with H,[48] Sn,[90–92,195] Al,[15,38,106,107] Ga,[127] Ni[12] and P.[90] P would normally be considered a p-type dopant, but was measured to be n-type in this case.[90] Hole type carriers are more difficult to introduce to materials such as ZnO[6,109] but p-type carriers have been introduced by nitrogen capping nanoparticles or films,[32,90,191] or doping with Cu,[88] As[102–104] or P.[196] It has been observed that the dopant ions are often not sufficient in themselves to produce the carrier concentrations observed[194] and additional carriers are present which are thought to originate from oxygen defiency, Zn interstitials or hydrogen donors.[4] Thus structural defects are intrinsically linked to carrier concentration and attempts to increase the latter necessarily induce the former.[4,108] It has been suggested that it would be naïve to attribute to carrier concentration, magnetic effects that may be more properly associated with defects.[4] Experiments to distinguish between the former and the latter have been attempted but have been limited by the fact that not all studies of the effect of carrier concentration on FM properties measure the carrier concentration. As Behan *et al.*, have shown,[15] introducing only a small number of carriers can actually decrease the resistivity and the magnetisation and it is only when the carrier concentration reaches extremely high levels that room temperature magnetisation is restored, Fig. 10.6. Even in this regime, the scatter in the experimental data and the sensitivity of the properties to the growth conditions implies that other effects must also play a role.[15,31,88,107] Nonetheless, evidence is emerging of carrier mediated ferromagnetism in some of these materials. The observation of such a carrier mediated mechanism and particularly the observation of a spin polarised transport current would be critical for these material's use in spintronics applications where spin injection is required.

10.3.2.1. *Theoretical review*

It is beyond the scope of this review to cover basic concepts associated with Double Exchange, RKKY etc. and the reader is referred to more theoretical reviews of the subject for further information.[5] There are many theoretical models which are based on a need to introduce carriers into a DMO in order that the DMO is FM and not AFM or spin-glass.[6,96,101,167] Hole type carriers are favoured in many studies.[6,9,37,150,167] Risbud *et al.*, found that in Co doped ZnO, the FM state was only stabilised by hole doping (introduced by Zn intersitials) while electron doping (oxygen vacancies) led to an AFM ground state.[167] Carrier based FM exchange dependent on electron doping has also been proposed[96,101] where the ferromagnetism was attributed to a Double Exchange mechanism. Lee *et al.*, even suggested that the RKKY was inappropriate in these materials[101] and suggested a possible cause for the large variability in FM properties. They proposed that to observe FM in Co doped ZnO, a high concentration of Co is needed. Increasing Co content decreases the electron-Co ratio. The FM properties are therefore a fine balance between these two effects. The authors suggested that this, rather than a defect mediated mechanism[4] was the cause of the large amount of variability in the reported experimental results.

Other studies have focussed on the means by which carriers can be introduced to DMOs. It has been thought that oxygen vacancies produce shallow donor defects in the host material that provide n-type carriers. There are many experimental studies in which non-magnetic films are subjected to a reducing anneal (to introduce n-type carriers) after which the films become conducting and ferromagnetic.[36,125] On annealing in an oxygen atmosphere (decreasing the free n-type carrier concentration), the films lose their ferromagnetism and become highly resistive thus the ferromagnetism is claimed to be linked to carrier concentration.[192] However, recent calculations have shown that in materials such as ZnO, the oxygen vacancy is a deep donor, not a shallow donor.[166,197–200] Janotti *et al.*, proposed that, rather than the oxygen vacancies being themselves responsible for the introduction of n-type carriers, instead, the oxgen vacancies introduced by the reducing anneal, led to highly reactive, free Zn bonds, which chemically bond to hydrogen.[166,197,198] Their ab-initio calculations of H substitution into ZnO showed that H substitution would form a shallow donor site in both ZnO and MgO.[166,197,198] Thus hydrogen substitution could explain the carrier concentration dependence on partial pressure of oxygen during growth. These theoretical calculations have recently received experimental support by optical transmission measurements.[199]

10.3.2.2. *Experimental results*

Hole doping

Initially, the results of Kittilsveld *et al.*, on carrier mediated FM exchange in Mn or Co doped ZnO[191] appeared to agree with the theoretical predictions of Dietl *et al.*[9] In this study, nitrogen capping was used to introduce p-type carriers into TM doped ZnO. For ZnO thin films doped with Mn, Kittilsveld *et al.*, found that nitrogen capping (p-type doping) was a necessary condition for observing room temperature ferromagnetism. Conversely, Co doped ZnO showed the opposite behaviour: room temperature ferromagnetism was only observed in the presence of n-type carriers. There are now several studies which have indicated that p-type carriers are necessary for ferromagnetism to be observed in conducting Mn doped ZnO thin films[104,126,196] and that there is a correlation between the FM and the carrier concentration.[38,59,104,125,126,146,191,195,196] Lee *et al.*, varied the p-type carrier concentration in Mn doped ZnO by As implantation to obtain carrier concentrations between 1 x 10^{17} cm^{-3} and 2.7 x 10^{18} cm^{-3}.[102,103] They found a direct correlation between carrier concentration and room temperature magnetism suggesting a carrier mediated exchange. However Hong *et al.*, used Cu doping at 0% and 5% to introduce hole carriers into Mn doped ZnO thin films grown by PLD.[88] They found that although hole doping had some effect on the room temperature magnetic behaviour, the greater effect was obtained from defects eg. oxygen vacancies, hence it was defects rather than carrier concentration that enhanced the ferromagnetic exchange.

Nonetheless, although these studies have indicated that carrier mediated ferromagnetism is possible, perhaps even preferred in Mn doped ZnO, hole doping studies have also provided an insight into the time dependence of the room temperature FM observed in these films. Wan[196] showed a clear correlation between the saturation magnetisation at room temperature and the carrier concentration (which was up to 2.4 x 10^{16} cm^{-3}) in P doped, p-type, Mn doped ZnO thin films. In this study they showed that a lower oxygen environment was preferable for the observation of ferromagnetism at room temperature as was a lower growth temperature, although the lower growth temperature resulted in a poorer structural quality. The maximum saturation magnetisation observed was 0.35 μ_B/Mn at room temperature. Over ten days, the films were re-measured and the saturation magnetisation and carrier concentration were shown to decrease while the resistivity, increased. After ten days, the films were no longer ferromagnetic. A similar time dependent behaviour was observed in Mn doped Indium tin oxide (ITO)[11] and in Ni implanted ZnO, where the FM was attributed to defects.[133]

Electron doping

In order to make meaningful statements concerning the relation between carrier density and magnetic behaviour it is instructive, indeed necessary, to measure both. Claims in the literature regarding carrier concentration are frequently made based on resistivity behaviour alone. Moreover, as has been pointed out by Pan *et al.*,[4] the ratio of the carrier density to the dopant density can give important information concerning the defect density. A measure of the carrier concentration is therefore of extreme importance. N-type carriers can easily be introduced to materials such as ZnO by a reducing anneal or using low oxygen partial pressure during film growth.[15,125,192,194] To introduce enough carriers into the lattice for FM behaviour to be re-introduced, co-doping is required.[15,107,194] It has been shown that the n-type carrier concentration produced by doping with Al produces more carriers than would be expected from ionisation of the Al.[194] So even in this case, low oxygen partial pressures are required during the growth process in order to produce films with high carrier concentrations. At carrier concentrations between ~3 x 10^{18} cm^{-3} and 1 x 10^{20} cm^{-3}, the so-called intermediate regime, the relation between the n-type carrier concentration and the magnetisation is non-trivial[15,90,91,194] and necessitates accurate determination of the carrier concentration. At n-type carrier concentrations of $<10^{20}$ cm^{-3}, the magnetisation may even decrease with increasing doping as shown by Ivill *et al.*, in Sn doped Mn:ZnO.[90,91] The results in this regime have been explained in terms of the bound magnetic polaron model.[15,90,91] In the theory, the ferromagnetic ordering between Mn moments is mediated by holes. Adding Sn to the ZnO in order to dope electrons into the structure cancels the free holes that can participate in this mediated exchange and thereby reduces the ferromagnetic exchange.[90–92] Similar behaviour was found by Behan *et al.*[15] whereby, in a certain region, introducing n-type carriers into the material acts against the BMP ferromagnetic mechanism and decreased the observed saturation magnetisation at room temperature. The ferromagnetic state could be induced at low temperature by freezing out the carriers as shown by the blue data points in Fig. 10.6.

At high carrier concentrations ($>10^{20}$ cm^{-3}), room temperature FM is restored in thin films of ZnO suggesting that a different, carrier mediated FM exchange mechanism is at work.[15] Several authors have indicated that at high carrier concentrations, a FM phase is stable in n-type films[15,38,125,126,195,201] both for Co doped and Mn doped ZnO. Defects need to be considered too and in a study of codoping ZnO with Al and Co, Liu *et al.*[107] showed that while the saturation magnetisation did increase with carrier concentration (up to 2.2×10^{20} cm^{-3}), the defects, as measured by the ratio of carrier concentration to Al concentration also correlated with the saturation magnetization.[107] The authors suggested that the ferromagnetic ordering was correlated with the structural defects rather than the electron concentration. The situation was quite different for Behan *et al.*, who suggested that the dependence of the saturation magnetisation at room temperature on the ratio of the free carriers to the TM dopants was reminiscent of that observed in GaMnAs.[202] Although the scatter in the data indicated that growth conditions were critical in the formation of a room temperature FM state, the clear indication was of a carrier mediated, rather than defect assisted mechanism. An intrinsic carrier mediated mechanism was also suggested by Yang *et al.*,[59] in Mn doped ZnO films codoped with Ga. In this case, the carrier concentration was very high and the saturation magnetisation was directly correlated with the carrier concentration. Intriguingly, for films of such a high carrier concentration, the saturation magnetisation shows very little temperature dependence between 4.2 K and 300 K, suggesting a very high T_c (Fig. 10.8).

A further problem when measuring co-doped films, or films with a high level of oxygen defects is the tendency to phase segregation.[11,203] In these studies, although there was an inverse correlation between the magnetic behaviour of the films and the resistivity, this was due to secondary phase formation rather than inherent ferromagnetism.

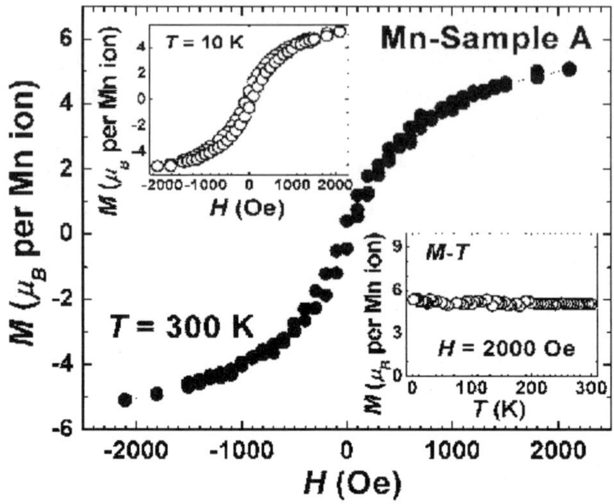

Fig. 10.8. Magnetisation loops at 300 K and 10 K in Mn doped ZnO co-doped with Ga, inset shows a temperature independent saturation magnetisation, Reprinted with permission from Ref. 59 (Copyright (2008), American Institute of Physics).

Measurements of spin polarisation, can these materials be used as spin injectors?

One key question which will determine the suitability of these materials for applications purposes will be whether the transport current in the conducting DMO materials is spin polarised.[1] There are various ways of determining the spin polarisation of the transport current. Xu *et al.*, measured the magnetoresistance of tunnel junctions made of Zn-CoO and inferred a spin polarised current was injected from the Co doped ZnO electrode.[204] While Ji *et al.*, found that the current that was spin injected from Co doped ZnO was polarised by 7% at room temperature.[24,205] These devices suggest that, under certain conditions, Co doped into ZnO carries a spin polarised current. A more direct method of measuring the transport spin polarisation is to use point contact Andreev reflection.[206–208] Point contact Andreev reflection (PCAR) spectroscopy uses the fact that in a spin polarised material, the conductance across a junction between a sharpened superconducting tip and the material is completely suppressed at voltages below the superconducting gap voltage.[206] The degree of suppression of the conductance gives an indication of the transport spin polarisation in any given material. For DMS, DMO and DMD research, the technique suffers from two major drawbacks. Firstly the fact that it requires a voltage drop across the contact between the superconductor tip and the film, means that it is restricted to highly conducting films. Secondly, the use of a superconducting tip necessitates that the measurements are conducted below the superconducting transition temperature of the tip material. Typically tips are Nb, Sn or Pb superconductors which means the measurements have to be taken below 10 K. Nonetheless, the ability of the technique to measure the actual transport spin polarisation and not the spin polarisation of the density of states has meant that it has been employed for a number of DMS materials. A spin polarised transport current of 52 ± 3% was observed in InMnSb but not in InBeSb.[209] Similarly, in GaMnSb a spin polarisation of 57 ± 5% was observed.[210] Spin polarisation has also been detected using the Andreev technique in GaMnAs.[211] Very recently, point contact results have been obtained on highly n-type ZnO codoped with Mn.[60] In this technique, due to the requirement for the film to have a low resistivity, PCAR measurements are restricted to films in the high carrier regime of Fig. 10.6. Nonetheless, in a highly conducting film (carrier concentration = 1.32×10^{21} cm^{-3}) and co-doped with Al and Mn, it was shown that this material can support a spin polarised current at 4.2 K.[60] A transport spin polarisation of up to 55% was observed in contacts in which a surface barrier had been punctured with the tip, Fig. 10.9.

In this figure, the quantity "1/Rn", is a measure of the pressure between the superconducting point contact and the DMS film. At low pressures the observed transport spin polarisation is low but as the tip

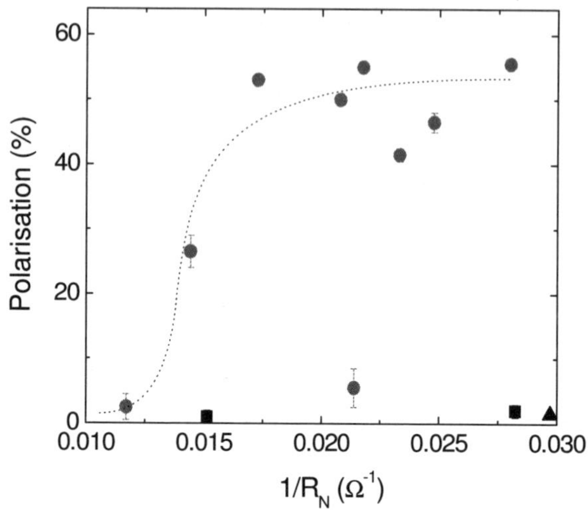

Fig. 10.9. Polarisation as a function of tip pressure (1/Rn) in 2% Mn, 1% Al doped ZnO (●) and 1% Al doped ZnO (●, ▲), the line is a guide for the eye.

pressure increases (and a surface barrier is punctured), the polarisation observed increases up to 55 ± 1%. In contrast, a non-magnetic film of ZnO that had been doped only with Al showed no dependence of the polarisation on tip pressure. These results suggest that the bulk of the thin film of Mn and Al co-doped ZnO supports a spin polarised transport current and therefore suggests a carrier mediated FM exchange mechanism operates in the high carrier concentration regime. The observation of a spin polarised transport current in highly n-type Mn doped ZnO opens this material up as a potential spin injector for spintronics devices, especially owing to the ease with which ZnO can be incorporated into the existing semiconducting architecture.

10.4. Devices Already Made with DMS DMO and DMD Materials

It has long been an aim of the spintronics research community to obtain spin injection into semiconducting materials.[8] In order to use DMS and

DMO in device structures, it is desirable that they can act as spin injectors and that they have a long spin diffusion length. Recent results have suggested that spin polarised transport is supported in highly n-type ZnO co-doped with a TM ion.[25,26,60] While Ghosh *et al.*, showed that the spin lifetime of carriers in Ga doped (n-type) ZnO was long even at room temperature.[25,26] The spin lifetime was also shown to increase with increasing electric field (Fig. 10.10) which has important consequences for devices made out of ZnO as it would allow spin manipulation by electrical fields rather than magnetic fields.[26]

The ability to make magnetic tunnel junctions (MTJs) with room temperature DMO materials is also being investigated.[24,115,204,205] Such devices have been very successful using low temperature DMS materials such as (Ga,Mn)As and reviews on spintronics devices based on these materials are already appearing.[8] Nonetheless, it is preferable technologically for devices to operate at, or close to room temperature. Using dc and rf magnetron sputtering Song *et al.*, grew all-oxide

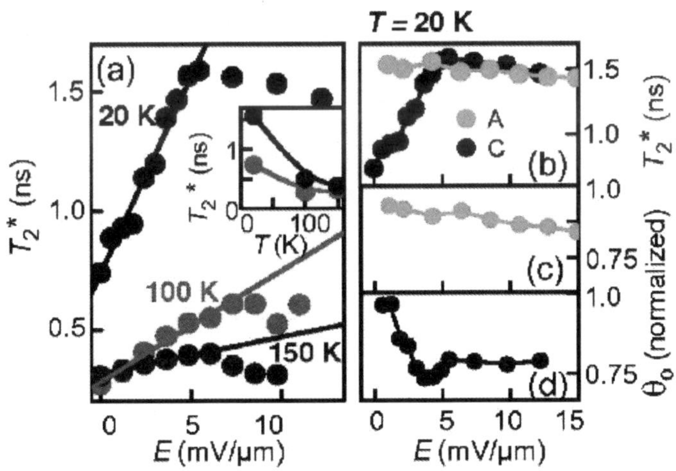

Fig. 10.10. Measurement of spin lifetime as a function of electric field and temperature in Ga doped ZnO, reprinted with permission from Ref. 26 (Copyright (2008), American Institute of Physics).

heterostructures of (Co doped ZnO)/ZnO/(Co doped ZnO) in order to obtain a tunnel magnetoresistance through the ZnO barrier, Fig. 10.11.[115] Song *et al.*, obtained a TMR of 20.8% at 4 K and in 2T. Although a TMR of 180% was observed at room temperature in Fe/MgO/Fe single crystal structures,[212] the results on ZnO structures are promising. Song *et al.*, attributed the large TMR observed in their structures to interface control which led to improved the temperature dependence of the TMR and indeed they were still able to observe a TMR at 300 K, although the magnitude of this effect had diminished to 0.35%. Xu *et al.* recently used MTJs to show that spin polarised transport existed and could be injected from Co doped ZnO at 5 K[204] in agreement with PCAR results on n-type Mn doped ZnO. Room temperature spin polarised injection into ZnO has been theoretically predicted[213] and experimentally measured[24,205] via spin valves made with Co doped ZnO as one of the electrodes. Ji *et al.*, showed that at 300 K the spin injection from Co doped ZnO into ZnO thin films was of the order of 7% (compared with 11.7% at 90 K).[24,205] There are a large number of devices now made with DMO based structures,[24,121,204,205,214,215] suggesting that there is hope for significant improvement in room temperature spintronics devices utilising these DMO materials.

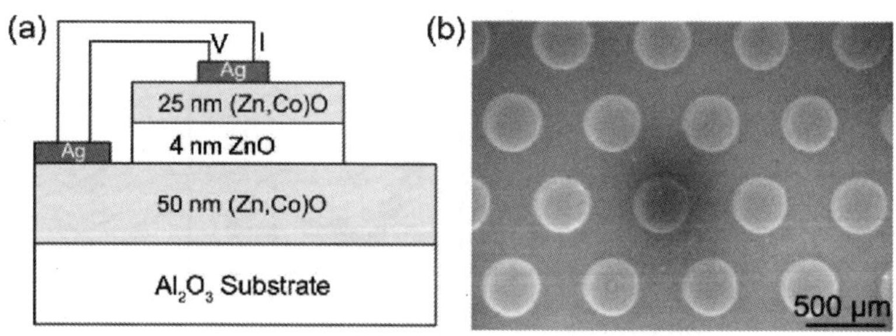

Fig. 10.11. Device structures made out of an all ZnO epitaxial heterostructure, (a) schematic, (b) SEM image of an array of MTJs. Reprinted with permission from Ref. 115 (Copyright (2007) American Institute of Physics).

ZnO itself also has applications in photovoltaics,[216] UV LEDs,[217] piezoelectrics[218] and electrochemical biosensors.[219] The crossover between the use of ZnO as a piezoelectric and doping it to be a ferromagnetic DMD suggests that doped ZnO and similar DMDs may play a role in the emerging field of multiferroics.[220] Indeed, claims of multiferroic behaviour (the coexistence of both ferroelectric and antiferromagnetic or ferromagnetic behaviour) have already been made by Lin *et al.*, in Li and Co co-doped ZnO[105] and Yang *et al.*, in Cr doped ZnO.[221]

10.5. Outlook

Although there has been a considerable amount of debate, there is growing evidence to support the argument that subtle changes in growth conditions can lead to large changes in magnetic behaviour in DMS and DMO based materials. The role played by defects in mediating the ferromagnetism is gradually being established, both when the material is insulating and when it contains free carriers, although much work remains to be done. There are strong suggestions that two types of FM exchange exist in these materials as a function of carrier concentration. At low and intermediate carrier concentrations, some form of BMP model appears to adequately explain the experimental results, in combination with the effects on the BMP model of defects. At high carrier concentrations, a carrier mediated FM exchange appears likely although the role of growth conditions and defects in this regime still needs to be established. In particular, spin polarised transport current has been observed in highly n-type FM ZnO both by direct measurements with point contact Andreev reflection and by inference from magnetic tunnel junction structures. These results mean that this material has enormous potential in spintronic devices. Even in the DMD regime, the combination of ferromagnetic behaviour with optical devices and/or piezoelectric properties means that these DMD materials also have high device potential. Nonetheless, measurements with a variety of characterisation techniques, especially including measuring the number and type of free carriers (in dilute magnetic conducting materials) will be invaluable in further elucidating the nature of the ferromagnetic exchange in these fascinating systems and in making it controllable so that it can be more easily incorporated into device structures.

Acknowledgements

The author is grateful for helpful discussions with colleagues and in particular to K. Morrison, W. R. Branford and L. F. Cohen for careful reading of the manuscript.

References

1. S. J. Pearton, C. R. Abernathy, D. P. Norton, et al., Mater. Sci. Eng.: R: Reports 40 (2003) 137.
2. R. Janisch, P. Gopal, and N. A. Spaldin, J. Phys.: Condens. Matter 17 (2005) R657.
3. J. M. D. Coey, Curr. Opin. Sol. St. 10 (2006) 83.
4. F. Pan, C. Song, X. J. Liu, et al., Mater. Sci. Eng.: R: Reports 62 (2008) 1.
5. T. Dietl, J. Phys.: Condens. Matter 19 (2007) 165204.
6. N. A. Spaldin, Phys. Rev. B 69 (2004) 125201.
7. A. M. Nazmul, S. Sugahara, and M. Tanaka, Phys. Rev. B 67 (2003) 241308.
8. C. K. P. Gould, G. Schmidt, L. W. Molenkamp, Adv. Mater. 19 (2007) 323.
9. T. Dietl, H. Ohno, F. Matsukura, et al., Sci. 287 (2000) 1019.
10. D. W. Abraham, M. M. Frank, S. Guha, Appl. Phys. Lett. 87 (2005) 252502.
11. M. Venkatesan, P. Stamenov, L. S. Dorneles, et al., Appl. Phys. Lett. 90 (2007) 242508.
12. L. M. Sandratskii, P. Bruno, and S. Mirbt, Phys. Rev. B (Condens. Matter Mater. Phys.) 71 (2005) 045210.
13. S. A. Chambers, T. Droubay, C. M. Wang, et al., Appl. Phys. Lett. 82 (2003) 1257.
14. A. J. Behan, J. R. Neal, R. M. Ibrahim, et al., J. Magn. Magn. Mater. 310 (2007) 2158.

15. A. J. Behan, A. Mokhtari, H. J. Blythe, et al., Phys. Rev. Lett. 100 (2008) 047206.
16. F. Bondino, K. B. Garg, E. Magnano, et al., J. Phys.: Condens. Matter 20 (2008) 275205.
17. E. Cespedes, G. R. Castro, et al., J. Phys.: Condens. Matter 20 (2008) 095207.
18. D. Chakraborti, S. Ramachandran, G. Trichy, et al., J. Appl. Phys. 101 (2007) 053918.
19. W. M. Chen, I. A. Buyanova, A. Murayama, et al., Appl. Phys. Lett. 92 (2008) 092103.
20. H. Frenzel, H. V. Wenckstern, A. Weber, et al., Physics of Semiconductors, Pts A and B, AIP Conference Proceedings, 893 (2007) 301.
21. H. Frenzel, H. von Wenckstern, A. Weber, et al., Phys. Rev. B 76 (2007) 035214.
22. H. Frenzel, A. Lajn, M. Brandt, et al., Appl. Phys. Lett. 92 (2008) 92108.
23. T. Fukumura, Z. Jin, M. Kawasaki, et al., Appl. Phys. Lett. 78 (2001) 958.
24. J. Gang, S.-S. Yan, Y.-X. Chen, et al., Chin. Phys. Lett. 23 (2006) 446.
25. S. Ghosh, V. Sih, W. H. Lau, et al., Appl. Phys. Lett. 86 (2005) 232507.
26. S. Ghosh, D. W. Steuerman, B. Maertz, et al., Appl. Phys. Lett. 92 (2008) 162109.
27. S.-J. Han, J. W. Song, C.-H. Yang, et al., Appl. Phys. Lett. 81 (2002) 4212.
28. S.-J. Han, T.-H. Jang, Y. B. Kim, et al., Appl. Phys. Lett. 83 (2003) 920.
29. Y. Ishida, J. I. Hwang, M. Kobayashi, et al., Phys. B-Condens. Matter 351 (2004) 304.
30. Z. Jin, T. Fukumura, M. Kawasaki, et al., Appl. Phys. Lett. 78 (2001) 3824.
31. T. C. Kaspar, T. Droubay, S. M. Heald, et al., New J. Phys. 10, (2008) 055010.
32. K. R. Kittilstved and D. R. Gamelin, J. Appl. Phys. 99 (2006) 08M112.

33. M. Kobayashi, Y. Ishida, J. I. Hwang, et al., Phys. Rev. B 72 (2005) 201201.
34. S. Kolesnik, B. Dabrowski, and J. Mais, J. Appl. Phys. 95 (2004) 2582.
35. H.-J. Lee, S.-Y. Jeong, C. R. Cho, et al., Appl. Phys. Lett. 81 (2002) 4020.
36. X.-C. Liu, E.-W. Shi, Z.-Z. Chen, et al., J. Phys.: Condens. Matter 20 (2008) 025208.
37. D. Maouche, P. Ruterana, and L. Louail, Phys. Lett. A 365 (2007) 231.
38. J. R. Neal, A. J. Behan, R. M. Ibrahim, et al., Phys. Rev. Lett. 96 (2006) 197208.
39. K. Nielsen, S. Bauer, M. Lubbe, et al., Phys. Status Solidi a-Appl. Mater. Sci. 203 (2006) 3581.
40. K. Potzger, S. Zhou, H. Reuther, et al., Appl. Phys. Lett. 88 (2006) 052508.
41. W. Prellier, A. Fouchet, B. Mercey, et al., Appl. Phys. Lett. 82 (2003) 3490.
42. A. Quesada, M. A. García, J. de la Venta, et al., The European Phys. J. B - Condens. Matter and Complex Systems 59 (2007) 457.
43. S. Ramachandran, A. Tiwari, J. Narayan, et al., Appl. Phys. Lett. 87 (2005) 172502.
44. S. Ramachandran, J. Narayan, and J. T. Prater, Appl. Phys. Lett. 88 (2006) 242503.
45. S. Ramachandran, J. T. Prater, N. Sudhakar, et al., Sol. Stat. Comm. 145 (2008) 18.
46. C. N. R. Rao and F. L. Deepak, J. Mater. Chem. 15 (2005) 573.
47. S. Riyadi, Muafif, A. A. Nugroho, et al., J. Phys.-Condens. Matter 19 (2007) 476214.
48. B. K. Roberts, A. B. Pakhomov, and K. M. Krishnan, J. Appl. Phys. 103 (2008) 07D133.
49. B. K. Roberts, A. B. Pakhomov, P. Voll, et al., Appl. Phys. Lett. 92 (2008) 162511.
50. K. Rode, A. Anane, R. Mattana, et al., J. Appl. Phys. 93 (2003) 7676.
51. P. Sharma, A. Gupta, K. V. Rao, et al., Nat. Mater. 2 (2003) 673.

52. P. Singh, Deepak, R. N. Goyal, et al., J. Phys.: Condens. Matter 20 (2008) 315005.
53. J. S. Thakur, G. W. Auner, V. M. Naik, et al., J. Appl. Phys. 102 (2007) 093904.
54. T. Tietze, M. Gacic, G. Schütz, et al., New J. Phys. 10 (2008) 055009.
55. K. Ueda, H. Tabata, and T. Kawai, Appl. Phys. Lett. 79 (2001) 988.
56. L. H. Van, J. Ding, M. H. Hong, et al., Surf. Rev. Lett. 15 (2008) 81.
57. X. J. Wang, I. A. Buyanova, W. M. Chen, et al., J. Appl. Phys. 103 (2008) 023712.
58. L. Yan, C. K. Ong, and X. S. Rao, J. Appl. Phys. 96 (2004) 508.
59. Z. Yang, J. L. Liu, M. Biasini, et al., Appl. Phys. Lett. 92 (2008) 042111.
60. K. A. Yates, A. J. Behan, J. R. Neal, et al., Phys. Rev. Lett. submitted (2008).
61. Y.-Z. Yoo, T. Fukumura, Z. Jin, et al., J. Appl. Phys. 90 (2001) 4246.
62. Y. Zheng, J. C. Boulliard, D. Demaille, et al., J. Cryst. Growth 274 (2005) 156.
63. N. H. Hong, N. Poirot, and J. Sakai, Phys. Rev. B (Condens. Matter Mater. Phys.) 77 (2008) 033205.
64. J. M. D. Coey, M. Venkatesan, P. Stamenov, et al., Phys. Rev. B (Condens. Matter Mater. Phys.) 72 (2005) 024450.
65. Y. Liu, Z. Lockman, A. Aziz, et al., J. Phys.: Condens. Matter 20 (2008) 165201.
66. S. Ghoshal and P. S. A. Kumar, J. Phys.: Condens. Matter 20 (2008) 192201.
67. M. Venkatesan, C. B. Fitzgerald, and J. M. D. Coey, Nat. 430 (2004) 630.
68. N. H. Hong, J. Sakai, N. Poirot, et al., Phys. Rev. B (Condens. Matter Mater. Phys.) 73 (2006) 132404.
69. D. M. Edwards and M. I. Katsnelson, J. Phys.: Condens. Matter 18, (2006) 7209.
70. S. Zhou, K. Potzger, G. Talut, et al., in Proceedings of the 52nd Annual Conference on Magnetism and Magnetic Materials (AIP, 2008), Vol. 103, p. 07D530.

71. F. Golmar, A. M. M. Navarro, C. E. R. Torres, et al., Appl. Phys. Lett. 92 (2008) 262503.
72. Y. Belghazi, G. Schmerber, S. Colis, et al., Appl. Phys. Lett. 89 (2006) 122504.
73. J. Alaria, H. Bieber, S. Colis, et al., Appl. Phys. Lett. 88 (2006) 112503.
74. K. Ando, Sci. 312 (2006) 1883.
75. K. Ando, H. Saito, Z. Jin, et al., Appl. Phys. Lett. 78 (2001) 2700.
76. S. Banerjee, K. Rajendran, N. Gayathri, et al., J. Appl. Phys. 104 (2008) 043913.
77. A. Barla, G. Schmerber, E. Beaurepaire, et al., Phys. Rev. B (Condens. Matter Mater. Phys.) 76 (2007) 125201.
78. K. P. Bhatti, S. Kundu, S. Chaudhary, et al., J. Phys. D: Appl. Phys. 39 (2006) 4909.
79. K. S. Burch, D. B. Shrekenhamer, E. J. Singley, et al., Phys. Rev. Lett. 97 (2006) 087208.
80. I. A. Buyanova, W. M. Chen, M. P. Ivill, et al., J. Vac. Sci. Technol. B: Microelectron. Nanometer Struct. 24 (2006) 259.
81. S. A. Chambers, Surf. Sci. Reports 61 (2006) 345.
82. S. A. Chambers, S. Thevuthasan, R. F. C. Farrow, et al., Appl. Phys. Lett. 79 (2001) 3467.
83. G. Cohen, V. Fleurov, and K. Kikoin, in 10th Joint MMM/Intermag Conference (AIP, Baltimore, Maryland (USA), 2007), Vol. 101, p. 09H106.
84. D. R. G. D. A. Schwartz, Adv. Mater. 16 (2004) 2115.
85. P. Dutta, M. S. Seehra, Y. Zhang, et al., J. Appl. Phys. 103 (2008) 07D104.
86. T. Fukumura, H. Toyosaki, K. Ueno, et al., New J. Phys. 10 (2008) 55018.
87. M. Gacic, G. Jakob, C. Herbort, et al., Phys. Rev. B (Condens. Matter Mater. Phys.) 75 (2007) 205206.
88. N. H. Hong, V. Brize, and J. Sakai, Appl. Phys. Lett. 86 (2005) 082505.
89. G. J. Huang, J. B. Wang, X. L. Zhong, et al., J. Mater. Sci. 42 (2007) 6464.
90. M. Ivill, S. J. Pearton, Y. W. Heo, et al., J. Appl. Phys. 101 (2007) 123909.

91. M. Ivill, S. J. Pearton, D. P. Norton, et al., J. Appl. Phys. 97 (2005) 053904.
92. M. Ivill, S. J. Pearton, S. Rawal, et al., New J. Phys. 10 (2008) 10.
93. N. K. J. L. MacManus-Driscoll, Y. Liu, M. E. Vickers, Adv. Mater. 19 (2007) 2925.
94. O. D. Jayakumar, I. K. Gopalakrishnan, K. Shashikala, et al., Appl. Phys. Lett. 89 (2006) 202507.
95. G.-H. Ji, Z.-B. Gu, M.-H. Lu, et al., J. Phys.: Condens. Matter 20 (2008) 425207.
96. E.-J. Kan, L.-F. Yuan, and J. Yang, J. Appl. Phys. 102 (2007) 033915.
97. T. C. Kaspar, S. M. Heald, C. M. Wang, et al., Phys. Rev. Lett. 95 (2005) 217203.
98. R. J. Kennedy, P. A. Stampe, E. Hu, et al., Appl. Phys. Lett. 84 (2004) 2832.
99. J.-Y. Kim, J.-H. Park, B.-G. Park, et al., Phys. Rev. Lett. 90 (2003) 017401.
100. S. Kolesnik, B. Dabrowski, and J. Mais, J. Supercond. 15 (2002) 251.
101. E.-C. Lee and K. J. Chang, Phys. Rev. B 69 (2004) 085205.
102. S. Lee, D. Y. Kim, Y. Shon, et al., Appl. Phys. Lett. 89 (2006) 022120.
103. S. Lee, Y. Shon, T. W. Kang, et al., Appl. Phys. Lett. 93 (2008) 022113.
104. S. W. Lim, M. C. Jeong, M. H. Ham, et al., Japn. J. Appl. Phys. Part 2-Lett. 43 (2004) L280.
105. Y.-H. Lin, M. Ying, M. Li, et al., Appl. Phys. Lett. 90 (2007) 222110.
106. X.-C. Liu, E.-W. Shi, Z.-Z. Chen, et al., Appl. Phys. Lett. 88 (2006) 252503.
107. X.-C. Liu, E.-W. Shi, Z.-Z. Chen, et al., J. Cryst. Growth 307 (2007) 14.
108. X. J. Liu, C. Song, F. Zeng, et al., J. Phys.: Condens. Matter 19 (2007) 296208.
109. D. C. Look, D. C. Reynolds, C. W. Litton, et al., Appl. Phys. Lett. 81 (2002) 1830.

110. A. Ney, R. Rajaram, S. S. P. Parkin, et al., Phys. Rev. B (Condens. Matter Mater. Phys.) 76 (2007) 035205.
111. H. H. Nguyen, W. Prellier, J. Sakai, et al., (AIP, 2004), Vol. 95, p. 7378.
112. U. Ozgur, Y. I. Alivov, C. Liu, et al., J. Appl. Phys. 98 (2005) 041301.
113. S. Ramachandran, A. Tiwari, and J. Narayan, Appl. Phys. Lett. 84 (2004) 5255.
114. S. R. Shinde, S. B. Ogale, S. Das Sarma, et al., Phys. Rev. B 67 (2003) 115211.
115. C. Song, X. J. Liu, F. Zeng, et al., Appl. Phys. Lett. 91 (2007) 042106.
116. P. A. Stampe, R. J. Kennedy, Y. Xin, et al., J. Appl. Phys. 93 (2003) 7864.
117. C. Sudakar, P. Kharel, G. Lawes, et al., in 10th Joint MMM/Intermag Conference (AIP, Baltimore, Maryland (USA), 2007), Vol. 101, p. 09H118.
118. M. Tay, Y. Wu, G. C. Han, et al., J. Appl. Phys. 100 (2006) 063910.
119. N. Theodoropoulou, V. Misra, J. Philip, et al., J. Magn. Magn. Mater. 300 (2006) 407.
120. M. Tortosa, M. Mollar, B. Mari, et al., J. Appl. Phys. 104 (2008) 033901.
121. H. Toyosaki, T. Fukumura, K. Ueno, et al., (AIP, 2006), Vol. 99, p. 08M102.
122. S. C. Wi, J. S. Kang, J. H. Kim, et al., Appl. Phys. Lett. 84 (2004) 4233.
123. A. Wojcik, M. Godlewski, E. Guziewicz, et al., Appl. Phys. Lett. 90 (2007) 082502.
124. H. Y. Xu, Y. C. Liu, C. S. Xu, et al., Appl. Phys. Lett. 88 (2006) 242502.
125. X. H. Xu, H. J. Blythe, M. Ziese, et al., New J. Phys. 8 (2006) 135.
126. W. Yan, Z. Sun, Q. Liu, et al., Appl. Phys. Lett. 90 (2007) 242509.
127. S. G. Yang, T. Li, B. X. Gu, et al., Appl. Phys. Lett. 83 (2003) 3746.
128. S. G. Yang, A. B. Pakhomov, S. T. Hung, et al., IEEE Trans. Magn. 38 (2002) 2877.

129. S. Zhou, K. Potzger, G. Zhang, et al., J. Appl. Phys. 100 (2006) 114304.
130. S. H. Xiong and Y. W. Du, Phys. Lett. A 372 (2008) 2114.
131. M. B. Knickelbein, Phys. Rev. Lett. 86 (2001) 5255-5257.
132. M. R. Pederson, F. Reuse, and S. N. Khanna, Phys. Rev. B 58 (1998) 5632-5636.
133. S. Zhou, K. Potzger, J. von Borany, et al., Phys. Rev. B (Condens. Matter Mater. Phys.) 77 (2008) 035209.
134. O. Kazakova, R. Morgunov, J. Kulkarni, et al., Phys. Rev. B (Condens. Matter Mater. Phys.) 77 (2008) 235317.
135. C. Song, K. W. Geng, F. Zeng, et al., Phys. Rev. B (Condens. Matter Mater. Phys.) 73 (2006) 024405.
136. J. Robin, Ann. Chim. 10 (1955) 400.
137. S. R. Shinde, S. B. Ogale, J. S. Higgins, et al., Phys. Rev. Lett. 92 (2004) 166601.
138. H. Kim, Phys. Stat. Sol. 241 (2004) 1553.
139. K. Samanta, P. Bhattacharya, R. S. Katiyar, et al., Phys. Rev. B (Condens. Matter Mater. Phys.) 73 (2006) 245213.
140. P. Dutta, M. S. Seehra, S. Thota, et al., J. Phys.: Condens. Matter 20 (2008) 015218.
141. P. Bobadova-Parvanova, K. A. Jackson, S. Srinivas, et al., J. Chem. Phys. 122 (2005) 014310.
142. D. C. Kundaliya, S. B. Ogale, S. E. Lofland, et al., Nat. Mater. 3 (2004) 709.
143. B. K. Rao and P. Jena, Phys. Rev. Lett. 89 (2002) 185504.
144. M. Zajac, J. Gosk, E. Grzanka, et al., J. Appl. Phys. 93 (2003) 4715.
145. W. J. Takei, R. R. Heikes, and G. Shirane, Phys. Rev. 125 (1962) 1893 LP.
146. M. Venkatesan, R. D. Gunning, P. Stamenov, et al., in Proceedings of the 52nd Annual Conference on Magnetism and Magnetic Materials (AIP, Tampa, Florida (USA), 2008), Vol. 103, p. 07D135.
147. X. Wang, J. Xu, X. Yu, et al., Appl. Phys. Lett. 91 (2007) 031908.
148. J. H. Kim, H. Kim, D. Kim, et al., Phys. B-Condens. Matter 327 (2003) 304.
149. H. Harima, J. Phys.: Condens. Matter 16 (2004) S5653.
150. T. Dietl, Phys. E-Low-Dimensional Sys. & Nanostruct. 35 (2006) 293.

151. K. Samanta, P. Bhattacharya, and R. S. Katiyar, Phys. Rev. B (Condens. Matter Mater. Phys.) 75 (2007) 035208.
152. C. Sudakar, P. Kharel, G. Lawes, et al., J. Phys.: Condens. Matter 19 (2007) 026212.
153. A. Ney, R. Rajaram, T. Kammermeier, et al., J. Phys.: Condens. Matter 20 (2008) 285222.
154. A. Espinosa, E. Cespedes, C. Prieto, et al., in Proceedings of the 52nd Annual Conference on Magnetism and Magnetic Materials (AIP, Tampa, Florida (USA), 2008), Vol. 103, p. 07D129.
155. C. Song, F. Zeng, K. W. Geng, et al., Phys. Rev. B (Condens. Matter Mater. Phys.) 76 (2007) 045215.
156. C. Song, S. N. Pan, X. J. Liu, et al., J. Phys.: Condens. Matter 19 (2007) 176229.
157. M. A. Garcia, M. L. Ruiz-Gonzalez, A. Quesada, et al., Phys. Rev. Lett. 94 (2005) 217206.
158. H. S. Hsu, J. C. A. Huang, S. F. Chen, et al., Appl. Phys. Lett. 90 (2007) 102506.
159. Q. Y. Xu, H. Schmidt, S. Q. Zhou, et al., Appl. Phys. Lett. 92 (2008) 082508.
160. J. M. D. Coey, M. Venkatesan, and C. B. Fitzgerald, Nat. Mater. 4 (2005) 173.
161. K. Kikoin and V. Fleurov, Phys. Rev. B (Condens. Matter Mater. Phys.) 74 (2006) 174407.
162. S. Halm, P. E. Hohage, E. Neshataeva, et al., Phys. Stat. Solidi a-Appl. Mater. Sci. 204 (2007) 191.
163. H. Toyosaki, T. Fukumura, Y. Yamada, et al., Nat. Mater. 3 (2004) 221.
164. Z. L. Lu, G. Q. Yan, S. Wang, et al., J. Appl. Phys. 104 (2008) 033919.
165. T. Schallenberg and H. Munekata, Appl. Phys. Lett. 89 (2006) 042507.
166. A. Janotti and C. G. Van de Walle, Nat. Mater. 6 (2007) 44.
167. A. S. Risbud, N. A. Spaldin, Z. Q. Chen, et al., Phys. Rev. B 68 (2003) 205202.
168. J. B. Torrance, M. W. Shafer, and T. R. McGuire, Phys. Rev. Lett. 29 (1972) 1168 LP.
169. D. E. Angelescu and R. N. Bhatt, Phys. Rev. B 65 (2002) 075211.

170. A. C. Durst, R. N. Bhatt, and P. A. Wolff, Phys. Rev. B 65 (2002) 235205.
171. B. W. Wessels, New J. Phys. 10 (2008) 055008.
172. T. Chanier, I. Opahle, M. Sargolzaei, et al., Phys. Rev. Lett. 100 (2008) 026405.
173. S. W. Jung, S.-J. An, G.-C. Yi, et al., Appl. Phys. Lett. 80 (2002) 4561.
174. K. G. Roberts, M. Varela, S. Rashkeev, et al., Phys. Rev. B (Condens. Matter Mater. Phys.) 78 (2008) 014409.
175. R. Seshadri, Curr. Opin. Sol. Stat. Mater. Sci. 9 (2005) 1.
176. K. A. Griffin, M. Varela, S. J. Pennycook, et al., (AIP, 2006), Vol. 99, p. 08M114.
177. G. Weyer, H. P. Gunnlaugsson, R. Mantovan, et al., J. Appl. Phys. 102 (2007) 113915.
178. N. Khare, M. J. Kappers, M. Wei, et al., Adv. Mater. 18 (2006) 1449.
179. X. J. Liu, C. Song, F. Zeng, et al., J. Appl. Phys. 103 (2008) 093911.
180. L. B. Duan, G. H. Rao, Y. C. Wang, et al., J. Appl. Phys. 104 (2008) 013909.
181. J. Hu, Z. Zhang, M. Zhao, et al., Appl. Phys. Lett. 93 (2008) 192503.
182. C. Madhu, A. Sundaresan, and C. N. R. Rao, Phys. Rev. B (Condens. Matter Mater. Phys.) 77 (2008) 201306.
183. J. Hong, J. Appl. Phys. 103 (2008) 063907.
184. P. Dev, Y. Xue, and P. Zhang, Phys. Rev. Lett. 100 (2008) 117204.
185. D.-Y. Cho, J.-M. Lee, S.-J. Oh, et al., Phys. Rev. B (Condens. Matter Mater. Phys.) 76 (2007) 165411.
186. P. Larson and S. Satpathy, Phys. Rev. B (Condens. Matter Mater. Phys.) 76 (2007) 245205.
187. V. Pardo and W. E. Pickett, Phys. Rev. B (Condens. Matter Mater. Phys.) 78 (2008) 134427.
188. P. Stamenov, M. Venkatesan, L. S. Dorneles, et al., in 50th Annual Conference on Magnetism and Magnetic Materials (AIP, San Jose, California (USA), 2006), Vol. 99, p. 08M124.
189. K. Sato, Jpn. J. Appl. Phys. 39 (2000) L555.

190. A. F. Jalbout, H. Chen, and S. L. Whittenburg, Appl. Phys. Lett. 81 (2002) 2217.
191. K. R. Kittilstved, N. S. Norberg, and D. R. Gamelin, Phys. Rev. Lett. 94 (2005) 147209.
192. A. C. Tuan, J. D. Bryan, A. B. Pakhomov, et al., Phys. Rev. B 70 (2004) 054424.
193. Y. W. Heo, M. P. Ivill, K. Ip, et al., Appl. Phys. Lett. 84 (2004) 2292.
194. A. Mokhtari, New J. Phys. submitted (2008).
195. D. P. Norton, S. J. Pearton, A. F. Hebard, et al., Appl. Phys. Lett. 82 (2003) 239.
196. Q. Wan, Appl. Phys. Lett. 89 (2006) 082515.
197. A. Janotti and C. G. V. de Walle, Phys. Rev. B 75 (2007) 121201.
198. A. Janotti and C. G. V. de Walle, Phys. Rev. B 76 (2007) 165202.
199. F. A. Selim, M. H. Weber, D. Solodovnikov, et al., Phys. Rev. Lett. 99 (2007) 085502.
200. L. S. Vlasenko and G. D. Watkins, Phys. B-Condens. Matter 376 (2006) 677.
201. B. Yang, A. Kumar, P. Feng, et al., Appl. Phys. Lett. 92 (2008) 233112.
202. A. Chattopadhyay, S. Das Sarma, and A. J. Millis, Phys. Rev. Lett. 87 (2001) 227202.
203. S. Lee, H. S. Lee, S. J. Hwang, et al., J. Cryst. Growth 286 (2006) 223.
204. Q. Xu, L. Hartmann, S. Zhou, et al., Phys. Rev. Lett. 101 (2008) 076601.
205. G. Ji, S. S. Yan, Y. X. Chen, et al., J. Mater. Sci. Technol. 24 (2008) 415.
206. R. J. Soulen, J. M. Byers, M. S. Osofsky, et al., Sci. 282 (1998) 85.
207. G. E. Blonder, M. Tinkham, and T. M. Klapwijk, Phys. Rev. B 25 (1982) 4515 LP.
208. I. I. Mazin, A. A. Golubov, and B. Nadgorny, J. Appl. Phys. 89 (2001) 7576.
209. R. P. Panguluri, B. Nadgorny, T. Wojtowicz, et al., Appl. Phys. Lett. 84 (2004) 4947.
210. R. P. Panguluri, B. Nadgorny, T. Wojtowicz, et al., Appl. Phys. Lett. 91 (2007) 252502.

211. R. P. Panguluri, K. C. Ku, T. Wojtowicz, et al., Phys. Rev. B 72 (2005) 054510.
212. S. Yuasa, T. Nagahama, A. Fukushima, et al., Nat. Mater. 3 (2004) 868.
213. L. Zhang, N. Deng, M. Ren, et al., Superlatt. Microstruct. 42 (2007) 222.
214. W. G. Wang, C. Ni, T. Moriyama, et al., Appl. Phys. Lett. 88 (2006) 202501.
215. S. J. May, P. J. Phillips, and B. W. Wessels, J. Appl. Phys. 100 (2006) 053912.
216. Z.-M. Liao, J. Xu, J.-M. Zhang, et al., Appl. Phys. Lett. 93 (2008) 023111.
217. M. Yoshikawa, K. Inoue, T. Nakagawa, et al., Appl. Phys. Lett. 92 (2008) 113115.
218. F. J. von Preissig, H. Zeng, and E. S. Kim, Smart Mater. Struct. 7 (1998) 396.
219. X. Lu, H. Zhang, Y. Ni, et al., Biosens. Bioelectron. 24 (2008) 93.
220. W. Eerenstein, N. D. Mathur, and J. F. Scott, Nat. 442 (2006) 759.
221. Y. C. Yang, C. F. Zhong, X. H. Wang, et al., J. Appl. Phys. 104 (2008) 064102.

Chapter 11

MÖSSBAUER SPECTROSCOPY AND ITS APPLICATIONS IN SPINTRONICS

Saeed Kamali

Department of Applied Science, University of California
Davis, California 95616
Physical Biosciences Division, Lawrence Berkeley National Laboratory
Berkeley, California 94720
E-mail: skamali@ucdavis.edu

There are a number of spectroscopic techniques available for characterization of condensed matter. Among these methods, the so-called nuclear probe techniques provide information about the samples under study at atomic level. Mössbauer spectroscopy, due to a very high energy resolution, which is in the neV range, is an outstanding technique providing valuable information about the electrical, magnetic and structural properties of condensed materials.

11.1. Introduction

In this chapter, an unique method for characterization of nano-structures, which are the basic structures in spintronics, will be introduced. Mössbauer spectroscopy (MS) is an excellent probe technique for the characterization of magnetic and electronic properties of materials at atomic level. It provides (via the hyperfine interaction) local (atomistic) information about the magnitude and direction of the atomic moment, the atomic environment and its symmetry around the probed atom, and also about the s-electron density at the probed nucleus in the sample. A German Ph.D. student, Rudolf Mössbauer, in 1958–59 discovered the effect, which is based on recoilless emission and absorption of γ-rays by nuclei. It is worth to mention that this effect is intimately related to the theory for neutron capture by atoms in a crystal, which was developed by Lamb in 1930's.[1] Dicke[2] studied the theory of the effect in early 50's. The powerfulness of the effect is more obvious, when one realizes that it can be used in so different fields such as solid state physics, chemistry, geology and even in biology. The first experiment was done on ^{191}Ir, but later on different isotopes were used, among which ^{57}Fe is the most used isotope. Mössbauer was awarded the 1961 Nobel prize in physics for this discovery. By applying this technique, one is able to study structural, electrical, magnetic and dynamical properties of materials.

To be able to understand the information one can extract from the samples under study by Mössbauer spectroscopy, an introduction of this technique is necessary. In the following, we will give a brief description of the technique. Since the discovery of the technique, a number of excellent books has been published, which describe Mössbauer spectroscopy and its applications in different fields.[3–10] The Mössbauer community, consisting of several thousands of researchers in different fields gather together in the International Conference on the Applications of the Mössbauer Effect (ICAME) every second year and also in other international as well as regional conferences. The web site, belonging to the International Board on the Applications of the Mössbauer Effect (IBAME),[11] as well as other web sites,[12–14] provide valuable information concerning the technique and other activities.

The outline of this chapter will be as follows. First, we will describe the basics of the technique. Then, we will present some applications of the technique in studying different kinds of superlattices.

11.2. Mössbauer Spectroscopy: The Basics

The main feature of Mössbauer spectroscopy is emission of γ-rays from excited nuclei and absorption of these γ-rays by identical nuclei in an absorber, i.e. the sample, whose properties are supposed to be studied. The process is schematically shown in Fig. 11.1.

However, if we take into account the conservation of momentum and energy, there will be some recoil of both the emitting and absorbing nuclei in the emission as well as absorption processes. Hence, there will be some loss of energy. After some simple calculations, the expression for the recoil energy will be as follows:

$$E_R = \frac{{E_\gamma}^2}{Mc^2} \tag{11.1}$$

where E_γ is the γ energy, M is the mass of the nucleus and c is the velocity of light. In case of free atoms, the recoil energy is enough to destroy the

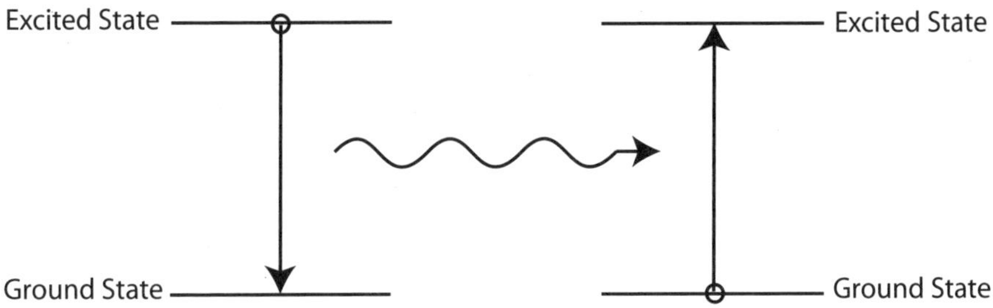

Fig. 11.1. Emission and absorption of γ-rays by identical nuclei.

resonance, i.e. no absorption will occur. It is why the free nuclei cannot experience the Mössbauer effect. On the other hand, in case of condensed matter, the nucleus in question is strongly bound to the crystal. In this case the whole crystal will be recoiled as a whole, and the mass in the Eq. (11.1) will be the mass of the whole crystal, hence the recoil energy can be ignored.

The so-called hyperfine interactions, which are the interactions between electronic and nuclear magnetic moments, not only change the energy of the nuclear states but also remove, partially or completely, the degeneracies of these states. The parameters, extracted from these interactions, give valuable information about oxidation states, bonding properties, covalency, electronegativity, structural and magnetic properties of the sample under study. The following three interactions are of paramount importance:

(1) Electric Monopole Interaction.
(2) Electric Quadrupole Interaction.
(3) Magnetic Dipole Interaction

It is worth mentioning that the dimension of all parameters extracted from these interactions is mm/s, because the vibrator carrying the radioactive source vibrates at such velocities to make resonances possible.

In most cases the radioactive sources emit a single line, i.e. the emitted γ-quanta have just one energy. On the other hand, due to the hyperfine interactions, the nuclei under study may have several energy transitions. In order to be able to study these transitions, one has to tune the energy of the γ-quanta to match these transitions. As is seen in Fig. 11.2, the radioactive source is put on a vibrator, hence the energy of the incoming γ-quanta is changed due to the Doppler effect. In this way, the whole energy range

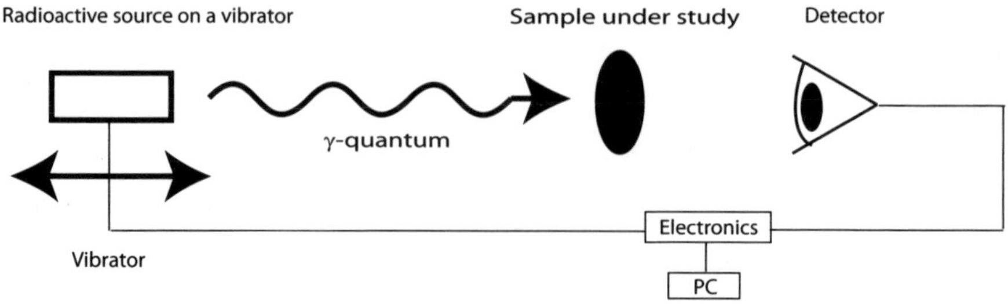

Fig. 11.2. A basic schematic picture of a Mössbauer set up in a transmission mode. By putting the radioactive source on a vibrator, vibrating with a constant acceleration, different velocities are achieved. Hence, due to the Doppler effect, a range of different energies is scanned.

under study can be scanned and all different transitions can be observed in a Mössbauer spectrum.

In the following parts, these interactions and the physical as well as chemical information extracted from each, are shortly described.

11.2.1. *Electric monopole interaction*

The simplest form of a Mössbauer spectrum is a single absorption line, also called singlet, where the changes in nuclear state energies are measured against a reference. There are two sources for this difference, isomer shift (IS) and second order Doppler shift (SOD).

11.2.1.1. *Isomer shift*

In a simple but real picture an atom consists of a nucleus with finite size and a positive charge immersed in a cloud of electrons. Hence, the electrostatic energy, due to the presence of the nucleus in the electrical potential, V(r), caused by the surrounding electrons, will be as follows:

$$E_{el} = \int_{all\ space} \rho_n V(r) d\tau \qquad (11.2)$$

where ρ_n is the charge density of the nucleus and $d\tau = dxdydz$ is the volume element.

Subtracting the overall potential, taking into account the fact that a nucleus has different radius in the excited state, R_e, and the ground state, R_g, then the isomer shift (IS), which is the difference between the nuclear energy levels in the emitter or the radioactive source (S) and the absorber (A), can be defined as:

$$IS = -S(Z)\left[\frac{2}{10\epsilon_0}Ze(R_e^2 - R_g^2)\right][\rho_e^S(0) - \rho_e^A(0)] \qquad (11.3)$$

where ρ_e is the electron density at the nucleus and ϵ_0 is the permeability constant. S(Z) is the relativistic correction factor due to the fact that in a relativistic calculation, in addition to s-electrons there is some probability for the presence of p-electrons in the nucleus. As is seen from Eq. (11.3), isomer shift is a function of electron concentration, then one can extract oxidation states, covalency, electronegativity and bonding properties of the probe atom by studying isomer shift.

11.2.1.2. *Second order Doppler shift*

Because the nuclei in a crystal are not frozen and vibrate around their equilibrium positions, the frequency of the emitted and absorbed γ-rays are changed due to the Doppler effect, which results in the change of γ-rays energy. At non-zero temperatures, the nuclei oscillate about their mean position in the crystal with a frequency of the order of 10^{12} Hz. Though, the averaged displacement during Mössbauer life time (140 ns) is then zero, the second term in the Doppler shift, which is non-zero gives rise to the second order Doppler shift (SOD) with energy change as:

$$\frac{\delta E_\gamma}{E_\gamma} = -\frac{\langle v^2 \rangle}{2c^2}\,. \qquad (11.4)$$

In the case of Fe, SOD = +0.07 mm/s as temperature is decreased by 100 K.[8]

11.2.1.3. *Centroid shift*

The measured energy shift, which is simply the position of the singlet relative to the reference is the sum of the mentioned quantities, i.e. the IS and SOD, and called the centroid shift, CS:

$$CS = IS + SOD\,. \qquad (11.5)$$

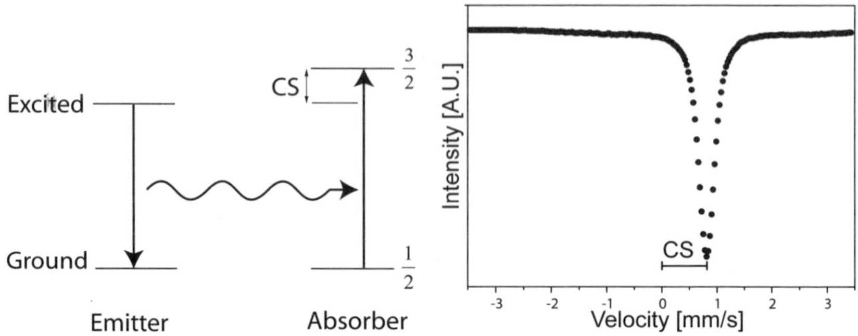

Fig. 11.3. The energy levels in the presence of just electric monopole interaction (left), which gives rise to a singlet (right).

The reference is generally α-Fe with CS positioned at zero mm/s. In Fig. 11.3 the Mössbauer spectrum, a singlet, is depicted together with the change in CS between the source and the absorber.

11.2.2. *Electrical quadrupole interaction*

A nucleus with non-spherical charge distribution, possesses a quadrupole moment, which can be described by a second rank tensor with nine elements. In a principal coordinate system, the off-diagonal elements of the tensor are all zero. The electrical quadrupole moment of a nucleus with a charge distribution of cylindrical symmetry and with the z-axis as the axis of quantization, can be expressed as:

$$Q = \frac{1}{e}\int \rho_n(r)(3z^2 - r^2)d\tau = \frac{1}{e}\int \rho_n(r)r^2(3\cos^2(\theta) - 1)d\tau \qquad (11.6)$$

where dτ is volume element and θ is the angle between the vector from the origin to a charge element in the nucleus, $\overrightarrow{r}$ and the symmetry axis z.

The degeneracy of the nuclear levels with spin $I > \frac{1}{2}$ of a nucleus possessing a quadrupole moment in a nonzero Electric Field Gradient (EFG), which is the second derivative of the potential caused by the surrounding electrons, will be lifted. This is shown in Fig. 11.4.

The EFG, described by a second rank tensor, can also be presented in a principal coordinate system, which results in diagonalizing the tensor. Due to Laplace's equation:

$$\sum V_{ii} = 0 \qquad (11.7)$$

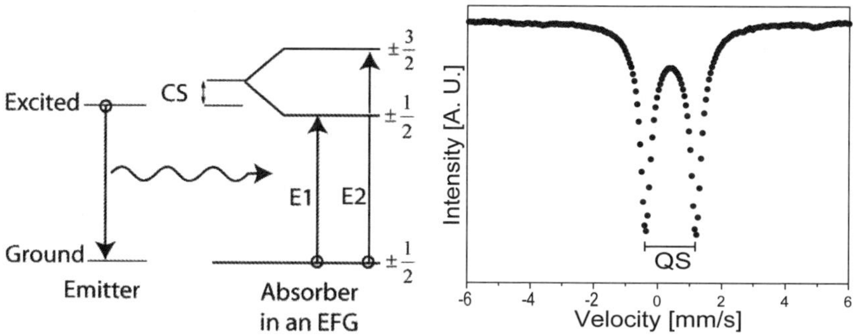

Fig. 11.4. The excited state with $I = \pm\frac{3}{2}$ level in presence of quadrupole splitting is split (left), which results in a doublet in the Mössbauer spectrum (right).

with i = x, y and z, there will remain only two independent elements; V_{zz} and η, which is the asymmetry parameter, defined as:

$$\eta = \frac{V_{yy} - V_{xx}}{V_{zz}} . \tag{11.8}$$

If the principal coordinate system is chosen in a way implying $V_{xx} \leq V_{yy} \leq V_{zz}$ then $0 \leq \eta \leq 1$.

The eigenvalues of the Hamiltonian of the quadrupole interaction will then be written as:

$$E_Q = \frac{eqV_{zz}}{4I(2I-1)}[3m_I^2 - I(I+1)]\left(1 + \frac{\eta^2}{3}\right)^{1/2} \tag{11.9}$$

where m_I is the nuclear magnetic spin quantum number with values ranging, -I, -I+1, ..., I-1, I, resulting in total (2I+1) different values. In the case of Fe, the ground state with $I = \frac{1}{2}$ will still be degenerate. On the other hand, the degeneracy of the excited state with $I = \frac{3}{2}$ will be lifted and will be split into $\pm\frac{1}{2}$ and $\pm\frac{3}{2}$. As an example, the so-called Quadrupole Splitting in an axially symmetric EFG, where $\eta = 0$, will be:

$$\Delta E_Q = E_Q\left(\pm\frac{3}{2}\right) - E_Q\left(\pm\frac{1}{2}\right) = \frac{eqV_{zz}}{2} . \tag{11.10}$$

11.2.3. *Magnetic hyperfine interaction*

A nucleus with spin quantum number $I > 0$ possesses a magnetic dipole moment $\overrightarrow{\mu}$:

$$\overrightarrow{\mu} = g_N \beta_N \overrightarrow{I} \tag{11.11}$$

where g_N is the nuclear Landé factor and $\beta_N = \frac{e\hbar}{2Mc}$, with M as nuclear mass, is the nuclear magneton. The Hamiltonian for the $\overrightarrow{\mu}$ in a magnetic field at the nucleus, $\overrightarrow{H}$, is given by:

$$H = -\overrightarrow{\mu}.\overrightarrow{H} . \tag{11.12}$$

The eigenvalues are given as:

$$E_M = -g_N \beta_N H m_I \tag{11.13}$$

where m_I is again the nuclear magnetic spin quantum number with values ranging, -I,-I+1,...,I-1,I, resulting in total (2I+1) values. The ground state is now split into two sublevels and the excited state into 4 levels as shown in Fig. 11.5.

Of the eight possible transitions, only six transitions are allowed for pure magnetic dipole transitions due to the selection rules, i.e. $\Delta I = 1$ and $\Delta m = 0, \pm 1$. Therefore, the transition from $\frac{3}{2}$ level in the excited state ($I = \frac{3}{2}$ and $m = \frac{3}{2}$) to -$\frac{1}{2}$ level in the ground state ($I = \frac{1}{2}$ and $m = -\frac{1}{2}$) and the transition from -$\frac{3}{2}$ level in the excited state ($I = \frac{3}{2}$ and $m = -\frac{3}{2}$) to $\frac{1}{2}$ level in the ground state ($I = \frac{1}{2}$ and $m = \frac{1}{2}$) are forbidden because $\Delta m = -2$ in the first case and $\Delta m = 2$ in the second case, respectively.

Though in some cases one can have external magnetic field during the measurement, the magnetic field at the nucleus originates mainly from the atomic electrons. There are different contributions to this field:

(1) The Fermi contact field H^c which is the result of the spin-polarization of inner filled s-shells, i.e. when the spin-up and spin-down electron

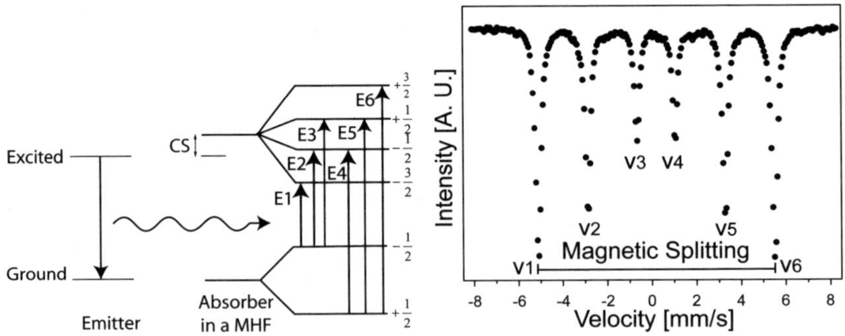

Fig. 11.5. The energy levels in presence of magnetic interaction (left) and the related Mössbauer spectrum.

densities are not equal. It is caused by spin-polarized partially filled outer d-shells.

(2) The orbital field H^L caused by orbital motion of valence electrons.
(3) The spin-dipolar field H^d, caused by atomic electron spins.

11.2.4. *Combined electric and magnetic hyperfine interaction*

Beside the electric monopole interaction, which is present in all cases, in many cases there is a combination of electric quadrupole and magnetic dipole interactions. Three cases are distinguished, $E_Q << E_M$, $E_Q >> E_M$ or $E_Q \approx E_M$. In the first cases, there is one dominant interaction and the other one is just a perturbation. When both interaction have equal strength, the full Hamiltonian should be used. Häggström[15] was the first, who solved this complicated case analytically. Due to our emphasis on magnetic materials, we consider here the case where $E_Q << E_M$, i.e. when the magnetic interaction is the main component and quadrupole interaction is just a perturbation. The EFG is assumed to be axially symmetric. The eigenvalues can be written as:

$$E_{M,Q} = -g_N \beta_N H m_I + (-1)^{|m_I|+\frac{1}{2}} \left(\frac{eQV_{zz}}{8} (3\cos^2\beta - 1) \right) \qquad (11.14)$$

where β is the angle between the axis of the magnetic field and the z-component of the the EFG's principal axis. If V_{zz} is positive then the $\pm\frac{3}{2}$ levels are shifted to higher energy by an amount which is equal to the quadrupole energy and the $\pm\frac{1}{2}$ levels are shifted to lower energy by the same amount. The shifts are reversed if V_{zz} is negative as can be seen in Fig. 11.6.

11.2.5. *Transmission vs. conversion electron Mössbauer spectroscopy*

In transmission Mössbauer spectroscopy, the absorber, i.e. the sample under study, will be placed between the radioactive source and the detector. The γ-rays, passed the absorber, will then be counted as a function of the velocity of the vibrator. In this way, the γ-rays, which are matched with a resonance energy, will be absorbed by the sample and the spectrum will consist of dips as for example shown in the right panel of Fig. 11.5. It is also possible to perform Mössbauer spectroscopy in a scattering geometry. In this mode, the excited Mössbauer nuclei in the sample will de-excite

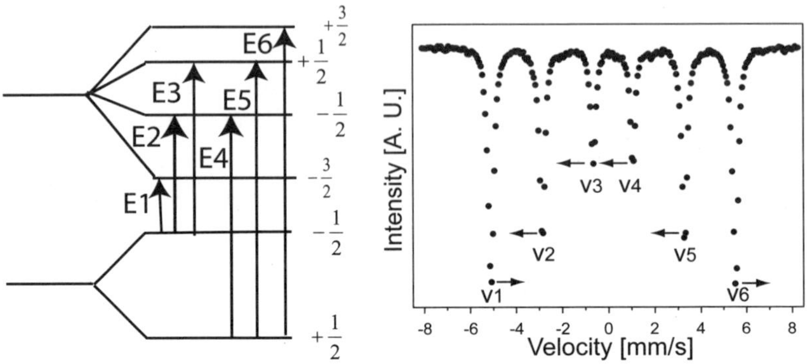

Fig. 11.6. The energy levels in the presence of magnetic interaction with quadrupole interaction as perturbation (left), which results in asymmetry of the line positions in the Mössbauer spectrum (right).

by emission of conversion electrons from K shell, hence, the name Conversion Electron Mössbauer Spectroscopy (CEMS). It is worth noting that the probability of de-excitation via this mode is over 90%. A conventional gas flow detector ($He + 10\% CH_4$), which is transparent to γ-rays and placed between the radioactive source and the sample, is then used to detect emitted electrons as a function of velocity. The spectra will then consist of peaks. Due to the scattering of the conversion electrons in the sample, only electrons from approximately 1000 Åunder the surface have enough energy to escape the surface and be detected by the detector. This makes CEMS a surface sensitive technique. Nevertheless, due to this fact that thin films and superlattices are always deposited on a thick substrate, the study of these systems, such as the examples presented in this chapter, are always performed in this mode.

11.2.6. *Relative intensities of resonance lines*

In spintronic devices, not only the magnetic moments, but also their directions play important roles. In Mössbauer spectroscopy, one can get information concerning the magnetic anisotropy of the sample under study by inspecting the line intensities. The relative intensities of the resonance lines are not only very sensitive to the properties of the nuclear transitions, e.g. spin and parity of the different states, but also to the geometrical configuration, i.e. they are functions of the angle between the direction of the photon and the quantization axis, which can be the magnetic field direction or the z-axis in the principal coordinate system of the EFG. They are also proportional to the Clebsch-Gordan coefficients. The relative intensity ratios between the six lines in a 1/2 - 3/2 transition are 3: x :1:1: x :3, where x can vary between 0 and 4. The ratio of the second (fifth) to the third (forth) is:

$$\frac{A_{2,5}}{A_{3,4}} = 4\frac{sin^2\theta}{1 + cos^2\theta} \tag{11.15}$$

As an example, in the case of powder sample, where all angles have the same probability, the ratios are 3:2:1:1:2:3. In the case of magnetic multilayers, if the magnetic field direction is in the plane, i.e. perpendicular to the direction of the photon, the ratios are 3:4:1:1:4:3, whereas in case of perpendicular magnetization, where the the magnetic field direction is perpendicular to the sample surface and parallel to the incoming γ-rays, the second and fifth lines vanish, i.e. the ratios will be 3:0:1:1:0:3. Thus, Mössbauer spectroscopy is a very good tool for studying the anisotropy in magnetic thin films and superlattices.

11.3. Superlattices, Thin Films

Thin films are defined as materials shrunk to nano scale in one dimension. These can be magnetic multilayers, which can be constructed artificially in the laboratory world by different techniques. Growing successive layers of different atomic species on a well characterized substrate will produce a multilayer. If layers of different elements are forced to have the same lattice constant one effectively has a single crystal and the multilayer is called a superlattice. The layers can be all magnetic, or magnetic layers can be separated by non-magnetic ones, the so-called spacer layers. The samples are not thermodynamically in their lowest energy state, but they are metastable. This is why the growing conditions are crucial factors for the resulting samples. Multilayers have differerent and unique properties compared to the bulk ones. They are, therefore, materials with properties that can be tailored to different applications. As an example one can mention growing an element in a crystallographic phase, which can not be found in nature because of thermodynamical laws. For example, body centered cubic (bcc) Co and Ni, which are hcp and fcc in bulk form, respectively. Other properties like, Magnetic Interlayer Coupling, which is the interaction between the magnetic layers via non-magnetic spacers, Giant Magnetic

Resistance (GMR) and Spin Polarized Tunneling (SPT) make such superlattices very nice devices in many sophisticated industrial applications. Another important task is Interface Magnetism because interfaces in multilayers have a very important role. In interfaces the symmetry is broken and the atomic environment is completely different from the interior part. It is then very important to understand the interactions occurring there. Again, because Mössbauer spectroscopy is a unique nuclear method, and furthermore, because it is sensitive enough to allow the study of monolayers by using isotopically enriched material, it may give in a fingerprint-type of way information on the presence of different magnetic and/or crystallographic phases in the sample. Moreover, because of the local character of the technique, it is a valuable tool to distinguish different properties at the different areas in a superlattice by introducing down to a monolayer of ^{57}Fe at those areas and measure the spectra. Here there are some studies described related to superlattices.

11.3.1. *Fe/Co superlattices*

11.3.1.1. *Magnetic hyperfine field*

The systems containing iron and cobalt are still the most interesting systems in materials research, especially in the field of applied magnetism. The very large magnetic moment per atom, and macroscopic magnetization, obtained in Fe-Co alloys with approximately 30 at% Co, at the maximum of the Slater-Pauling curve[16] are some of the reasons for the intense investigations of these systems. The potentiality of Fe-Co alloys in information storage applications has been studied in theoretical papers.[17] Bulk Fe-Co alloys have been studied for a long time and almost all their structural and magnetic properties are revealed by specially local-probe techniques.[18–25] Superlattices of these elements have also been studied intensively ever since the possibilities to grow such system by different deposition techniques have been available experimentally[26–29] as well as theoretically.[30,31] Mössbauer spectroscopy has also been a suitable technique to study such iron-based system,[32–37] because there is no need for introducing impurities in such system, due to the present of Fe, which is a Mössbauer isotope. In most of these works, experimental as well as theoretical, it was inferred that the Fe hyperfine field and the magnetic moment of Fe increase at the interfaces. In contrary to these studies, in a experiment performed with Nuclear Resonance Scattering (NRS), which is a synchrotron based Mössbauer spectroscopy technique, Lindgren *et al.*[33] reported that the B_{hf} is largest in the middle of the Fe layer in Fe(6 ML)/Co(3 ML) and Fe(5 ML)/Co(5 ML) supperlattices, where ML stands for monolayers. This was a puzzling result. This discrepancy was resolved partially in the study by Kamali *et al.*, where a systematic study was performed on such superlattices.[37] They measured a series of ^{57}Fe(x ML)/Co(7 ML) by means of Conversion Electron Mössbauer Spectroscopy (CEMS), where x was 2, 3, 5, 9 and 14 MLs. The CEMS together with the B_{hf} distributions are plotted in Fig. 11.7.

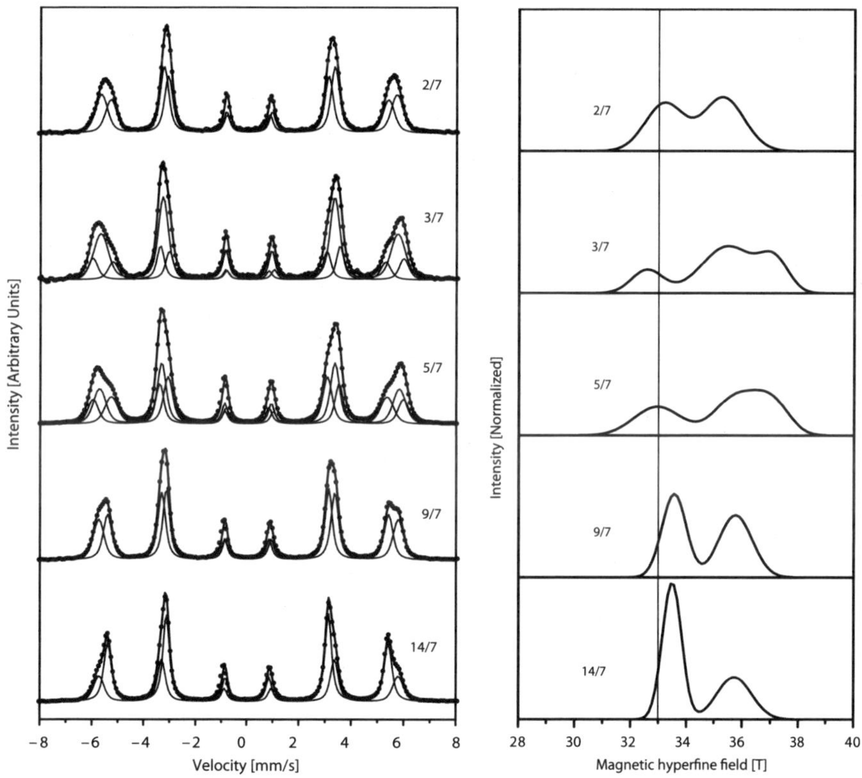

Fig. 11.7. In the left panel, CEM spectra of all Fe/Co samples denoted by x/y, where x and y stand for the nominal number of monolayers of Fe and Co, respectively, are illustrated. In the right panel, the correspondent Fe magnetic hyperfine field distributions are plotted. The line indicates the magnetic hyperfine field for Fe atoms in α-Fe, which is 33.0 T. Reprinted with permission from Kamali *et al.*,[37] J. Phys.: Condens. Matter 18 (2006) 5807. © 2006 IOP Publishing Ltd.

They concluded that although the B_{hf} and hence the magnetic moment of Fe atoms at interfaces increase compared with bulk Fe, the highest values are reached approximately in the third layer in the Fe layers due to intermixing of Fe and Co atoms with a lower Co concentration than 50% as in the first interface layer. The B_{hf} will then decrease for inner parts of the Fe layers and reaches the bulk value.

11.3.1.2. *Magnetic anisotropy energy*

Another important issue for such superlattices is the orientation of the B_{hf}. The out of plane orientation can be easily extracted by studying the intensity ratio of second (fifth) to third (forth) lines, in a measurement geometry, where the incoming γ-ray direction is perpendicular to the sample surface. However, the direction of the B_{hf} in plane cannot be resolved by this geometry since all directions have 90° angle with respect to the incoming γ-rays. Hägström *et al.*[36] by using a geometry similar to the one shown in Fig. 11.10, were able to extract information about the magnetic anisotropy in that system. They concluded that the magnetic moment of the system lay in the [110] directions.

11.3.2. *Fe/Cr*

During the last few decades, there has been a huge interest in investigating the Fe/Cr superlattices, due to the fundamental and practical interests, which will be described shortly. One of the interesting effects discovered first in Fe/Cr systems, which are artificial ferromagnet/antiferromagnet superlattices, was the existence of the antiferromagnetic coupling[38,39] between the Fe layers separated by Cr layers. Soon after this discovery, an important related phenomenon, the so-called giant magnetoresistance (GMR) effect was discovered both in Fe/Cr superlattices[40] and Fe/Cr/Fe trilayers.[41] In this phenomenon, electrons with different spin directions experience different potential barriers in passing from one ferromagnet to another via the spacer layer. In case of parallel alignment of the magnetic layers, spin-up electrons have high conductivity and spin-down electrons have low conductivity. In case of anti-parallel alignment, electrons with both spin directions will experience high barrier potential and the conductivity will be low. While in zero external magnetic field, the ferromagnetic Fe layers couple antiferromagnetically to each other, they will couple ferromagnetically under an external magnetic field with sufficient strength. GMR has many applications in spintronics, such as magnetic sensors and magnetic recording technology, and has been recognized as an important effect, which resulted in the awarding of the Nobel prize for physics to Fert and Gründberg in 2007. Fe/Cr was also one of the systems, in which the oscillatory magnetic exchange coupling was observed.[42] This effect has been discussed as a consequence of Ruderman-Kittel-Kasuya-Yosida (RKKY) interaction.[43]

Still the current investigations reveal a lot of surprising effects in magnetization of such multilayers. Nevertheless, the interface structure in these systems is of paramount role in determining magnetic properties in such systems. Moreover, other new phenomena with industrial applications, such as non-collinear magnetic exchange interaction between Fe layers,[44,45] depend on the defects, steps, roughness and intermixing at the interfaces. It is why the understanding of the interfaces is very important in designing the spintronic devices. Techniques such as X-ray diffraction are just able to give an average roughness in the whole sample and most of the other techniques may in the best case give indirect information on the atomic scale about the interfaces. Once again, due to the monolayer resolution, Mössbauer spectroscopy, has been a powerful technique to study interfaces in Fe/Cr superlattices.[46–55] Landes *et al.*[46,47] studied the interfaces in Fe/Cr superlattices by introducing 2 ML ^{57}Fe at the interfaces and successively moving them from interfaces to the Fe layers. In such a way, an oscillatory behavior of the B_{hf} was demonstrated. It means that, while the B_{hf} on the Fe is reduced drastically at the interface, there are some increases at the second and the third layers of Fe inside the Fe layer, before the B_{hf} reaches the Fe bulk value for Fe layers very far from interfaces. Furthermore, in these studies, by measuring the B_{hf} as a function of temperature, a $T^{3/2}$ spin wave law was obtained:

$$B_{hf}(T) = B_{hf}(0)(1 - bT^{3/2}) \tag{11.16}$$

where b, the spin wave parameter and the slope of the function, is inversely proportional to the exchange interaction, J, i.e. the higher the value of b the weaker the coupling at the interface. The strength of this parameter at the interface compared to the bulk is as high as double.[56,57]

In analogy with alloys of Fe and Cr,[58] the effect of Cr atoms on the Fe B_{hf} in the interfaces has been described phenomenologically by Klinkhammer *et al.*:[50]

$$B_{hf} = B_{hf}(bulk) + n_1 B_1 + n_2 B_2 + \Delta B + \Delta B_{i=2} \tag{11.17}$$

where n_1 and n_2 are the number of the nearest neighbour (nn) and next nearest neighbour (nnn) Cr atoms and B_1 and B_2 are the contributions from one Cr atom in the first and second shell on the Fe B_{hf}. Starting with Fe B_{hf}(bulk)=-33.3 T, the fitted parameters were B_1=2.5 T and B_2=2.05 T, which means that the B_{hf} is decreased as the number of nn and nnn Cr is increased. The constant term, ΔB=-1.75 T, which is related to the broken translational symmetry at the interface, is in accordance with the observed enhancement of the B_{hf}. The last term $\Delta B_{i=2}$=-1.2 T is related to the B_{hf} at the second Fe layer at the interface, where n_1=0 and n_2=1. This term vanishes for all other Fe sites. This procedure is in accordance with the oscillatory behaviour of the B_{hf} at the interface as mentioned before. It is worth to mention that the oscillatory behaviour of the B_{hf} has been observed only in molecular beam epitaxially growth samples, where smooth interfaces exist, and not for sputtered samples. These experimental results have been combined with band structure calculations[59] to give a clear picture of such systems. Kazansky and Uzdin,[60] using a method of self-consistent calculation of magnetic moments, studied the magnetic structure of Fe/Cr interfaces with different degrees of disordered roughness. Uzdin *et al.*[54] performed CEMS measurements on Fe/Cr superlattices with ^{57}Fe ML at the different interfaces with the following compositions:

- MgO/Cr(50 Å)/[^{57}Fe(3 ML)/^{nat}Fe(8 ML)/^{57}Fe(3 ML)/Cr(8 ML)] × 10 (sample 1)
- MgO/Cr(50 Å)/[^{57}Fe(0.7 ML)/^{nat}Fe(8 ML)/Cr(8 ML)] × 40 (sample 2)
- MgO/Cr(50 Å)/[^{57}Fe(0.7 ML)/^{nat}Fe(8 ML)/Cr(8 ML)] × 200 (sample 3)
- MgO/Cr(50 Å)/[^{57}Fe(0.7 ML)/^{56}Fe(8 ML)/Cr(8 ML)] × 200 (sample 4)

The CEMS spectra together with the B_{hf} distributions are plotted in Fig. 11.8.

As can be seen, there are six distinct maxima, located at B_{hf}=33.1, 30.6, 28.0, 25.2, 22.7, and 19.6 T for sample 1, and with very small deviations for samples 2-4. There is an additional peak at 16.9 T for samples 2-4 only, which has been explained due to the thin ^{57}Fe probe layer in these samples. The highest value (33.1 T in sample 1 and 33.2 T in other samples) emanates probably from the bulk-like environment and, in agreement with Eq. (11.17), from the Fe atoms in the second layer under the surface, due to the existence of a distribution of B_{hf}, which includes higher values than the bulk value. Other values are attributed to Fe atoms with different numbers

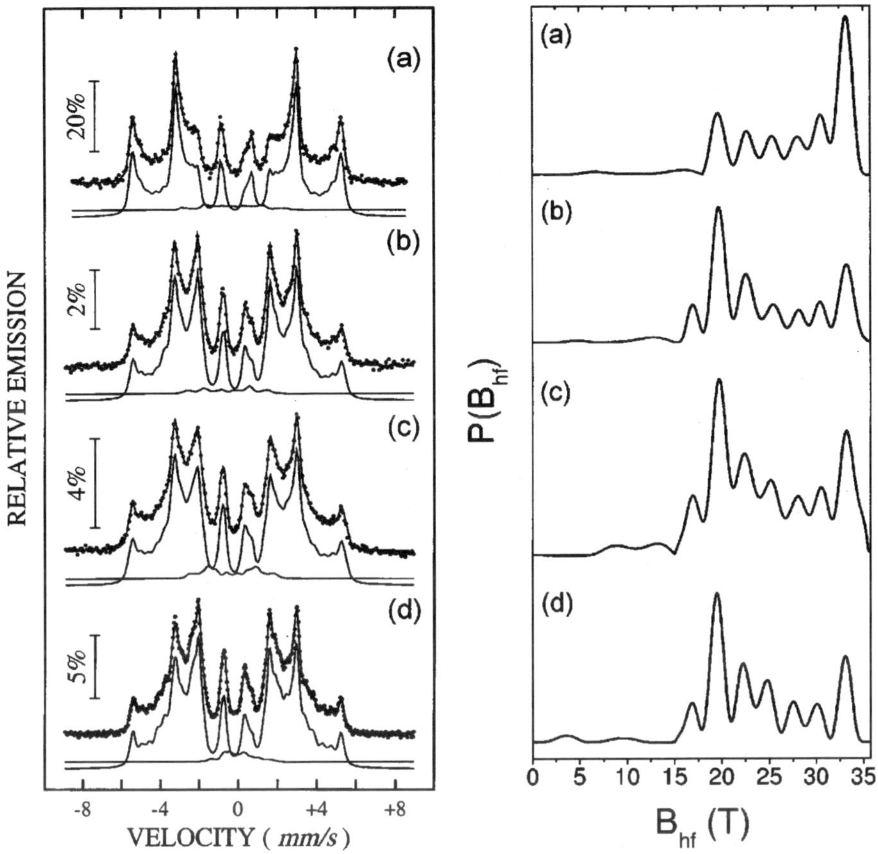

Fig. 11.8. (Left) CEMS spectra of all Fe/Cr measured at room temperature. (a) Sample 1, (b) Sample 2, (c) Sample 3, and (d) Sample 4. (Right) Magnetic hyperfine field distributions of the same sample in the same order as in the left panel. Reprinted with permission from Uzdin *et al.*,[54] Phys. Rev. B 63 (2001) 1044071. © 2001 The American Physical Society.

of Cr as nn and nnn. It is worth to mention that according to Landes *et al.*,[46,47] the 19.6 T should be attributed to a flat surface, i.e. n_1=4 and n_1=1, while according to Klinkhammer *et al.*,[50] the value 22.7 T should be attributed to the flat surface. Furthermore, Uzdin *et al.*[54] used these B_{hf} values and the results from the self-consistent calculation of magnetic moments for the same superlattice structures, to correlate the magnetic moments to magnetic hyperfine fields. In the later study,[55] by using the same procedure, they investigated the differences between Fe lower and upper interfaces, as a function of annealing temperature.

11.3.3. *Fe/V superlattices*

Although the interesting effects discussed in the previous section, such as the existence of the antiferromagnetic coupling,[38,39] giant magnetoresistance (GMR) effect[40,41] and the oscillatory magnetic exchange coupling,[42] were all discovered in Fe/Cr systems, the same phenomena also exist in systems of magnetic layers separated by nonmagnetic metals in form of superlattices such as Fe/V.[61,62] It is also possible to tune interlayer exchange coupling by hydrogen absorption in V layers.[63,64] In the periodic table, V stands between the non-magnetic and magnetic elements in the 3d transition metals. Though it is a non-magnetic element, when dissolved in Fe it gets an induced magnetic moment.[65] In case of Fe/V thin films[66] or superlattices,[67] the induced magnetic moment in the V layers at the interface couple antiferromagnetically to Fe magnetic moments. The exchange coupling in Fe/V and Fe/VCr superlattice was studied using first principle method.[68,69] In a later study, Schwickert *et al.* combined augmented spherical wave method calculations with the experimental techniques to investigate magnetic properties of Fe/V systems with interdiffused interfaces.[61] They found that V magnetic moment in an interdiffused interface is enhanced compared with an ideal interface. They also showed that the induced V magnetic moments, which couple ferromagneticlly to each other, decay monotonically away from Fe interface. The decaying extent, which is 4 ML (6 Å), was termed as a "transient ferromagnetic" state by them. Furthermore, they observed three antiferromagnetic coupling peaks for V thickness at 22, 32 and 42 Å. They expected to observe another peak at 12 Å V thickness, but they explained that the probable reason not to observe it was due to the transient ferromagnetic behavior of V. In an ideal Fe/V superlattice, an increase of Fe magnetic moment compared to the bulk value, i.e. 2.2 μ_B was predicted by Niklasson *et al.*[30] This is in accordance with the results from dilute FeV alloys.[70] In the latter study,[70] the authors suggested that substitution of every nn and nnn Fe atom by a V atom will decrease B_{hf} by 3 T. Tage *et al.*,[71] by applying first principle theory, were able to show that by geometrical tuning of superlattices with appropriate choice of material, interesting magnetic configuration can be produced. They demonstrated that the magnetization moment in every second layer in Fe/V/Co superlattice can be stabilized in an out of plane direction and perpendicular to the magnetization direction of the adjacent magnetic layers. A magnetic configuration that can be used in sensor applications.

As it has been mentioned before, the interface configurations have crucial role in the above mentioned effects and hence in the construction of the spintronic devices. Then it is an interesting subject to study interfaces in Fe/V superlattices to understand how the V atoms influence the neighboring Fe atoms at the interfaces and deeper in the Fe layer. CEMS has been extensively used to study Fe/V superlattices. It is obvious that B_{hf} and the magnetic moments of Fe atom decreases as the surrounding Fe atoms are replaced by non-magnetic atoms such as V. By depositing a monolayer of ^{57}Fe at Fe(10 ML)/V(10 ML) interface and at different distances from the interface, using 15 T/μ_B conversion factor, the average magnetic moment of Fe was estimated to be 1.1 μ_B at the interface, while it was increased gradually to 2.2 μ_B, the characteristic value for bulk Fe, for Fe atoms deep inside the Fe layer.[72] In a series of studies, Wäppling *et al.* used CEMS to investigate Fe hyperfine field distribution in Fe(x ML)/V(y ML) superlattices as function of x and y.[73–77] Some spectra from these studies are presented in the left panels in Fig. 11.9. B_{hf} distributions from these spectra and some other related B_{hf} distributions are indicated in right panels in Fig. 11.9.

The samples with Fe layer less than 5 ML showed no magnetic splitting down to 133 K. The average magnetic hyperfine field was strongly dependent on the Fe layer thickness. Nevertheless, the bulk value, i.e. 33 T was not found even in a 10 ML thick Fe layer at room temperature, which according to their argument, was due to the lower Curie temperature compared to bulk value. Furthermore, they concluded that Fe B_{hf} depends not only on the Fe layer thickness but also on the V layer thickness. For a fixed Fe layer thickness, the increase of V layer thickness caused the decrease of B_{hf}. They explained that interlayer coupling, which is stronger for the thinner spacer layer, contributes to the magnetization of the non-magnetic part, hence there will be positive influence on the magnetic hyperfine field distribution. By depositing ^{57}Fe at the different interfaces, i.e. Fe on V and V on Fe, they[74] studied different interfaces in Fe/V superlattices. They showed that the average B_{hf} is higher for the upper interface of Fe than the lower interface, which means that intermixing is less in the Fe upper interface, and thus, that the Fe layer is smoother than the V layer. They also studied the interface quality as a function of growth temperature.[78] While the XRD data suggested that the temperature growth is optimum for $T_g = 300-330$°C, the CEM spectra were almost identical for

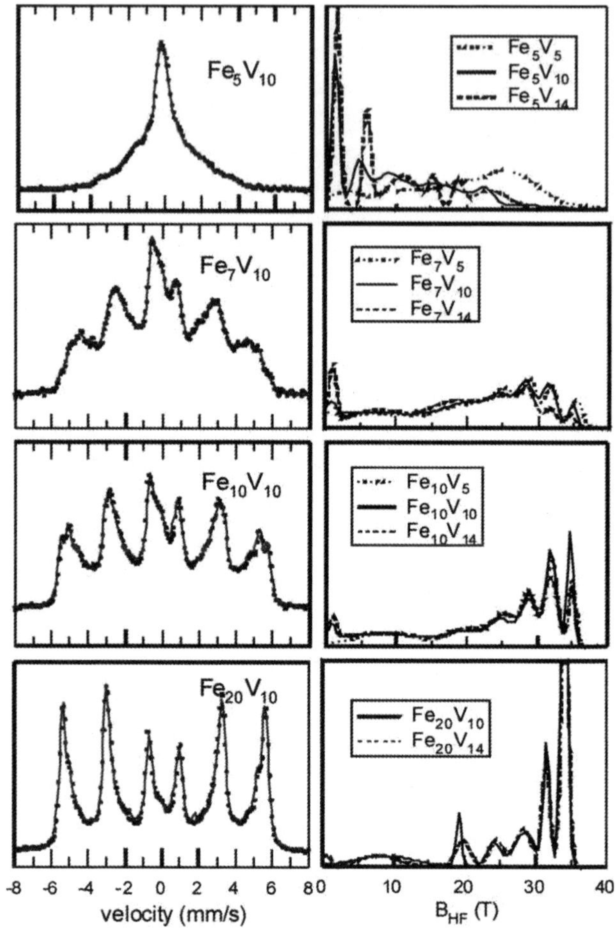

Fig. 11.9. (Left) CEMS spectra from Fe/V superlattices with varying Fe and V thicknesses. (Right) Magnetic hyperfine field distributions for the samples depicted in the left panels and some other related superlattices. Reprinted with permission from Nordsröm *et al.*,[76] Hyperfine Interact. 141/142 (2002) 465. © 2002 Kluwer Academic Publishers.

$T_g = 230-330$°C. This means that the short-range order is similar in the samples grown in this range and the growth temperature has minor effect if it is kept under the optimum temperature, i.e. 330°C.

In a more recent study, Uzdin and Häggström[79] used Mössbauer spectroscopy together with the theoretical modeling and self-consistent calculations of magnetic moments to investigate the magnetic structure of Fe/V systems with different intermixings at the interface. They used the so called floating model[55] to describe the intermixing at the interface. In this model, it is supposed that during the growth a fraction, ψ, of the atoms of the deposited monolayer exchanges position with the atoms in the "substrate", i.e. the layer under. The extreme values, ψ=0 and ψ=1 mean ideal interface and complete interchange of the two monolayers in question, respectively. By using B_{hf} distribution, it is then possible to extract a value for the intermixing parameter, ψ. Furthermore, they found a linear relationship between calculated Fe magnetic moment and Fe B_{hf} from CEMS, with the proportionality factor of 15 T/μ_B.

11.3.4. *Exchange spring magnets*

Magnetization reversal process is one of the fundamental questions of magnetism. The evolution of a magnetic structure as a response of the system on an external magnetic field can lead to a rich variety of behaviours at the nanoscale. The investigation of such processes is important to elucidate the internal nature of the magnetic interactions, as well as their modification upon confined geometry. It is quite relevant also in technological applications for the improvement of the functionality of magnetic memory media and spintronic devices. The development of nanotechnology has pushed the study and utilization of non-collinear magnetic ordering in single magnetic elements. One of such systems where non-collinear magnetic structure can be reversibly varied by magnetic field is the system consisting of exchange coupled of hard and soft magnetic phases, the so-called spring magnets. In spring magnets the hard and the soft magnetic bilayers or multilayers are coupled to each other ferromagnetically by an exchange field, H_{ex}, which is a function of magnetic parameters and the geometry of the soft layers. For external applied magnetic field, H_{app}, less than H_{ex} antiparallel to the magnetization direction of the system, the magnetic moment of the soft layers will remain parallel to the magnetic moment of the hard layers. As the H_{app} is increased behind this limit, the magnetization reversal occurs via the development of a non-collinear spiral structure in the soft material, but if the field is then removed, the spiral structure will spring back to the original direction, i.e. the easy axis of the hard magnet, hence the system is called spring magnet.[80] For further increase of H_{app}, the magnetization of the hard layer is irreversibly switched at the so-called nucleation field, H_{nu}, which depends on magnetic parameters and the geometry of both hard and soft layers. The variation of the external magnetic field leads to the reversible change of the magnetic moment direction mainly in the soft magnetic layers. Goto *et al.* were the first who studied the magnetization reversal process in the soft layer. Later on, Skomski and Coey,[81] and Coey[82] studied the exchange coupling theoretically. A model of one-dimensional chain of classical spins[83] were used to simulate characteristic behaviour of such systems. If one divide the whole magnetic system, i.e. Fe layer plus FePt layer, into N sublayers, the total energy of this system can be written as:

$$E = -\sum_{i=1}^{N-1} \frac{A_{i,i+1}}{d^2} cos(\phi_i - \phi_{i+1}) - \sum_{i=1}^{N} K_i cos^2(\phi_i) - \sum_{i=1}^{N} HM_i cos(\phi_i - \phi_H) \tag{11.18}$$

which is the sum of the magnetic exchange and anisotropy energies and the dipolar interaction with the external field. Here $A_{i,i+1}$ is the exchange constant between sublayers i and i+1, $d = (d_{soft} + d_{hard})/N$ is the sublayer thickness, ϕ is the angle between the sublayer magnetization relative to the remanent magnetization of the hard layer and H is the external magnetic field. K_i and M_i are the the anisotropy constant and magnetization of the ith sublayer, respectively. To achieve the equilibrium spin configuration, one has to minimize the energy with respect to each of ϕ_i. The combination of the high magnetization of the soft phase with the high coercivity of the hard one, makes these systems suitable for building permanent magnets, with large energy product (magnetic saturations times coercivity).[84,85] An energy product as high as 1090 kJ/m^3 is predicted to be achieved by suitable nanostructured composites such as $Sm_2Fe_{17}N_3/Fe_{65}Co_{35}$. Exchange spring media are also promising systems for magnetic recording media.[86–90] Various experimental methods have been used for the investigation of magnetic ordering in spring magnets, which in turn became a model system for benchmarking different methods of non-collinear magnetism study. Large efforts were undertaken to characterize the angular distribution of magnetic moments directions across the soft magnetic layer. For example, Kuncser *et al.*[91] used angle dependent Mössbauer spectroscopy to study the spin structure of Sm-Co(10nm)/Fe(15nm) bilayers. Two samples were studied. In the first sample (A), the whole Fe layer was homogeneously enriched in ^{57}Fe, whereas in the second sample (B) 2 nm layer of ^{57}Fe was deposited at the interface. It means that, while the Mössbauer signal is from the whole Fe layer in the first sample, it reveals just the spin direction at the interface in the second sample. The first measurements in zero external field and with perpendicular geometry, i.e. when γ-rays have the same direction as the normal direction of the samples, the 3:4:1:1:4:3 line ratios of the

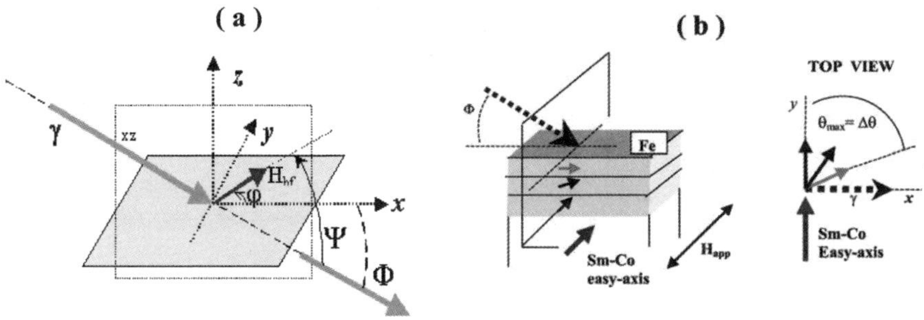

Fig. 11.10. (a) Schematic geometrical arrangement of the CEMS measurement. The external magnetic field is varied along y axis, which is the easy axis of the hard magnetic layer. (b) Schematic representation of a spin spiral distribution in the plane of the soft magnetic layer. Reprinted with permission from Kuncser *et al.*,[91] Phys. Rev. B 68 (2003) 0644161. © 2003 the American Physical Society.

Mössbauer spectra, revealed that all spins lay in the plane of the samples. With the geometry depicted in Fig. 11.10, the second line intensity in the Mössbauer spectra is changed as the applied magnetic field is increased.

Figure 11.11 shows the CEMS spectra of both samples as a function of the external magnetic field. The intensity ratio R_{23} in both samples undergoes characteristic changes as the external magnetic field is varied. The intensity ratio has a minimum value of about 1.5 at a field intensity of 0.4 T.

Because the Fe spin moments in an applied external field will form a spiral structure, the angle between the incident γ-ray and the spin moment will not be constant but will have a certain angular distribution $P(\phi)$ with:

$$\int_0^{2\pi} P(\varphi)d\varphi = 1 \tag{11.19}$$

The intensity ratio will then be written as:

$$R_{23} = 4\int_0^{2\pi} \frac{1 - cos^2\phi\ cos^2\varphi}{1 + cos^2\phi\ cos^2\varphi} P(\varphi)d\varphi \tag{11.20}$$

Taking into account that the incoming γ-rays are perpendicular to the easy axis of the hard magnetic layer and also that the conversion electrons from different depths have different probabilities to escape the sample surface, Eq. (11.20) can be expressed as:

$$R_{23} = \frac{4}{0.67\cdot\Delta\theta}\int_{\theta_0}^{\theta_0+\Delta\theta} \frac{1 - cos^2\phi\ sin^2\theta}{1 + cos^2\phi\ sin^2\theta} \times (0.60 + \frac{0.14\cdot(\theta-\theta_0)}{\Delta\theta})d\theta \tag{11.21}$$

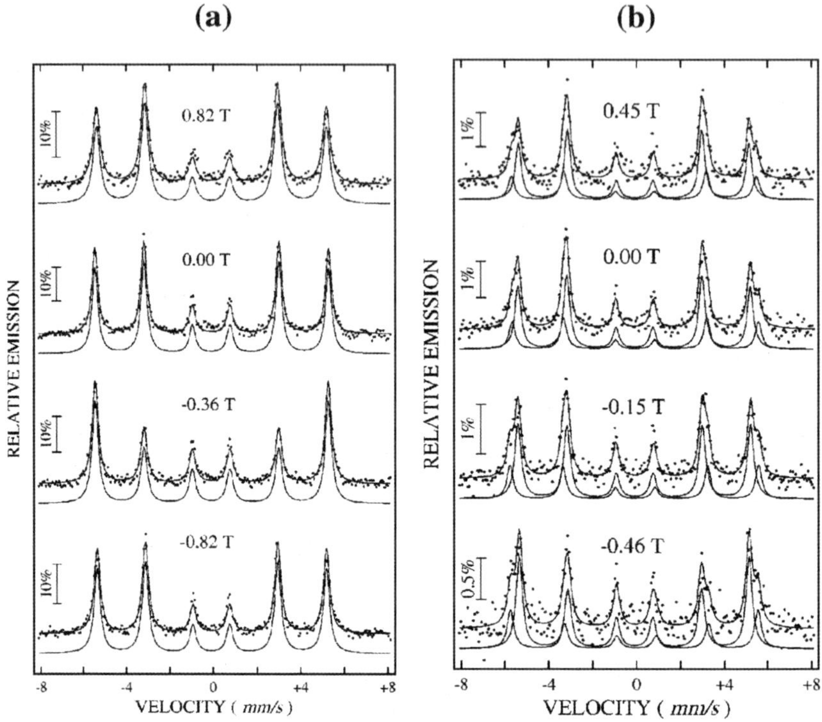

Fig. 11.11. CEMS spectra at room temperature as a function of external magnetic field. (a) Sample A with an incident angle of γ radiation, ϕ, of 20° (b) Sample B with an incident angle of γ-rays, ϕ, of 30°. Reprinted with permission from Kuncser *et al.*,[91] Phys. Rev. B 68 (2003) 0644161. © 2003 The American Physical Society.

where θ_0 is the angle between the magnetic direction of the first Fe layer and the easy axis of the hard layer, and $\Delta\theta$ is the moment aperture, i.e. the angle between the Fe moment at the interface and the Fe magnetic moment at the surface of the layer. Now, by having intensity ratios, R_{23} from a series of measurements, one is able to extract the distribution parameters, which has been the aim of this study. In combination with the intensity ratio from the second sample, which gave the angle between the Fe moment and the easy axis of the hard magnetic layer at the interface, they were able to infer that there exists a uniform spin spiral structure in the Fe layer. Furthermore, they concluded that for the highest value of the applied magnetic field before the irreversible switching of the hard magnet occurs, the Fe moment direction reaches a maximum of 40° relative the easy axis of

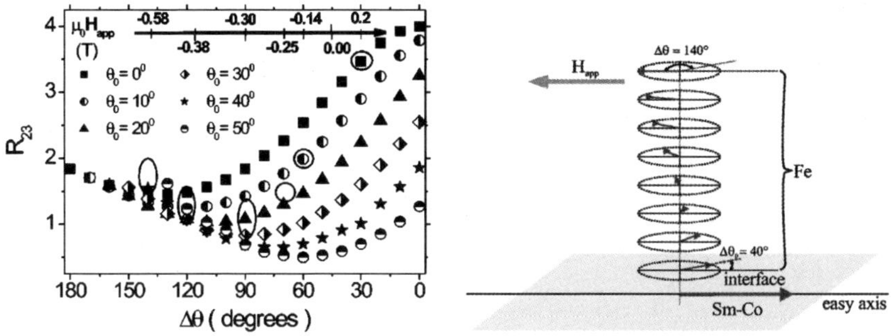

Fig. 11.12. (Left) Calculated intensity ratios R_{23} according to Eq. (11.21) for sample A as a function of moment aperture, $\Delta\theta$, with θ_0 as a parameter. Large circles indicate experimental data points and their corresponding applied fields values for positive fields and negative fields < 0.25 T and ellipses indicate the same for negative fields > 0.30 T. (Right) Schematic illustration of the spiral structure of spins in the soft magnetic layer in case of an applied magnetic field of 0.58 T. Reprinted with permission from Kuncser *et al.*,[91] Phys. Rev. B 68 (2003) 0644161. © 2003 the American Physical Society.

the Sm-Co layer and the moment aperture was 140°. These are summarized in Fig. 11.12.

For utilization of exchange spring magnets in electronic applications it is very important to know how the properties of the system depend on the characteristics of the hard and the soft magnetic layers. One of the ways which allows tuning the exchange spring behaviour is the variation of soft layer composition. Recently, a method to improve the exchange-spring nanocomposite permanent magnets was suggested.[92–94] In this method, by annealing or depositing Sm-Co/Fe bilayers at high temperatures induced intermixing at the interface occurs, which results in an improvement of the energy product. This result was in contrast to the earlier belief that an ideal interface should give the optimal exchange coupling. The hysteresis loop becomes more single phase-like, though the magnetization remains fully reversible. The interpretation of this phenomenon is based on a simple phenomenological model, which considers the spring magnet as one-dimensional chain of classical spins.[83] Micromagnetic simulation requires the knowledge about the concentration profile near the interface as well as the dependence on parameters (uniaxial anisotropy, saturation magnetization and exchange constant) and on the concentration in alloy. It cannot give information about real magnetic structure at the atomic scale. Only recently an atomic scale quantum-mechanical theory based on a non-collinear Tight-Binding formalism for self-consistent calculations of electronic and magnetic structure of d-metallic systems in an external magnetic field has been developed.[95,96] The theory was applied to the description of the magnetic reversal process in exchange spring magnets with Fe as the soft magnetic material. The calculations reproduce all the main features of the exchange spring behavior. The theory is suitable for the study of more complex systems with interface roughness, interdiffusion and non-coplanar magnetic ordering. This approach can be used for the interpretation of magnetization reversal processes in systems with different soft layer composition. Based on the theoretical approach, magnetic structures measured by Mössbauer spectroscopy in external magnetic field can be interpreted to give a complete picture, which will be highly appreciated not only by the scientists in this field, but also by the industrial research centers, which can use these results in very sophisticated devices.

Acknowledgments

I sincerely thank my colleagues, L. Häggström, V. Uzdin, H. Raanaei, E. Goikolea, J. Åkerman, for suggestions and comments, and especially R. Wäppling for reading the manuscript. Special thanks go to my colleagues, whose permissions for reusing their results, make this work possible. I would also like to thank my colleagues at the Mössbauer Effect Data Center[13] (MEDC), J. Stevens, N. Hall and A. Khasanov, for a kind collaboration. Japanese society for promotion of science (JSPS) is acknowledged for financial support.

References

1. W. E. J. Lamb, *Phys. Rev.* **55**, 190, (1939).
2. R. H. Dicke, *Phys. Rev.* **89**, 472, (1953).
3. H. Frauenfelder, *The Mössbauer Effect.* (W. A. Benjamin, Inc, New York, 1963).
4. G. K. Wertheim, *Mössbauer Effect: Principles and Applications.* (Academic Press, New York, 1964).
5. V. I. Goldanskii and R. H. Herber, *Chemical Applications of Mössbauer Spectroscopy.* (Academic Press, New York, 1968).
6. N. N. Greenwood and T. C. Gibb, *Mössbauer Spectroscopy.* (Chapman and Hall, London, 1971).
7. L. May, *An Introduction to Mössbauer Spectroscopy.* (Plenum, New York, 1971).
8. T. C. Gibb, *Principles of Mössbauer Spectroscopy.* (Chapman and Hall, London, 1977).

9. P. Gutlich, R. Link, and A. Trautwein, *Mössbauer Spectroscopy and Transition Metal Chemistry.* (Springer-Verlag, Berlin, 1978).
10. G. Long, Ed., *Mössbauer Spectroscopy Applied to Inorganic Chemistry, Volume I.* (Plenum Press, New York and London, 1984).
11. `http://www.ibame.org/index.html`.
12. `http://www.mossbauer.org/index.html`.
13. `http://orgs.unca.edu/medc/`.
14. `http://pecbip2.univ-lemans.fr/~moss/webibame/`.
15. L. Häggström. Report UUIP-851, University of Uppsala, Sweeden. Technical report, Physics Institute, (1974).
16. C. Kittel, *Introduction to Solid State Physics.* (John Wiley & Sons, Inc., 1996).
17. T. Burkert, L. Nordström, O. Eriksson, and O. Heinonen, *Phys. Rev. Lett.* **93**(2), 027203, (2004).
18. M. F. Collins and J. B. Forsyth, *Phil. Mag.* **8**, 401, (1963).
19. C. E. Johnson, M. S. Ridout, and T. E. Cranshaw, *Proc. Phys. Soc.* **81**, 1079, (1963).
20. T. R. McGuire and D. I. Bardos, *J. Appl. Phys.* **40**, 1371, (1969).
21. I. Vincze, I. A. Campbell, and A. J. Meyer, *Solid State Commun.* **15**, 1495, (1974).
22. A. Narayanasamy, T. Nagarajan, P. Muthukumarasamy, and T. S. Radhakrishnan, *J. Phys. F: Metal Phys.* **9**(11), 2261, (1979).
23. B. deMayo, *Phys. Rev. B.* **24**, 6503, (1981).
24. B. Fultz, *Hyperfine Interact.* **41**, 607, (1988).
25. H. H. Hamdeh, B. Fultz, and D. H. Pearson, *Phys. Rev. B.* **39**, 11233, (1989).
26. J. Dekoster, E. Jedryka, M. Wójcik, and G. Langouche, *J. Magn. Magn. Mater.* **126**, 12, (1993).
27. P. Blomqvist, R. Wäppling, A. Broddefalk, P. Nordblad, S. G. E. Velthuis, and G. P. Felcher, *J. Magn. Magn. Mater.* **248**, 75, (2002).
28. P. Blomqvist and R. Wäppling, *J. Cryst. Growth.* **252**, 120, (2003).
29. M. Björck, G. Andersson, B. Lindgren, R. Wäppling, V. Stanciu, and P. Nordblad, *J. Magn. Magn. Mater.* **284**, 273, (2004).
30. A. M. N. Niklasson, B. Johansson, and H. L. Skriver, *Phys. Rev. B.* **59**(9), 6373, (1999).
31. O. Eriksson, L. Bergkvist, E. Holmström, A. Bergman, O. LeBacq, S. Frota-Pessoa, B. Hjövarsson, and L. Nordström, *J. Phys.: Condens. Mater.* **15**(5), 599, (2003).
32. J. Dekoster, H. Bemelmans, S. Degroote, J. DeWachter, E. Jedryka, R. Moons, A. Vantomme, and G. Langouche, *Hyperfine Interact.* **95**, 191, (1995).
33. B. Lindgren, M. Andreeva, L. Häggström, V. S. B. Kalska, A. Chumakov, O. Leupold, and R. Rüffer, *Hyperfine Interact.* **136/137**, 439, (2001).
34. B. Kalska, P. Blomquist, L. Häggström, and R. Wäppling, *J. Phys.: Condens. Mater.* **13**, 2963, (2001).
35. B. Kalska, P. Blomquist, L. Häggström, and R. Wäppling, *J. Magn. Magn. Mater.* **226-230**, 1773, (2001).

36. L. Häggström, B. Kalska, E. Nordström, P. Blomqvist, and R. Wäppling, *J. Alloys and Compounds.* **347**, 252, (2002).
37. S. Kamali-M, A. Bergman, G. Andersson, V. Stanciu, and L. Häggström, *J. Phys.: Condens. Mater.* **18**, 5807, (2006).
38. P. Grünberg, R. Schreiber, Y. Pang, M. B. Brodsky, and H. Sowers, *Phys. Rev. Lett.* **57**(19), 2442, (1986).
39. C. Carbone and S. F. Alvarado, *Phys. Rev. B.* **36**(4), 2433, (1987).
40. M. N. Baibich, J. M. Broto, A. Fert, F. N. V. Dau, F. Petroff, P. Eitenne, G. Creuzet, A. Friederich, and J. Chazelas, *Phys. Rev. Lett.* **61**(21), 2472, (1988).
41. G. Binasch, P. Gründberg, F. Saurenbach, and W. Zinn, *Phys. Rev. B.* **39** (7), 4828, (1989).
42. S. S. P. Parkin, N. More, and K. P. Roche, *Phys. Rev. Lett.* **64**(19), 2304, (1990).
43. M. A. Ruderman and C. Kittel, *Phys. Rev.* **96**, 99, (1954).
44. J. C. Slonczewski, *J. Magn. Magn. Mater.* **150**, 13, (1995).
45. A. Schreyer, J. F. Ankner, T. Zeidler, H. Zabel, C. F. Majkrzak, M. Schäfer, and P. Grünberg, *Europhys. Lett.* **32**, 595, (1995).
46. J. Landes, C. Sauer, R. A. Brand, W. Zinn, and Z. Kajcsos, *Hyperfine Interact.* **57**, 1941, (1990).
47. J. Landes, C. Sauer, R. A. Brand, W. Zinn, S. Mantl, and Z. Kajcsos, *J. Magn. Magn. Mater.* **86**, 71, (1990).
48. D. L. Williamson, B. M. Lairson, A. P. Payne, N. M. Rensing, and B. M. Clemens, *Hyperfine Interact.* **92**, 1271, (1994).
49. J. Zukrowski, G. Liu, H. Fritzsche, and U. Gradmann, *J. Magn. Magn. Mater.* **145**, 57, (1995).
50. F. Klinkhammer, C. Sauer, E. Y. Tsymbal, S. Handschuh, Q. Leng, and W. Zinn, *J. Magn. Magn. Mater.* **161**, 49, (1996).
51. R. Schad, P. Belien, G. Verbanck, K. Temst, V. V. Moshchalkov, Y. Bruynseraede, B. Bahr, J. Falta, J. Dekoster, and G. Langouche, *Europhys. Lett.* **44**, 379, (1998).
52. P. B. R. Schad, G. Verbanck, K. Temst, H. Fischer, S. Lefebvre, M. Bessiere, D. Bahr, J. Falta, J. Dekoster, L. Langouche, V. V. Moshchalkov, and Y. Bruynseraede, *J. Magn. Magn. Mater.* **198-199**, 104, (1999).
53. M. Kopcewicz, T. Lucinski, F. Stobiecki, and G. Reiss, *J. Appl. Phys.* **85**(8), 5039, (1999).
54. V. M. Uzdin, *Phys. Rev. B.* **63**, 104407, (2001).
55. V. Uzdin, W. Keune, and M. Walterfang, *J. Magn. Magn. Mater.* **240**, 504, (2002).
56. C. T. Rado and T. R. Weertman, *J. Phys. Chem. Solids.* **11**, 315, (1959).
57. D. L. Mills and A. A. Maradudin, *J. Phys. Chem. Solids.* **28**, 1855, (1967).
58. S. M. Dubiel and J. Zukrowski, *J. Magn. Magn. Mater.* **23**, 214, (1981).
59. A. J. Freeman and C. L. Fu, *J. Appl. Phys.* **61**, 3356, (1987).
60. A. K. Kazansky and V. M. Uzdin, *Phys. Rev. B.* **52**, 9477, (1995).
61. M. M. Schwickert, R. Coehoorn, M. A. Tomaz, E. Mayo, D. Lederman, and W. L. O'Brien, *Phys. Rev. B.* **57**(21), 13681, (1998).

62. R. M. A. Broddefalk, P. Nordblad, P. Blomqvist, R. Wäppling, J. Lu, and E. Olsson, *Phys. Rev. B.* **65**, 214430, (2002).
63. B. Hjörvarsson, J. A. Dura, P. Isberg, T. Watanabe, T. J. Udovic, G. Andersson, and C. F. Majkrzak, *Phys. Rev. Lett.* **79**(5), 901, (1997).
64. V. Leiner, K. Westerholt, A. M. B. H. Zabel, and B. Hjörvarsson, *Phys. Rev. Lett.* **91**(3), 037202, (2003).
65. I. Mirebeau, G. Parette, and J. W. Cable, *J. Phys. F: Met. Phys.* **17**, 191, (1987).
66. T. G. Walker and H. Hopster, *Phys. Rev. B.* **49**(11), 7687, (1994).
67. M. A. Tomaz, W. J. A. Jr., W. L. O'Brien, and G. R. Harp, *J. Phys.: Condens. Matter.* **9**, L179, (1997).
68. M. van Schifgaarde and F. Herman, *Phys. Rev. Lett.* **71**(12), 1923, (1993).
69. M. van Schifgaarde, F. Herman, S. P. Parkin, and J. Kudrnovský, *Phys. Rev. Lett.* **74**(20), 4063, (1995).
70. J. C. Krause, J. Schaf, and J. M. I. da Costa, *Phys. Rev. B.* **61**(9), 6196, (2000).
71. A. Tage, L. Nordström, P. James, B. Johansson, and O. Eriksson, *Nature.* **406**, 280, (2000).
72. L. Häggström, P. Isberg, M. Wikner, B. Degroote, and R. Wäppling, *Hyperfine Interact.* **C 3**, 401, (1998).
73. B. Kalska, L. Häggström, P. Blomquist, and R. Wäppling, *J. Phys.: Condens. Matter.* **12**, 539, (2000).
74. B. Kalska, P. Blomquist, L. Häggström, and R. Wäppling, *Europhys. Lett.* **53**(3), 395, (2001).
75. B. Kalska, L. Häggström, P. Blomquist, and R. Wäppling, *J. Magn. Magn. Mater.* **226-230**, 1782, (2001).
76. E. Nordström, B. Kalska, L. Häggström, P. Blomqvist, and R. Wäppling, *Hyperfine Interact.* **141/142**, 465, (2002).
77. E. Nordström, B. Kalska, L. Häggström, P. Blomqvist, and R. Wäppling, *J. Alloys Comp.* **348**, 208, (2003).
78. G. Andersson, E. Nordström, and R. Wäppling, *Europhys. Lett.* **60**(5), 731, (2002).
79. V. M. Uzdin and L. Häggström, *Phys. Rev. B.* **72**, 024407, (2005).
80. E. F. Kneller and R. Hawig, *IEEE Trans. Magn.* **27**(4), 3588, (1991).
81. R. Skomski and J. M. D. Coey, *Phys. Rev. B.* **48**(21), 15812, (1993).
82. J. M. D. Coey, *Solid State Commun.* **102**, 101, (1997).
83. E. E. Fullerton, J. S. Jiang, M. Grimsditch, C. H. Sowers, and S. D. Bader, *Phys. Rev. B.* **58**(18), 12193, (1998).
84. R. Skomski, *J. Appl. Phys.* **76**(10), 7059, (1994).
85. S. D. Bader, *Rev. Mod. Phys.* **78**, 1, (2006).
86. R. H. Victora and X. Shen, *IEEE Trans. Magn.* **41**(2), 537, (2005).
87. R. H. Victora and X. Shen, *IEEE Trans. Magn.* **41**(10), 2828, (2005).
88. D. Suess, T. Schrefl, S. Fähler, M. Kirschner, G. Hrkac, F. Dorfbauer, and J. Fidler, *Appl. Phys. Lett.* **89**, 0125041, (2006).
89. D. Suess, *Appl. Phys. Lett.* **89**, 1131051, (2006).
90. J. P. Wang, W. Shen, and J. Bai, *IEEE Trans. Magn.* **41**(10), 3181, (2005).

91. V. E. Kuncser, M. Doi, W. Keune, M. Askin, H. Spies, J. S. Jiang, A. Inomata, and S. D. Bader, *Phys. Rev. B.* **68**, 064416, (2003).
92. J. S. Jiang, J. E. Pearson, Z. Y. Liu, B. Kabius, S. Trasobares, D. J. Miller, S. D. Bader, D. R. Lee, D. Haskel, G. Srajer, and J. P. Liu, *Appl. Phys. Lett.* **85**(22), 5293, (2004).
93. J. S. Jiang, J. E. Pearson, Z. Y. Liu, B. Kabius, S. Trasobares, D. J. Miller, S. D. Bader, D. R. Lee, D. Haskel, G. Srajer, and J. P. Liu, *J. Appl. Phys.* **97**, 10K3111, (2005).
94. Y. Choi, J. S. Jiang, Y. Ding, R. A. Rosenberg, J. E. Pearson, S. D. Bader, A. Zambano, M. Murakami, I. Takeuchi, Z. L. Wang, and J. P. Liu, *Phys. Rev. B.* **75**, 104432, (2007).
95. V. M. Uzdin and A. Vega, *Phys. Rev. B.* **77**, 134446, (2008).
96. V. M. Uzdin and A. Vega, *Nanotechnol.* **19**, 315401, (2008).

Chapter 12

NUCLEAR RESONANCE SCATTERING AND ITS APPLICATIONS IN SPINTRONICS

Saeed Kamali

Department of Applied Science, University of California
Davis, California 95616
Physical Biosciences Division, Lawrence Berkeley National Laboratory
Berkeley, California 94720
E-mail: skamali@ucdavis.edu

Parallel to the development of synchrotron radiation facilities, in addition to the use of such radiation in conventional techniques, new methods have been developed. One of these methods is synchrotron radiation based Mössbauer spectroscopy, which is also called Nuclear Resonance Scattering. Nuclear Resonance Scattering is now a very powerful technique for characterization of nanostructured materials.

12.1. Introduction

Nuclear Resonance Scattering (NRS) is the Synchrotron Radiation based Mössbauer spectroscopy. As a continuation of the previous chapter, NRS will be described here. NRS is now a routine technique for studying many different kinds of samples, especially in extreme conditions such as high pressure. By introducing some examples, the use of NRS in spintronics will be presented.

Because the source for this technique is synchrotron radiation, a short description of it will be given before introducing the technique and its applications.

12.2. Synchrotron Radiation

Synchrotron radiation (SR), which once was an unwanted radiation for high energy physicists, now has become a standard source for almost all conventional techniques, especially in the third generation synchrotron facilities, where insertion devices such as undulators and wigglers are used. Its applications cover all natural sciences, ranging from material science to biophysics. Synchrotron radiation, which is based on the emission of electromagnetic radiation by relativistic electrons in magnetic fields, has several advantages over conventional X-ray sources, some of which are listed below:

- Brilliance: It is brighter than the conventional sources by several orders of magnitude. Shorter times are needed for measurements, especially in the case of the nanostructures.
- Polarization: Different kinds of polarizations. Linearly or circularly polarized can be arranged thanks to the conservation of angular momentum law. It also results in an increase in sensitivity and a decrease in background noise.
- Collimation: The beam can be collimated to sub-micron scale.
- Continuous Spectrum: From infra red to hard X-rays are produced.
- Pulsation: It can be powerful in dynamical studies.

The photon energy from such a source ranges from a few tens eV up to several hundreds keV. This means that all the conventional techniques based on the interaction of X-rays with different kinds of materials can be performed using such radiation. Parallel to the development of Synchrotron radiation, the interest for nanotechnology and its applications has increased drastically during the last decade. Fabrication of all kinds of nanostructured materials is now possible by various techniques due to the recent progress in this field. Due to the increase of the ratio of surface atoms with respect to the total number of atoms of the particle and also, to the changes of the electronic structure of nanomaterials, there are significant differences between properties of such materials and bulk properties. Moreover, there are some phenomena, which are characteristic just for nanostructured materials. Such phenomena make nanostructured materials applicable for many devices. Then, understanding the physical properties and electronic structure of nanomaterials is of crucial importance.

Synchrotron radiation, due to its novel properties, is a proper source for studying nanostructured materials. The development of different synchrotron based techniques enables us to perform measurements, which have not been possible before. There are now several synchrotron radiation facilities all over the world, among them SPring-8[1] in Japan is the biggest with a circumference of 1436 m and an electron energy of 8 GeV. Advanced Photon Source (APS)[2] in USA with 7 GeV electron energy and European

Synchrotron Radiation Facility (ESRF)[3] in France with 6 GeV are the other largest synchrotron facilities in the world.

For a detailed description of synchrotron radiation and its properties, the interested readers are referred to a number of excellent books in this field.[4–6]

12.3. Nuclear Resonance Scattering

In 1974 Ruby[7] suggested to used synchrotron radiation as the source for Mössbauer spectroscopy. Gerdau *et al.*[8] were the first who monochromatized the SR down to 10^{-8} eV at an energy of 14.4 keV by double nuclear Bragg diffraction from yttrium iron garnet (YIP) single crystal enriched with ^{57}Fe. Later on, they performed the first NRS experiment on YIG sample in 1985 at HASYLAB (Hamburger Synchrotron Strahlungslabor) and observed the so-called quantum beats in the time domain.[9] Since then, NRS has been developed very rapidly to establish itself as a powerful technique. One can refer to some interesting articles and reviews[10–19] and books[20–22] for a detailed description of the technique.

In addition to coherent elastic NRS, i.e. a spectroscopy based on analyzing beat patterns in the temporal evolution of the nuclear decay, there is also inelastic NRS (coherent and incoherent), which is a suitable technique for studying phonon density of states. This technique was performed for the first time by Seto *et al.*[23] Here we will just consider coherent elastic NRS, denoted by NRS from now on. For studying inelastic NRS one can, for example, refer to the interesting book by Röhlsberger.[22]

A main component in NRS is the detector. The basic requirements for a detector in NRS experiments are, for example, high quantum efficiency, high counting rate, low background noise and high time resolution. A detector fulfilling these criteria is the Avalanche Photodiode (APD) detector.[24,25] APDs are made of thin silicon wafer with a depletion region of 50-100 mm and are the most efficient detectors in nuclear resonance scattering experiments due to their better count-rate capability (up to few GHz), lower noise (around 0.01 Hz) and small size (0.5 cm^3). APDs have an effective gain in the range of 100-1000 (below the breakthrough voltage), and thus, low energy photons (below 10 keV) are also detectable, which is an important task in inelastic nuclear scattering experiments.

Due to the broad bandwidth of the synchrotron radiation, which currently can be reduced to sub-meV in the best case, all nuclear states are excited simultaneously. After de-excitation of these states, which occurs

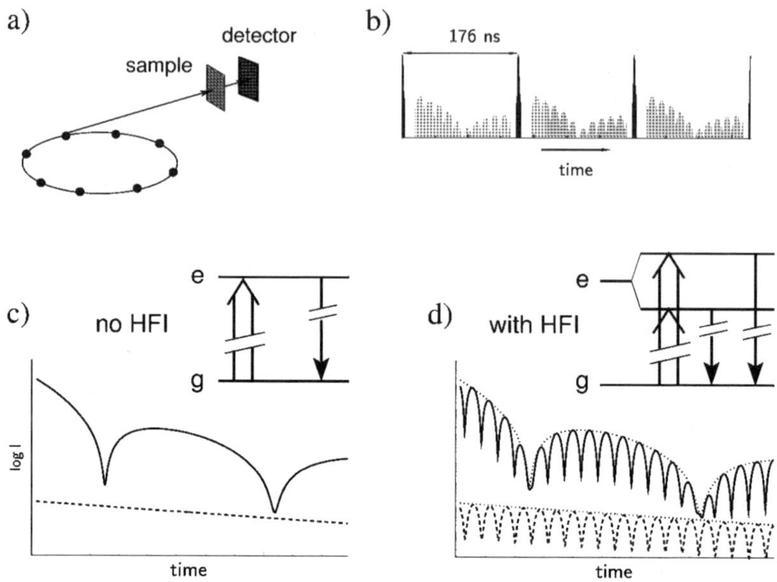

Fig. 12.1. (a) A schematic experimental set-up for NRS. Black dots present the electron bunches in the storage ring. (b) Time structure of the detected radiation. The black peaks correspond to the exciting pulses with 176 ns in between in 16-bunch mode at ESRF. (c) NRS time patterns for nuclei with just one transition. (d) NRS time patterns for the case, where there are two nuclear transitions. The forward scattering process from two close lying nuclear levels gives rise to the modulation of the time pattern, which is due to the emission of γ-rays with slightly different energies. The dashed lines in (c) and (d) correspond to the response of a single nucleus, and the full lines result from multiple scattering in a thick sample. The dotted lines in (d) correspond to the intensity pattern of the unsplit nuclear levels that determine the envelope of the quantum beat pattern. Reprinted with permission from Leupold,[18] Hyperfine Interact. 144/145 (2002) 21. © 2002 Kluwer Academic Publishers.

after e.g. 100 ns in case of ^{57}Fe, the emitted photons, which have different energies will interfere as time goes on, resulting in the quantum beats. Figure 12.1 shows a schematic set-up, the synchrotron pulses and some time spectra for absent and present of hyperfine interactions, respectively.

In Fig. 12.1(a), a very schematic experimental set-up for NRS, in a transmission mode, is depicted. Figure 12.1(b) shows the time structure in the experiment. Figures 12.1(c) and (d) show that, depending on the absence or the existence of hyperfine interactions, the time spectra will undergo characteristic changes. In case of a single nucleus with no hyperfine interactions in a very thin sample, where the probability of multiple scattering is negligible, the response will be an exponential decay, as indicated by the dashed line in Fig. 12.1(c). On the other hand, for thick samples, according to quantum mechanics, there is the possibility that the incoming photon will interact with more than one nucleus, the so-called multiple scattering, the result of which is the modulation of the time pattern depicted by the full line in Fig. 12.1(c). In case of existence of hyperfine interactions, there will be more than one transition in a single nucleus. The emitted γ-rays from these transitions have slightly different energies, the interference of which results in the modulation of the time pattern as indicated by the dashed line in Fig. 12.1(d). Again, in the thick samples, where the multiple scattering exists, there will be response from different nuclei, which results in the time pattern modulation indicated by full line in Fig. 12.1(d).

Furthermore, because the γ-rays are polarized in NRS, depending on the direction of the magnetic hyperfine field, different transitions can be excited, which reflect themselves in the time spectra as indicated in Fig. 12.2.

Figure 12.3 depicts in more detail the relative orientation of incidence of photons with $\vec{k}$ to a sample with magnetization direction $\vec{m}$.

Figure 12.4 indicates the Mössbauer transitions and the related energies in case of ^{57}Fe. The six allowed transitions decompose into three different groups depending on the polarization. ω's are angular frequencies of the different transitions, and F_{-1}, F_0 and F_{+1} are functions describing the scattering of radiation with left-circular, linear, and right-circular polarization, respectively.[22]

As an example related to magnetism, Fig. 12.5 shows time spectra for different orientation of the magnetic hyperfine field for purely σ polarized incident radiation on a 2 nm thick ^{57}Fe film on a W substrate.

It should be emphasized that in contrast to conventional Mössbauer spectroscopy, which is performed in the energy domain, NRS is performed in the time domain. Fourier analysis will then be applied to extract the relevant magnetic and electronic parameters. Several computer based fitting programs have been developed during the last decade for analyzing the NRS spectra in reflection as well as transmission mode.[26–30]

One of the most interesting cases for application of NRS in spintronics is the so-called Grazing Incidence mode, the theory of which can be studied in a series of papers by Hannon *et al.*[31–35] Related to this mode, Grazing Incidence Mössbauer Spectroscopy (GIMS) or Grazing Incidence Reflectivity (GIR) mode has now become a standard technique to study thin films and superlattices.[36–39] In this technique, presented in a series of papers

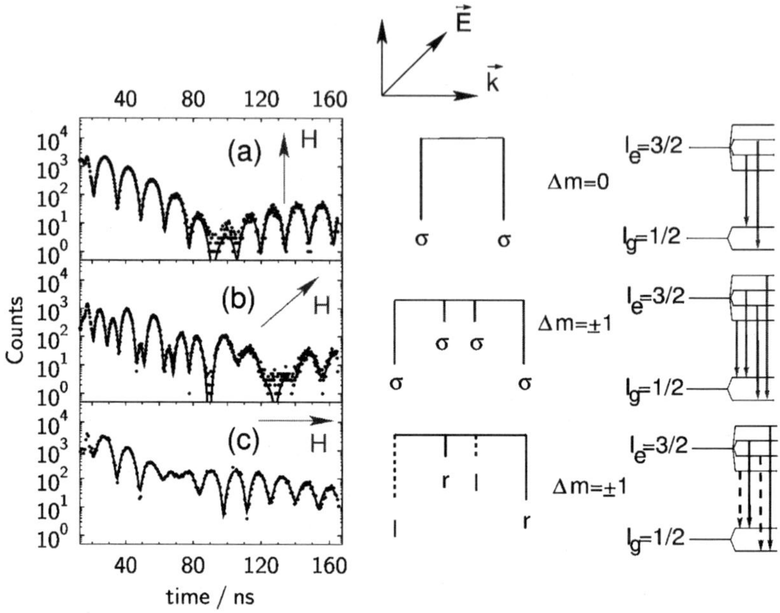

Fig. 12.2. NRS spectra of a 3 μm thin ^{57}Fe foil with different magnetization directions. E is the polarization of the incident photon (σ polarization) with propagation direction k. The corresponding transitions are also indicated in the right part of the figure. l and r denote left and right circular polarization, respectively. Reprinted with permission from Leupold,[18] Hyperfine Interact. 144/145 (2002) 21. © 2002 Kluwer Academic Publishers.

by Irkaev *et al.*[40–42] and Röhlsberger,[43] the incoming photons penetrate the sample with a very small angle relative to the surface of the sample. By successively increasing the angle, the photons penetrate more and more in the sample, so one is able to study different parts of the sample in a systematic way.

In addition to all Mössbauer spectroscopy applications, NRS can be used in other fields such as dynamical processes thanks to the very fast data acquisition.

NRS has been used in many kinds of superlattice systems, for example, the systems that we have already discussed in the previous chapter such as Fe/Co,[44] Fe/Cr[45–48] and Fe/V.[49] Presenting the results from all these and other interesting studies done by NRS will be beyond the scope of this chapter. In the following sections, just a few examples will be presented, so that the readers get some kind of feeling for the technique.

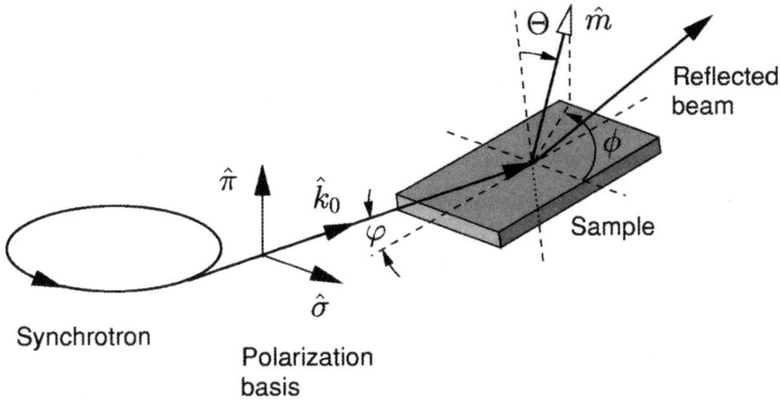

Fig. 12.3. Incidence of photons on a sample with magnetization direction m. Reprinted with permission from Röhlsberger *et al.*,[19] Phys. Rev. B 67 (2003) 245412. © 2003 The American Physical Society.

12.4. Exchange Spring Magnets

As an instructive example, which is also related to one of the former topics discussed earlier, the magnetic spin structure of exchange coupled thin films, studied by NRS will be presented. Röhlsberger *et al.*[50] did a direct measurement of the depth-dependent spin structure in a Exchange Spring Magnet consisting of $Fe_{55}Pt_{45}$ (denoted here as FePt) as the hard magnetic layer and Fe as the soft magnetic layer. The sample was prepared by rf magnetron sputtering in an Ar atmosphere. First a 30 nm FePt layer was deposited on a superpolished Si substrate. After annealing, a wedge-shaped Fe layer with a slope of 0.5 nm/mm was grown on the FePt, on which a 0.7 nm ^{57}Fe layer was deposited. Another layer of Fe was deposited with opposite slope. The sample was then coated by 3 nm Ag to prevent the oxidation. As seen in the right panel of Fig. 12.6, it was then possible to probe different depths of the sample just by exposing different parts of the sample to the SR. In the figure, the transverse displacement of the exposure of the SR relative to the edge of the sample, i.e. where the ^{57}Fe probe atoms are in contact with the FePt layer, Δx, and the depth position of them, D, are indicated. In the left panel of Fig. 12.6, the spin structure of the system is schematically shown after applying an external magnetic field perpendicular to the easy axis of the hard layer.

In this way, it has been possible to perform a series of NRS measurements on a single sample to extract magnetic properties through out the

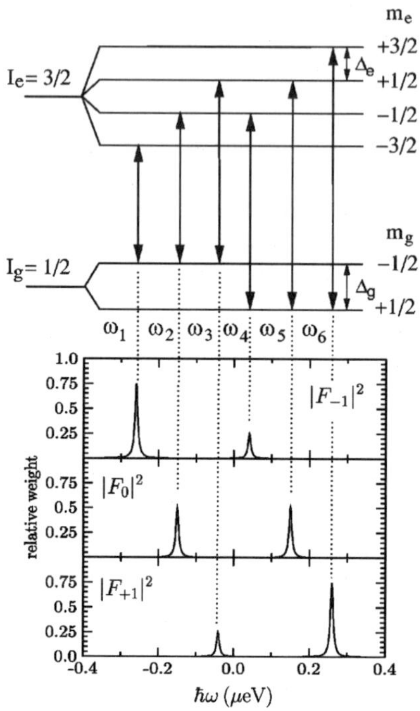

Fig. 12.4. The six dipole allowed Mössbauer transitions of ^{57}Fe decompose into three different polarization dependencies given by F_{-1}, F_0 and F_{+1}, which are functions of ω and describing the scattering of radiation with left-circular, linear, and right-circular polarization, respectively. Reprinted with permission from Röhlsberger *et al.*,[19] Phys. Rev. B 67 (2003) 245412. © 2003 The American Physical Society.

soft magnetic layer in form of the spin structure. Figure 12.7 shows the time spectra of the grazing incidence reflection. In the left panel, the time spectra at various depths under an applied external magnetic field of 160 mT are depicted. In the right panel, the time spectra for variation of the magnetic hyperfine field at the center of Fe layer as a function of external magnetic field are plotted.

As can be seen, both subspectra, as a function of the depth and as a function of the external magnetic field, undergo characteristic changes, which are directly related to the rotation of the magnetization. The result of the fitted spectra is summarized in Fig. 12.8.

The characteristic behaviour of such systems was simulated using a model of one-dimensional chain of classical spins.[51] In this model, where the whole magnetic system, i.e. soft magnetic layer plus hard magnetic

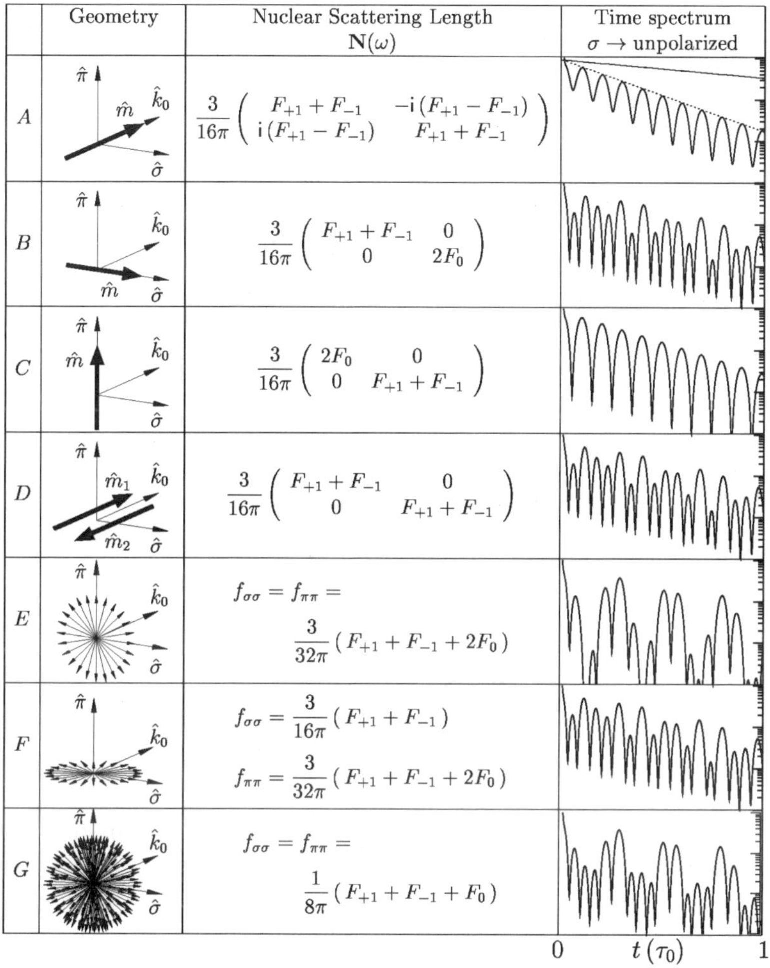

Fig. 12.5. Calculated time spectra for a 2 nm thick ^{57}Fe film on a W substrate for different orientations of the magnetic hyperfine field for purely σ polarized incident radiation. Reprinted with permission from Röhlsberger *et al.*,[19] Phys. Rev. B 67 (2003) 245412. © 2003 The American Physical Society.

layer, is divided into N sublayers, the total energy of the system can be expressed as:

$$E = -\sum_{i=1}^{N-1} \frac{A_{i,i+1}}{d^2} cos(\phi_i - \phi_{i+1}) - \sum_{i=1}^{N} K_i cos^2(\phi_i) - \sum_{i=1}^{N} HM_i cos(\phi_i - \phi_H)\,. \tag{12.1}$$

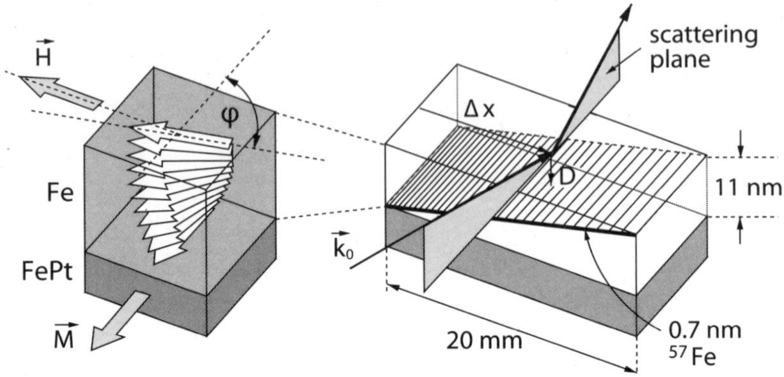

Fig. 12.6. Left: schematic spin structure in an exchange spring bilayer consisting of a soft-magnetic layer (Fe) on a hard-magnetic layer (FePt) with uniaxial anisotropy after applying an external field H perpendicular to the remanent magnetization M of the hard magnetic layer. Right: The different components of the sample and the direction of the SR are shown. By adjusting the transverse displacement, Δx, it is possible to probe the magnetic properties of the different depths, D, of the sample. Reprinted with permission from Röhlsberger *et al.*,[50] Phys. Rev. Lett. 89 (2002) 237201. © 2002 The American Physical Society.

Equation (12.1) was used to simulate such characteristic behaviour. The parameters used for FePt layer in the simulation were $K_h = 4.0 \cdot 10^7$ erg/cm^3, $M_h = 1100$ emu/cm^3 and $A_h = 4.0 \cdot 10^{-6}$ erg/cm. The parameters used for Fe layer were $K_s = 1.0 \cdot 10^3$ erg/cm^3, $M_s = 1900$ emu/cm^3 and $A_s = 1.0 \cdot 10^{-6}$ erg/cm. The solid lines in Fig. 12.8 are the results from these simulations. The measured data at the top of the Fe layer deviate from the simulation. The authors explained it as the diffusion of the oxygen through the capping layer, which results in the formation of an Fe oxide phase with a reduced exchange coupling. By letting this parameter to be adjusted during the simulation, a reduced exchange coupling of $A_s = 3.0 \cdot 10^{-7}$ erg/cm for top 3 nm Fe layer was achieved, which results in the dashed lines.

12.5. Magnetic Tunnel Junctions

The development of the applied spintronic technology is one of the hottest scientific fields in the world. One result of this technology is Magnetoresistive Random Access Memory (MRAM) which will have a crucial role in industry. MRAM is slated for commercial introduction in the very near

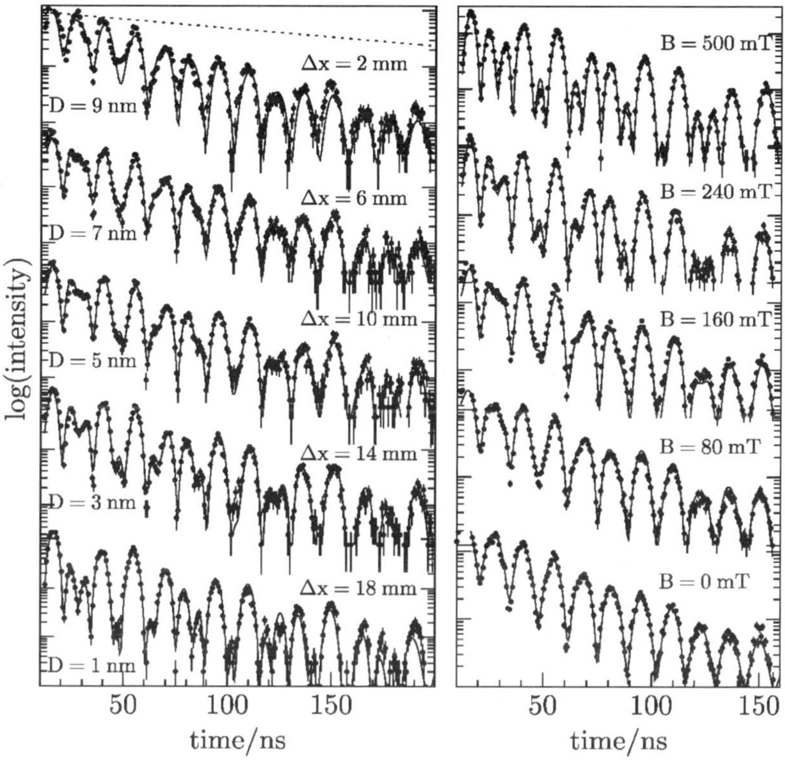

Fig. 12.7. Time spectra of Fe/FePt exchange spring bilayer, depicted in Fig. 12.6, in a grazing incidence reflection geometry. Left: As a function of different depths under influence of a perpendicular magnetic field of B=160 mT. Right: variation of the time spectra at the center of the Fe layer as a function of external magnetic field. In both cases the time spectra undergo characteristic changes, which reflect the rotation of the magnetization direction. The dots are the measured data and the solid lines are theoretical simulations according to Eq. (12.1). Reprinted with permission from Röhlsberger *et al.*,[50] Phys. Rev. Lett. 89 (2002) 237201. © 2002 The American Physical Society.

future and will greatly simplify memory use in many applications thanks to its unique attributes such as non-volatility, fast read and write, and virtually unlimited read and write cycles.[52,53] The core element of the MRAM bit cell is the Magnetic Tunnel Junction (MTJ), which in its simplest form consists of two ferromagnetic films separated by a very thin (10–15 Å) dielectric tunneling barrier.[54,55] The information is stored in the relative orientation of the magnetization of the two ferromagnetic layers [parallel(P)/antiparallel(AP)]. The MTJ typically has a Tunneling Magnetoresistance (TMR) greater than 40%, i.e. the AP state has a 40% greater

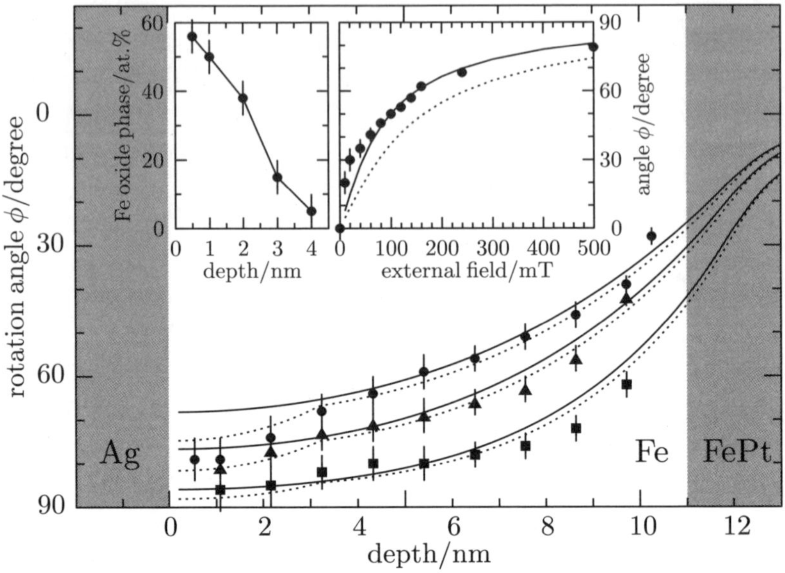

Fig. 12.8. The spin rotation as a function of depth in the Fe layer for different external magnetic fields of 160 mT (circles), 240 mT (triangles) and 500 mT (squares). Solid lines are simulations according to the model described by Eq. (12.1). The dashed lines are simulations with a reduced exchange coupling in the first 3 nm of the top Fe layer due to the possible oxidation. This oxidation phase of Fe is depicted in the left inset. The rotation angle of the magnetization in the center of the Fe layer is plotted as a function of the external magnetic field in the right inset. Solid and dashed lines are simulation for the mentioned model and with a reduced exchange coupling, respectively. Reprinted with permission from Röhlsberger *et al.*,[50] Phys. Rev. Lett. 89 (2002) 237201. © 2002 The American Physical Society.

resistance than the P state. The memory information is read out by applying a voltage on the MTJ and measuring the resulting current. Figure 12.9 shows schematic representation of tunneling and MTJ's basic components.

For MRAM and also the read heads of the hard disk drives (HDD) currently in use, a TMR ratio of about 70% is sufficient, but for next generation of spintronic devices such as high density MRAM and ultra density HDD, much higher TMR ratios are necessary. In 2001, Butler *et al.*[57] and Mathon and Umerski[58] used first-principle calculations to study the mechanism of TMR in epitaxial Fe/MgO/Fe MTJs. A TMR ratio of over 1000% was then predicted. In 2007, Lee *et al.*[59] reported a TMR ratio as high as 500% at room temperature and of 1010% at 5 K for a $[Co_{25}Fe_{75}]_{80}B_{20}$(4 nm)/MgO(2.1 nm)/$[Co_{25}Fe_{75}]_{80}B_{20}$(4.3 nm) MTJ.

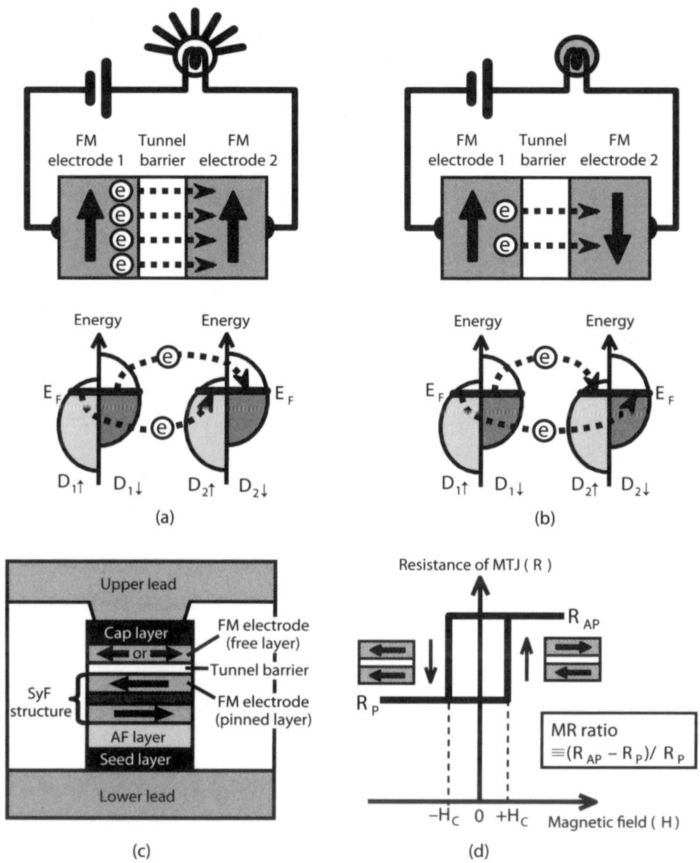

Fig. 12.9. Schematic illustration of the different components of a MTJ and of the TMR effect. (a) In the parallel alignment of the magnetizations of the two electrodes (P state), there is a low tunneling resistance, due to the matching of the density of states (DOSs) at the Fermi level, E_F, for spin-up, $D_{1\uparrow}$, $D_{2\uparrow}$, and spin-down, $D_{1\downarrow}$, $D_{2\downarrow}$, respectively, in the electrodes 1 and 2. (b) When the magnetizations of the electrodes are aligned antiparallel, the tunneling resistance is high, because the DOSs are not matched. (c) Different components of a typical MTJ. (d) A magnetoresistance curve of a MTJ, where H_C denotes the coercivity field for the free layer of MTJ. The inset shows the definition of MR ratio. Reprinted with permission from Yuasa,[56] J. Phys. Soc. Jpn. 77 (2008) 031001. © 2008 The Physical Society of Japan.

The tunneling barrier is typically formed by first depositing a metal film (e.g. Al or Mg) and subsequently oxidizing the metal in an oxygen plasma. Proper oxidation of the MTJ is critical for achieving high TMR, and in this sense optimal oxidation is typically defined as the condition for which TMR is maximized.[60] Under-oxidation leaves unoxidized metal at the bottom electrode, and over-oxidation results in a low-quality magnetic oxide added to the tunneling barrier. Both scenarios reduce the spin polarization at the bottom interface and consequently reduce the TMR. In order to achieve a completely oxidized tunneling barrier it is favorable to over-oxidize the MTJ at the oxidation step, continue with the deposition of the top layer, and finally anneal the MTJ material stack to reduce the magnetic oxide back to metal and increase the oxidation level of the tunneling barrier.

The quality of the two interfaces to the tunneling barrier (top and bottom) is also critical for achieving good long-term reliability of the MTJ.[61–63] Any remaining magnetic oxide is of low quality, prone to both gradual deterioration and dielectric breakdown, and is consequently detrimental to the reliability of the MTJ. It has been argued that it is primarily oxidized Fe that governs the poor reliability of the magnetic oxide.

By depositing an Al or Mg wedge followed by a single oxidation one will be able to probe an entire range of oxidation conditions from severely under-oxidized via optimal to severely over-oxidized. Since Fe plays a key role in MRAM technology as one of the main elements in both the top and bottom magnetic electrode of the Magnetic Tunnel Junctions (MTJ), and is believed to be the most critical element for achieving both high Tunneling Magnetoresistance and long-term reliability, it will be extremely valuable to use Mössbauer Spectroscopy to extract as much information as possible about the nature of Fe in MTJs. While the MTJ oxidation process and the composition at the bottom interface has been studied in the past using various spectroscopic techniques, no attempt has been made to use Mössbauer spectroscopy (or actually NRS) for this purpose, although Mössbauer spectroscopy, as mentioned before, is a unique method to provide atomistic information about the magnitude of the Fe atomic moment, the environment around Fe nuclei at the atomic level and also the symmetry around the Fe atom. Valuable information, which can be used to design such devices in an optimum way. Through dusting of the interfaces with ^{57}Fe and using NRS, Kamali *el al.*[64] were the first to study the magnetic environment of the ^{57}Fe atoms and determined how the oxidation state of Fe evolves with the number of annealing steps, both at the top and the bottom interface, and how the local interface quality correlates with the overall MTJ quality.

They proposed fabrication of trilayers of the typical compounds used in MRAM devices, CoFeB for the top and bottom ferromagnetic electrodes, and MgO for the tunneling barriers. In addition they deposited probing layers of ^{57}Fe and used nuclear resonance scattering of this layer to determine the oxidation state of Fe at the two interfaces as a function of oxidation conditions and annealing times. It is worth mentioning that MTJs used in spintronic devices, are trilayers and not multilayers, so there will be just a few atomic ^{57}Fe layers in such samples, and conventional Mössbauer spectroscopy, even with the most strongest sources, will be insufficient for getting enough statistic. On the other hand, the high brilliance of synchrotron radiation facilities makes it possible to perform NRS on samples with very low quantity of ^{57}Fe.

12.6. Conclusions

Synchrotron based Mössbauer Spectroscopy, the so-called Nuclear Resonance Scattering (NRS), is a very sensitive probe technique to study the magnetic state and environment of Fe atoms and other Mössbauer isotopes. Thanks to the very high photon intensity available at third generation synchrotron facilities, NRS and other Mössbauer related techniques have already had and will have a significant scientific contribution to the active field of spintronic materials and devices.

Acknowledgments

I sincerely thank my colleagues, L. Häggström, R. Wäppling, V. Uzdin, M. Andreeva, Y. Yoda, Y. Sakurai, E. Goikolea and J. Åkerman for suggestions and comments. I would especially like to thank R. Röhlsberger for the permission of reusing his work on Exchange Spring Magnets and also for reading the manuscript. I would also like to thank my colleagues at the Mössbauer Effect Data Center[65] (MEDC), J. Stevens, N. Hall and A. Khasanov, for a kind collaboration. Japanese society for promotion of science (JSPS) is acknowledged for financial support.

References

1. `http://www.spring8.or.jp`.
2. `http://www.aps.or`.
3. `http://www.ESRF.fr`.
4. H. Winick, *Synchrotron Radiation Sources: A Primer, Vol 1.* (World Scientific Publishing Co Pte Ltd, Singapore, 1995).
5. D. Attwood, *Soft X-Rays and Extreme Ultraviolet Radiation: Principles and Applications.* (Cambridge University Press, Cambridge CB2 2RU, UK, 1999).

6. J. Als-Nielsen, *Elements of Modern X-ray Physics.* (John Wiley & Sons Ltd, Chichester West Sussex England, 2001).
7. S. L. Ruby, *J. de Physique.* **35**(C6), 209, (1974).
8. E. Gerdau, R. Rüffer, H. Winkler, W. Tolksdorf, C. P. Klages, and J. P. Hannon, *Phys. Rev. Lett.* **54**(8), 835, (1985).
9. E. Gerdau, R. Rüffer, R. Hollatz, and J. P. Hannon, *Phys. Rev. Lett.* **57**, 1141, (1986).
10. E. Gerdau, *Hyperfine Interact.* **90**, 301, (1994).
11. W. Sturhahn and E. Gerdau, *Phys. Rev. B.* **49**(14), 9285, (1994).
12. G. V. Smirnov, *Hyperfine Interact.* **97/98**, 551, (1996).
13. G. V. Smirnov, *Hyperfine Interact.* **123/124**, 31, (1999).
14. Y. V. Shvyd'ko, *Hyperfine Interact.* **123/124**, 275, (1999).
15. Y. V. Shvyd'ko, *Phys. Rev. B.* **59**(14), 9132, (1999).
16. G. R. HOY, *Hyperfine Interact.* **135**, 191, (2001).
17. R. Rüffer, *Hyperfine Interact.* **141/142**, 83, (2002).
18. O. Leupold, *Hyperfine Interact.* **144/145**, 21, (2002).
19. R. Röhlsberger, J. Bansmann, V. Senz, K. L. Jonas, A. Bettac, K. H. Meiwes-Broer, and O. Leupold, *Phys. Rev. B.* **67**, 245412, (2003).
20. E. Gerdau and H. de Waard, Eds., *Nuclear Resonant Scattering of Synchrotron Radiation (Part A).* (Hyperfine Interact, The Netherlands, 1999).
21. E. Gerdau and H. de Waard, Eds., *Nuclear Resonant Scattering of Synchrotron Radiation (Part B).* (Hyperfine Interact, The Netherlands, 2000).
22. R. Röhlsberger, *Nuclear Condensed Matter Physics with Synchrotron Radiation: Basic Principles, Methodology and Applications.* (Springer-Verlag, Berlin Heidelberg Germany, 2004).
23. M. Seto, Y. Yoda, S. Kikuta, X. W. Zhang, and M. Ando, *Phys. Rev. Lett.* **74**(19), 3828, (1995).
24. S. Kishimoto, *Nucl. Instr. Meth. Res. A.* **309**, 603, (1991).
25. A. Q. R. Baron and S. L. Ruby, *Nucl. Instr. Meth. Res. A.* **343**, 517, (1993).
26. W. Sturhahn, *Hyperfine Interact.* **125**, 149, (2000).
27. Y. V. Shvyd'ko, *Hyperfine Interact.* **125**, 173, (2000).
28. M. Haas, E. Realo, H. Winkler, W. Meyer-Klaucke, and A. Trautwein, *Hyperfine Interact.* **125**, 189, (2000).
29. H. Spiering, L. Déak, and L. Bottýan, *Hyperfine Interact.* **125**, 197, (2000).
30. M. Andreeva, B. Lindgren, and V. Panchuk. REFTIM, Version 6.1. `http://www.esrf.fr/computing/scientific/REFTIM/MAIN.htm`.
31. J. Hannon, G. Trammell, M. Mueller, E. Gerdau, H. Winkler, and R. Rüffer, *Phys. Rev. Lett.* **43**(9), 636, (1979).
32. J. Hannon, N. V. Hung, G. Trammell, E. Gerdau, M. Mueller, R. Rüffer, and H. Winkler, *Phys. Rev. B.* **32**(8), 5068, (1985).
33. J. Hannon, N. V. Hung, G. Trammell, E. Gerdau, M. Mueller, R. Rüffer, and H. Winkler, *Phys. Rev. B.* **32**(8), 5081, (1985).
34. J. Hannon, G. Trammell, M. Mueller, E. Gerdau, R. Rüffer, and H. Winkler, *Phys. Rev. B.* **32**(10), 6363, (1985).
35. J. Hannon, G. Trammell, M. Mueller, E. Gerdau, R. Rüffer, and H. Winkler, *Phys. Rev. B.* **32**(10), 6374, (1985).

36. A. I. Chumakov, L. Niesen, D. L. Nagy, and E. E. Alp, *Hyperfine Interact.* **123/124**, 427, (1999).
37. D. L. Nagy, L. Bottyán, L. Deák, E. Szilágyi, H. Spiering, J. Dekoster, and G. Langouche, *Hyperfine Interact.* **126**, 353, (2000).
38. M. A. Andreeva, *Hyperfine Interact.* **156/157**, 595, (2004).
39. M. A. Andreeva and B. Lindgren, *Phys. Rev. B.* **72**, 125422, (2005).
40. S. M. Irkaev, M. A. Andreeva, V. G. Semenov, G. N. Belozerskii, and O. V. Grishin, *Nucl. Instr. and Meth. B.* **74**, 545, (1993).
41. S. M. Irkaev, M. A. Andreeva, V. G. Semenov, G. N. Belozerskii, and O. V. Grishin, *Nucl. Instr. and Meth. B.* **74**, 554, (1993).
42. S. M. Irkaev, M. A. Andreeva, V. G. Semenov, G. N. Belozerskii, and O. V. Grishin, *Nucl. Instr. and Meth. B.* **103**, 351, (1995).
43. R. Röhlsberger, *Hyperfine Interact.* **123/124**, 301, (1999).
44. B. Lindgren, M. Andreeva, L. Häggström, V. S. B. Kalska, A. Chumakov, O. Leupold, and R. Rüffer, *Hyperfine Interact.* **136/137**, 439, (2001).
45. T. S. Toellner, W. Sturhahn, R. Röhlsberger, E. E. Alp, C. H. Sowers, and E. E. Fullerton, *Phys. Rev. Lett.* **74**(17), 3475, (1995).
46. T. Ruckert, W. Keune, W. Sturhahn, M. Hu, J. Sutter, T. Toellner, and E. Alp, *Hyperfine Interact.* **126**, 363, (2000).
47. M. A. Andreeva, V. G. Semenov, B. Lindgren, L. Häggström, B. Kalska, A. I. Chumakov, O. Leupold, R. Rüffer, K. A. Prokhorov, and N. N. Salashchenko, *Hyperfine Interact.* **141/142**, 119, (2002).
48. C. L'abbé, J. Meersschaut, W. Sturhahn, J. S. Jiang, T. S. Toellner, E. E. Alp, and S. D. Bader, *Phys. Rev. Lett.* **93**(3), 037201, (2004).
49. B. Kalska, L. Häggström, B. Lindgren, P. Blomquist, R. Wäppling, M. A. Andreena, Y. V. Nikitenko, V. V. Proglyado, V. L. Aksenov, V. G. Semenov, A. I. Chumakov, O. Leupold, and R. Rüffer, *Hyperfine Interact.* **136/137**, 295, (2001).
50. R. Röhlsberger, H. Thomas, K. Schlage, E. Burkel, O. Leupold, and R. Rüffer, *Phys. Rev. Lett.* **89**(23), 237201, (2002).
51. E. E. Fullerton, J. S. Jiang, M. Grimsditch, C. H. Sowers, and S. D. Bader, *Phys. Rev. B.* **58**(18), 12193, (1998).
52. J. Åkerman, M. DeHerrera, M. Durlam, B. Engel, J. Janesky, F. Mancoff, J. Slaughter, and S. Tehrani, *Magnetic Tunnel Junction Based Magnetoresistive Random Access Memory*, In ed. M. Johnson, *Magnetoelectronics*, pp. 231–272. (Elsevier Ltd, 2004).
53. J. Åkerman, *Sci.* **308**, 508, (2005).
54. J. S. Moodera, L. R. Kinder, T. M. Wong, and R. Meservey, *Phys. Rev. Lett.* **74**, 3273, (1995).
55. T. Miyazaki and N. Tezuka, *J. Magn. Magn. Mater.* **139**(3), L231, (1995).
56. S. Yuasa, *J. Phys. Soc. Jpn.* **77**(3), 031001, (2008).
57. W. H. Butler, X. G. Zhang, and T. C. Schulthess, *Phys. Rev. B.* **63**, 054416, (2001).
58. J. Mathon and A. Umerski, *Phys. Rev. B.* **63**, 220403, (2001).
59. Y. M. Lee, J. Hayakawa, S. Ikeda, F. Matsukura, and H. Ohno, *Appl. Phys. Lett.* **90**, 212507, (2007).

60. S. Moodera, E. F. Gallagher, K. Robinson, and J. Nowak, *Appl. Phys. Lett.* **70**, 3050, (1997).
61. J. Åkerman, P. Brown, M. DeHerrera, M. Durlam, E. Fuchs, D. Gajewski, M. Griswold, J. Janesky, J. Nahas, and S. Tehrani, *IEEE Transactions on Device and Materials Reliability.* **4**(3), 428, (2004).
62. J. Åkerman, P. Brown, D. Gajewski, M. Griswold, J. Janesky, M. Martin, H. Mekonnen, J. Nahas, S. Pietambaram, J. Slaughter, and S. Tehrani, *IEEE International Reliability Physics Symposium.* **43**, 163, (2005).
63. S. Kim and B. Cho, *Appl. Phys. Lett.* **86**, 142106, (2005).
64. S. Kamali and *et al.*, Unpublished.
65. `http://orgs.unca.edu/medc/`.

Chapter 13

BIONANOMAGNETISM

Peter Svedlindh[*], Klas Gunnarsson[*], Mattias Strömberg[*] and Sven Oscarsson[†]

[*]*Department of Engineering Sciences, Uppsala University*
P.O. Box 534, SE-75121, Sweden
[†]*Department of Physics and Materials Science, Uppsala University*
P.O. Box 530, SE-75121 Uppsala, Sweden
E-mail: peter.svedlindh@angstrom.uu.se

Magnetic nanoparticles constantly find new applications in biotechnology and medicine. In all cases, the response of the nanoparticles to a magnetic field plays an essential role, either by creating local magnetic fields or by being attracted to magnetic field sources. This article will not attempt to review existing and emerging applications, but will instead focus on two methods with high potential for future bioassay applications. Firstly, we describe a new principle for DNA sequence detection using oligonucleotide probe-tagged magnetic beads. The target-DNA is recognised by hybridisation to a linear DNA-probe, forming upon ligation a circularised target-probe complex. Thereafter these complexes are volume-amplified to micrometer-sized DNA coils, which are detected by the addition of the probe-tagged beads. Upon incorporation of the beads, the complex magnetisation vs. frequency spectrum changes dramatically. Secondly, a novel method for controlled transport and release of proteins immobilised to magnetic beads is presented. Protein covered beads are added to a substrate containing micron-sized magnetic elements, which together with a rotating magnetic field control their movement. Adding a reductive agent, the immobilised proteins are momentarily released from the beads. Using this method, transportation and local release of sub-attomoles of any kind of molecule can be achieved.

13.1. Introduction

During the last decades, single domain magnetic nanoparticles, often referred to as magnetic beads, have found numerous applications in biosciences. A unique property of magnetic beads is their response to a magnetic field; magnetic beads acquire a sizeable field induced magnetic moment and are attracted to sources of magnetic field. In medicine, magnetic nanoparticles are used as magnetic resonance imaging (MRI) contrast agents.[1] MRI is based on the nuclear magnetic resonance signal from protons in e.g. tissue water. The imaging is carried out by detecting the relaxation processes of the proton spins after having momentarily being influenced by a transverse radio-frequency (RF) field; one distinguishes between a longitudinal relaxation process (T_1-decay) when the proton spins align with a static magnetic field and a transversal de-phasing relaxation process when the coherent spin precession of the proton spins is lost with the RF field switched off (T_2-decay). It is by now well established that the magnetic field from nanoparticles will have a strong effect on the T_2-decay,[2] the relaxation time of this decay is shortened, and can thus be used for MRI contrast enhancement. Antibody conjugated microparticles carrying iron-oxide nanoparticles have e.g. been used for *in vivo* MRI contrast enhancement, probing small-sized cancers[3] and acute brain inflammation.[4] Another interesting application of magnetic nanoparticles in medicine is in hyperthermia cancer therapy.[1,5] Here the therapy is e.g. achieved by using *in vivo* administrated stealth magnetic nanoparticles designed for selective uptake by tumour cells. Once inside a cell, under the influence of an AC magnetic field, energy is dissipated, either due to Néel relaxation of the particle magnetic moments or because of rotational Brownian motion of the nanoparticles.[1]

Array-based bioassay methods are powerful as tools for analyses of DNA, RNA, proteins and other biomolecules.[6] Today, fluorescence and chemiluminescence technologies are dominating the read-out of microarrays owing to their high sensitivity and multiplex capabilities. However, magnetic beads can be used as labels in bioassays and give alternatives for detection of biological processes. Magnetic bead based bioassays involve labelling probe biomolecules with magnetic beads and upon hybridisation with the target, either static or dynamic magnetic bead properties can be detected. Furthermore, magnetic biosensor schemes are either classified as substrate-based (sensor on chip) or substrate free. In the former biosensor category, if the target is present, probe functionalised magnetic beads bind directly to the sensor surface, thereby inducing a signal change. Examples of existing magnetic sensor based techniques are micro-Hall effect[7] and magnetoresistive micro-biosensor platforms,[8] and magnetic force microscopy.[9] One substrate-free method is the Brownian relaxation biosensor scheme,[10] which makes use of the decrease in the bead Brownian relaxation frequency caused by the hydrodynamic size increase upon probe-target hybridisation.

For bioseparation, the use of bioconjugated magnetic beads is a well-known and routinely used technique capable of specific separation of e.g. cells,[11,12] bacteria,[13] and proteins.[14] State-of-the-art procedures imply that millions of magnetic beads, each bead often carrying thousands of probe-target complexes, are separated simultaneously as a group.

A controlled transport of magnetic beads on microchips, on the single bead level, opens up new perspectives for novel instrumentation in bioseparation and bioassay technology, where magnetic beads with individually attached probe molecules, e.g. immunoglobulines, are allowed to continuously pass through a sample in a controlled and predictable way. Various methods for magnetic bead control have been published,[15–17] and we will here focus on one method that allows for both transportation and separation of magnetic beads on the single bead level.[15] In addition, we will describe a substrate-free biosensor method[18] combining the Brownian relaxation of magnetic nanobeads with a method to amplify the size of the target-probe complex, thus also amplifying the effect on the Brownian relaxation process. Combining the two here described methods[15,18] has to potential to yield a powerful all magnetic bioassay method.

13.2. Properties and Biofunctionalisation of Magnetic Beads

13.2.1. *Magnetic beads*

Today, a variety of magnetic beads are available where the surfaces of the beads have been functionalised in order to fit the needs of different biochemical applications.[19] The magnetic material used can be either ferromagnetic, a magnetic transition metal element or alloy,[20] or ferrimagnetic, e.g. magnetite (Fe_3O_4).[21] For temperatures below 350°C and particle sizes less than 500 nm, particles of magnetite by time oxidise into ferrimagnetic maghemite (γ-Fe_2O_3).[22,23] The vast majority of prepared magnetic bead systems make use of ferrimagnetic iron-oxides as magnetic material, although ferromagnetic transition metal nanoparticles offer higher saturation magnetisations. In the applications described here, magnetic nanoparticles are embedded in a non-magnetic host material, which defines the bead size and the surface to be functionalised. The magnetic material consists of grains of γ-Fe_2O_3, typically with a grain diameter somewhere in the range 5–15 nm, each grain being small enough to be in a single domain state. The grains either form a clustered core or are randomly dispersed in the pores of the non-magnetic matrix. Beads in suspension will exhibit a superparamagnetic behaviour, i.e. exhibiting a strong field induced magnetisation but showing negligible magnetic hysteresis. Important properties of the beads are e.g. the concentration of beads in solution, the content of magnetic material, and shape and size of the beads. In most applications, the beads need to be mono-dispersive, i.e. their diameter should be well defined.

In the experiments described here, the following types of spherical beads have been utilised. For all products, the carrying liquid is deionised water. In Section 13.3, so-called Nanomag® -D beads of diameters 40 nm, 80 nm, 130 nm and 250 nm, respectively, have been used.[24] In this type of bead, a cluster of maghemite nanoparticles form a core surrounded by a dextran hull. In Section 13.4, Dynabead® M-270[25] or the Micromer® -M[24] beads have been used, with diameters of 2.8 μm and 4.9 μm, respectively. In the former case, maghemite grains are randomly trapped inside the pores of a polystyrene matrix, while in the latter case, the nanoparticles are clustered to form the core in a latex casing.

13.2.2. *Biofunctionalisation of magnetic beads – the SPDP coupling chemistry*

Surface modification of magnetic beads is often necessary in order to achieve ferrofluidic stability, solubility in a certain solvent environment or in order to introduce molecular entities on the surface serving either as sites for further functionalisation or as molecular probes in biotechnological applications. Immobilisation of biomolecules onto the surface of magnetic beads is a process commonly denoted biofunctionalisation. There exist numerous publications on biofunctionalisation of magnetic beads in the scientific literature and by that also a large number of coupling chemistries. Just to mention a few examples, Ma *et al.*[26] synthesized bovine serum albumin protein coated magnetic microspheres, Aurich *et al.*[27] synthesised polyaspartate coated magnetic nanoparticles for biomedical applications and Gong *et al.*[28] performed immobilisation of monoclonal antibody against human α-fetoprotein on amino group functionalised silica-coated magnetic nanoparticles. In particular, conjugation of short single-stranded DNA molecules (oligonucleotides) onto magnetic beads have been performed by e.g. Lee *et al.*[29] and Kouassi and Irudayaraj.[30]

The *N*-succinimidyl 3-(2-pyridyldithio)propionate (SPDP) molecule is a heterobifunctional cross linker, which can be used to cross-link an amino group containing entity and an aliphatic thiol group containing entity through a disulphide bridge. The SPDP coupling chemistry, first demonstrated by Carlsson *et al.*[31] in 1978, is a reversible covalent immobilisation method since the disulphide bond can be reduced with dithiothreitol (DTT), thereby cleaving the bridge. Figure 13.1 taken from Strömberg *et al.*[32] schematically illustrates how thiolated single-stranded oligonucleotides can be conjugated to magnetic beads having amino groups on the bead surface. For cross-linking other entities, the procedure is analogous. In step (I), the primary amino groups on the bead surface are SPDP-activated by reaction with the *N*-hydroxysuccinimide

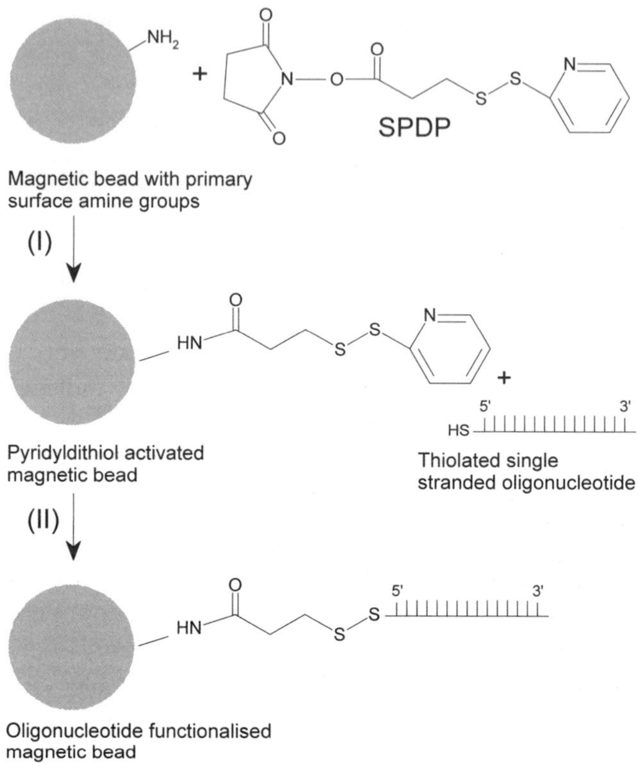

Fig. 13.1. Schematic illustration of covalent immobilisation of thiolated single-stranded oligonucleotides on magnetic beads with amino groups using SPDP coupling chemistry. In step (I), the *N*-hydroxysuccinimide ester group of the SPDP molecule reacts with the amino group on the bead and in step (II), the 2-pyridyl disulphide group of the SPDP molecule reacts with the oligonucleotide thiol group. Reprinted with permission from Ref. 32. Copyright (2007) IOP Publishing Ltd.

ester group of the SPDP molecule. In step (II), the oligonucleotide molecules attach to the bead by reaction between the oligonucleotide thiol group and the 2-pyridyl disulphide group of the SPDP molecule.

Briefly, the practical procedure begins with resuspension of the beads in PBS buffer (pH ~ 7) followed by addition of SPDP dissolved in dimethylsulphoxide (DMSO) and the mixture is incubated at room temperature during ~1 hr. After washing the beads by exchanging the solvent a number of times, DTT reduced thiolated oligonucleotides are added followed by incubation over night at room temperature. After washing, any remaining SPDP-activated amino groups are blocked by incubation with iodoacetamide at room temperature for ~1 hr, thereby alkylating these sites. After some final washing steps, the as-synthesised batch is obtained. In the following sections describing two bionanomagnetism applications, the SPDP-coupling chemistry was used for biofunctionalisation of amino group covered magnetic iron oxide beads with oligonucleotides (Section 13.3) and protein molecules (Section 13.4).

13.3. An Example of a Recently Developed Magnetic Biosensor Scheme — The Volume-Amplified Magnetic Nanobead Detection Assay

In this section, a magnetic biosensor principle recently developed by some of the authors is briefly described, namely the volume-amplified magnetic nanobead detection assay (VAM-NDA). This bioassay principle is a powerful combination between changes in dynamic magnetic properties of magnetic beads and the use of high performance molecular tools. However, we will begin with a brief theoretical background relevant for the VAM-NDA method, i.e. frequency dependent magnetic properties of magnetic beads and magnetic relaxation mechanisms.

13.3.1. *Dynamic magnetic properties and relaxation mechanisms of magnetic beads*

As mentioned in Section 13.2.1, the magnetic beads used in this work are composed of a clustered core consisting of single domain maghemite nanoparticles (each having a diameter ~15 nm) held together by a dextran casing, where the nanoparticle moments are thermally blocked relative to each other at room temperature.

According to the Debye theory,[33] the complex low-field susceptibility $\chi(\omega)$ for an ensemble of beads, reads

$$\chi(\omega) = m(\omega)/H_{AC} = (\chi_0 - \chi_\infty)/(1 + i\omega\tau) + \chi_\infty \tag{1}$$

where ω is the angular frequency of the applied AC magnetic field, χ_∞ is the high frequency susceptibility, χ_0 is the low field static susceptibility, $m(\omega) = m'(\omega) - im''(\omega)$ is the complex magnetisation,

H_{AC} is the amplitude of the applied AC magnetic field and τ is the characteristic relaxation time of the beads. The magnetic relaxation is governed by either of two relaxation mechanisms: The Néel relaxation,[34] where the magnetic moment rotates within the nanoparticles or the Brownian relaxation,[35] where the entire bead rotates in response to the AC magnetic field. In the Néel relaxation process, the magnetic moment needs to overcome an energy barrier $E_b = KV_p$ with the relaxation time expressed as[36]

$$\tau_N = \tau_0 \exp\left(E_b/\beta\right) \tag{2}$$

where K is the magnetic anisotropy constant, V_p is the nanoparticle volume, $\beta = kT$ is the thermal energy and τ_0 is a weakly temperature and volume dependent microscopic relaxation time. In the Brownian relaxation model, the characteristic relaxation time is instead given by

$$\tau_B = 3\eta V_B/\beta\,. \tag{3}$$

Here V_B is the hydrodynamic volume of the bead and η is the dynamic viscosity of the carrier liquid. Experimentally, in our case, the relaxation time is obtained from the Brownian relaxation frequency $f_B = \left(2\pi\tau_B\right)^{-1}$, which is the frequency characterising the position of the peak in the m'' vs. frequency spectrum.

13.3.2. *Brief overview of the volume-amplified nanobead detection assay*

The volume-amplified magnetic nanobead detection assay for detection of single-stranded DNA targets, first time shown by Strömberg *et al.*,[18] combines the Brownian relaxation biosensor principle[10] with high performance molecular tools, viz. padlock probe target recognition[37,38] and rolling circle amplification (RCA).[39,40] It is a non-optical magnetic bioassay principle and is also of lab-on-a-bead type. By this, the VAM-NDA method offers the possibility to design a new generation of diagnostic devices, mostly of point-of-care and over-the-counter type, at a substantially lower cost than those used today. Although the authors hitherto only have demonstrated detection of DNA targets, other kinds of biomolecules, viz. RNA and proteins, can be detected using the so-called proximity ligation technique[41,42] that produce RCA products as a consequence of coincident binding of two or more specific antibodies equipped with DNA strands.

Figure 13.2, taken from Ref. 18, schematically shows the VAM-NDA bioassay principle. A collection of single-stranded target-DNA molecules (black lines) is shown in the left part of the figure. Linear padlock probe molecules (light grey lines) designed to exactly match the target-DNA are added, which upon hybridisation with the target-DNA molecules form circularised probe-target complexes. The ends of the padlock probes are joined together by ligation. The addition of RCA polymerase initiates RCA, and the circularised padlock probes are amplified for a certain time (the RCA-time). After completion of RCA, the sample contains ~1 μm-sized (for a RCA time ~1 hr) random-coil single-stranded DNA molecules having a periodic sequence with the complement of the padlock probe as the repeating motif. The RCA-coils are detected by addition of single-stranded detection probe oligonucleotide functionalised magnetic beads (total hydrodynamic diameter ~150 nm), where the coupled oligonucleotides are complementary to a region of the repeating motif in the RCA-coils. The probe-tagged beads and the RCA-coils will – during their diffusive motion – approach each other and the probe-tagged beads are then incorporated into the coils by base-pair hybridisation (bead immobilisation). The hydrodynamic size of the immobilised beads is thus strongly increased, now essentially corresponding to the diameter of an RCA-coil. Non-immobilised beads exhibit an unaltered hydrodynamic diameter. When target-DNA is not present, no circularised padlock probes and hence no RCA-coils are formed. In this case, all probe-tagged beads remain free in solution. The lower right part of Fig. 13.2 shows the m'' vs. frequency spectrum for one positive sample and for a negative control sample. The negative control sample exhibits a well defined f_B value, and the peak magnitude gives a measure of the number of free probe-tagged beads. The positive sample has essentially two relaxation frequencies where the low-frequency peak (LFP) mainly corresponds to single RCA-coils with immobilised beads and the high-frequency peak (HFP) arises from probe-tagged beads that remain free in solution. The two relaxation events have been resolved using a Cole-Cole fitting procedure[43] and the bold black curves show the two contributions to the

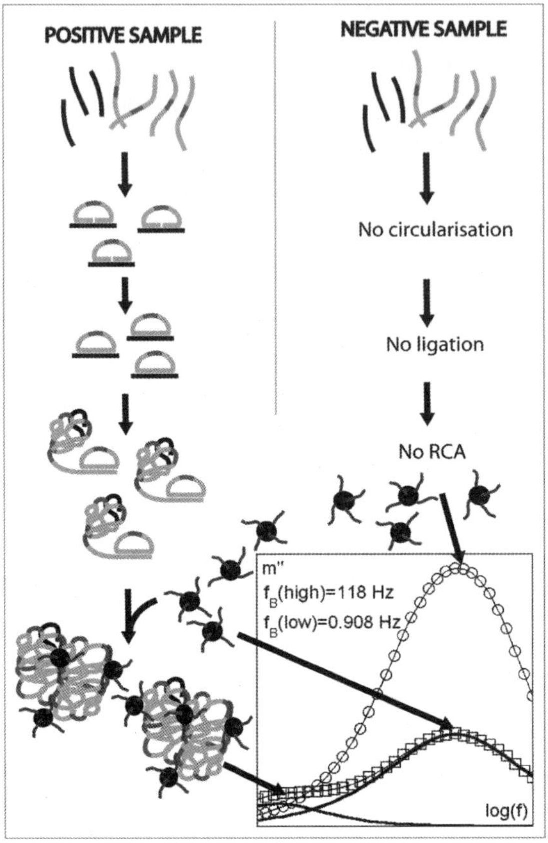

Fig. 13.2. Schematic illustration of the volume amplified magnetic nanobead detection assay. In a positive sample (left part), single-stranded target DNA molecules (black lines) are recognised by padlock probes (light grey lines), i.e. the 5′ and 3′ ends of the padlock probe hybridise to the target strand, thereby forming a circularised probe-target complex. After ligation and addition of RCA polymerases, each circularised complex is enzymatically amplified during a certain time (the RCA-time) to a macromolecular DNA coil (RCA-coil) having a periodic sequence with the complement of the padlock probe as the repeating motif. The RCA-coils are detected by adding detection probe oligonucleotide functionalised magnetic beads (filled black circles with dark grey lines) exhibiting Brownian relaxation behaviour where the oligonucleotides are complementary to part of the repeating sequence of the RCA-coils. Upon this, a certain amount of the probe-tagged beads bind to the RCA-coils (bead immobilisation) depending on the RCA-coil concentration. This in turn, compared to a negative control sample (right part), is manifested as a decreased m'' high frequency peak level (turn-off detection) and/or an increased m'' low frequency peak level (turn-on detection). Reprinted with permission from Ref. 18. Copyright (2008) American Chemical Society.

measured positive sample curve. The extracted peak frequency values at 310 K are also indicated. By measuring the *decrease* in the m'' HFP level relative to the corresponding level of the negative control sample, quantitative DNA target concentration analysis can be performed.

Figure 13.3, taken from Ref. 18, shows the outcome of a quantitative detection experiment of single-stranded target DNA using magnetic beads with a bare diameter of 130 nm. Samples were incubated for 30 min at 343 K, the probe-tagged bead concentration was 1 nM and the RCA-time was 1 hr. The complex magnetisation spectra were recorded at 310 K by a commercial SQUID magnetometer using an AC magnetic field amplitude of 2 Oe. In order to achieve similar bead sedimentation rates and chemical environment in all samples, the negative control sample contained 100 pM of RCA-coils having a sequence non-complementary to the detection probes on the beads.

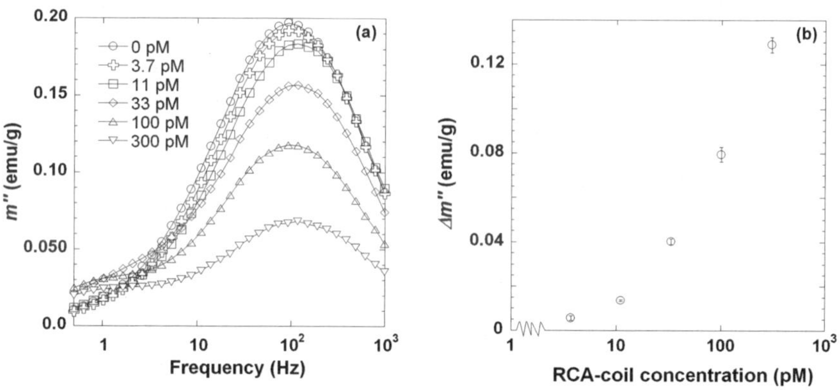

Fig. 13.3. Quantitative turn-off type detection of single-stranded target DNA using magnetic beads with a bare diameter of 130 nm. (a) shows imaginary part of the complex magnetisation (normalised with respect to the solid content of beads) vs. frequency at 310 K for samples with different RCA-coil concentrations (c). Samples were incubated for 30 min at 343 K, the probe-tagged bead concentration was 1 nM and the RCA-time was 1 hr. In order to achieve similar bead sedimentation rates and chemical environment in all samples, the negative control sample contained 100 pM of RCA-coils having a sequence non-complementary to the detection probes on the beads. (b) shows a more condensed representation of the data in panel (a), viz. $\Delta m''$, defined by $\Delta m''' = m''(0) - m''(c)$ at the high frequency peak vs. RCA-coil concentration. Reprinted with permission from Ref. 18. Copyright (2008) American Chemical Society.

A limit of detection of ~3 pM was achieved. In this case, the detection scheme is of *turn-off* type since it is based on the *decrease* in the m'' HFP level due to immobilisation of magnetic beads in the RCA-coils, i.e. the number of free beads *decreases* in the carrier liquid. On the other hand, *turn-on* type detection relates to the *increase* in the m'' LFP level when the number of beads immobilised in the coils *increases*. Apparently, from Fig. 13.3, only turn-off type detection is possible using 130 nm beads since the m'' LFP levels do not correlate well with RCA-coil concentration even though LFPs are clearly visible.

In Strömberg *et al.*,[44] turn-on and turn-off detection possibilities were investigated in more detail with respect to bead size. It was found that when performing quantitative analysis with 40 nm beads as in Fig. 13.3, both turn-off and turn-on type detection is possible. In the same paper, a similar investigation was also presented for 250 nm beads showing that even for RCA-coil concentrations as high as 300 pM, the observed LFPs were broad and low in magnitude. These observations suggest that the 40 nm beads preferably immobilise such that a collection of more or less well separated magnetic bead containing RCA-coils are formed (LFP located at ~1 Hz at 310 K for 1 hr RCA-time). This implies that the 40 nm beads preferably bind to the interior of the coils. On the other hand, the 250 nm beads immobilise in such a way that coil aggregates are formed with sizes much larger than the size of a single RCA-coil. This can only occur if the 250 nm beads cross-link coils by binding to the exterior of the coils. These coil aggregates exhibit characteristic frequencies below the low frequency limit (0.5 Hz) of the SQUID experiments. For 130 nm beads, the situation is intermediate although only turn-off detection is practically possible.

Also, in Ref. 44, it was found that for a given RCA-coil concentration, the ratio between the number of immobilised beads and the number of coils increased with decreasing bead size. This is reasonable since the available free space inside a coil is limited and a smaller bead size will therefore allow for a larger number of immobilised beads. It was also found that smaller bead size yields faster immobilisation kinetics. Furthermore, a larger number of detection probe oligonucleotides per bead and a lower concentration of probe-tagged beads were found to yield more efficient bead immobilisation.

On the one hand, according to the above discussion, from a bioassay sensitivity point of view, one should use a low concentration of beads and a small bead size. On the other hand, the magnetic moment per bead will be small and using a low concentration of beads will yield a small total magnetic moment of a sample. Since we have used a commercial SQUID magnetometer, with a comparably low pick-up coil filling factor, to record the complex magnetisation, we have not yet been able to improve the detection limit of a few pM of target DNA. However, as we will discuss further in Section 13.5, our intention is to replace the SQUID magnetometer with a miniaturised magnetic field sensor with size adapted to the size of the magnetic beads. This further development is expected to considerably improve the detection sensitivity of the VAM-NDA bioassay method.

Finally, as was pointed out in Refs. 18 and 44, the VAM-NDA method can also be used for turn-off type *multiplex analysis* by using different bead sizes for each type of target molecule such that the free bead relaxation peaks (HPFs) of the respective bead sizes are well separated in frequency. Presence of one kind of target results in a decreased m'' HFP level for the corresponding bead size, presence of two targets lead to decreased m'' HFP levels for the corresponding bead sizes, and so on and so forth. It should be mentioned that for each multiplexed sample, all contributions to the measured m'' profile have to be extracted using a Cole-Cole fit procedure in order to see how each m'' HFP level has changed comparing to the corresponding negative control levels. One also has to choose the largest bead size used in a multiplex analysis test such that its HFP does not interfere significantly with the frequency region where one finds the LFP for RCA-coils containing magnetic beads. A qualitative proof of duoplex analysis was recently presented in Ref. 44, where simultaneous detection of two kinds of bacterial DNA sequences, viz. *Vibrio cholerae* (250 nm beads) and *Vibrio vulnificus* (130 nm beads) is described.

13.4. Transportation and Release of Biomolecules Using Magnetic Beads

Functionalised magnetic beads used in biosciences are usually manipulated as large ensembles. A system that enables controlled transport and separation of beads one-by-one would open up a large number of possibilities in bioscience.[16,17]

In this section, a lab-on-a-chip method for single bead transport is presented,[15] making it possible to direct a particular bead into a specific position. Proteins immobilised to the surface of the beads can be successfully transported and released on a micrometer scale.[45] The essential part of the system consists of a substrate containing micron-sized elements made of a soft magnetic material, the so-called transport microchip. Exposed to an in-plane rotating magnetic field, an element is continuously magnetised and demagnetised and will thus exert an attractive force on beads varying in time as the magnetic field rotates.

The substrate of the bead transport microchip is a 5×5 mm^2 (100) silicon substrate with a native silicon oxide layer. The magnetic structures were fabricated by either e-beam or optical lithography. The pattern was written in a layer of positive or negative resist spin-coated onto the substrate. After developing the pattern, the magnetic material (Permalloy) was deposited by thermal evaporation and the procedure was completed by a lift-off process. The magnetic properties of the elements effectively depend on size and shape of the elements. In the case of Permalloy ellipses, the easy/hard axis of magnetisation is along the long/short axis of an element.

The magnetic domain structure of the elements has been investigated using magnetic force microscopy (MFM). Figure 13.4 shows the domain state of $2 \times 6 \times 0.1$ μm^3 ellipses in a weak in-plane magnetic field (field direction indicated by the arrow). In Fig. 13.4(a), the ellipses in the bottom row, having the field parallel to their long axes, are magnetised close to saturation. Consequently, the fringing magnetic field is large at the short ends of the ellipses. The ellipses in the top row, where the field is parallel to the short axis, show a distorted multi-domain structure. Even in this case, however, there is a weak net magnetisation in the direction of applied field. In Fig. 13.4(c), the situation is reversed compared to Fig. 13.4(a): Since the field now is parallel to the long axes of the top row ellipses, these are in a state close to saturation. On rotating the field, this domain state cycle will be repeated and the direction of net magnetisation will rotate in phase with the field rotation. The domain

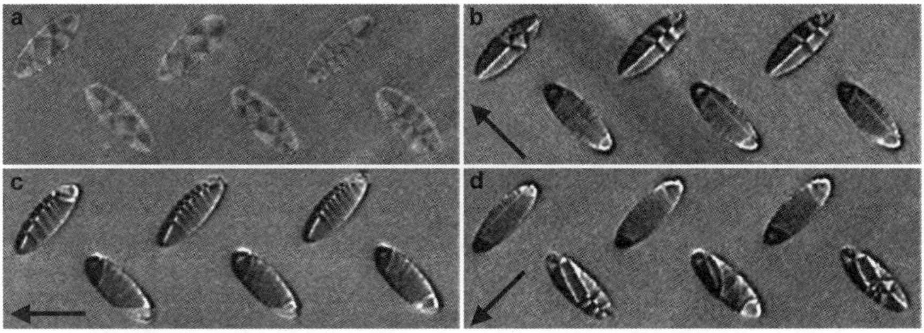

Fig. 13.4. MFM images of the magnetic domain structures of Permalloy ellipses $2 \times 6 \times 0.1\ \mu m^3$. (a) corresponds to demagnetised magnetic elements, while (b-d) give the domain states in an in-plane magnetic field of constant amplitude, 6.3 kA/m; arrows indicate the directions of the magnetic field. Reprinted with permission from Ref. 15. Copyright (2005) Wiley-VCH Verlag.

structures mapped by MFM are further verified by results from computer simulations using the OOMMF micromagnetic solver.[46] These simulations[15] indicate that the net magnetisation for a saturated ellipse is five times larger than for a multi-domain ellipse.

Also triangular Permalloy elements have been prepared. Exposed to an in-plane magnetic field, these elements show distorted multi-domain states with a net magnetisation in the direction of the field.[45] The maximum fringing magnetic field will appear at the apex of a triangle when the field is directed perpendicular to the base.

In conclusion, the net magnetisation of an element follows the direction of the applied field and the magnitude of magnetisation will depend on the direction of the magnetic field with respect to the element geometry.

In the experimental setup, the transport microchip is fixed inside a fluid cell and is observed using an optical microscope. Two mutually perpendicular electromagnets generate a uniform in-plane rotating magnetic field at the position of the elements. When the bead suspension is injected into the fluid cell, the fringing magnetic fields due to the Permalloy elements create attractive forces on close enough beads and these beads will thus move towards the elements. Finally, they will stick

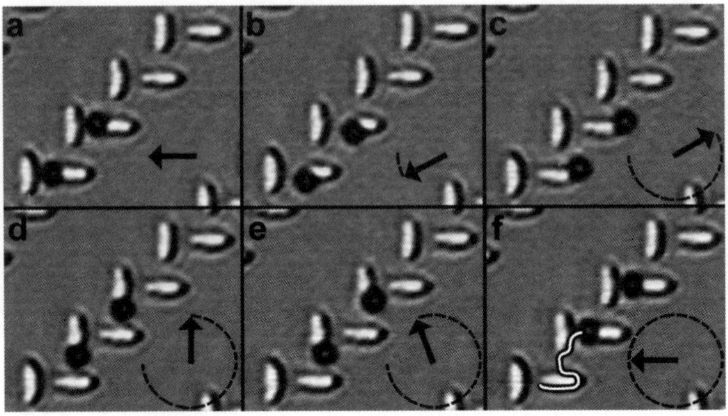

Fig. 13.5. Transport of magnetic beads (Dynabead M-270, diameter = 2.8 μm) on a staircase pattern of Permalloy ellipses ($2 \times 6 \times 0.1\ \mu m^3$) in an in-plane rotating magnetic field (6.3 kA/m). The arrows indicate the direction of the field. After one complete field revolution, the beads have moved one step in the pattern as indicated by the white curve in (f). Reprinted with permission from Ref. 15. Copyright (2005) Wiley-VCH Verlag.

to the rim of the elements at a point defined by the direction of the net magnetisation.

Figure 13.5 shows two beads (dark circles) being transported in a staircase pattern of Permalloy ellipses (white structures). The magnetic particles will move along the perimeter of the ellipses, following the rotation of the net magnetisation. In Fig. 13.5(d), the beads are attracted by the strong force from the apex of the now saturated vertical ellipses. Hence, the beads will jump to the neighbouring vertical ellipses. Following the pace of the rotating field, the beads can move at a speed in range 2–20 μm/s. The staircase pattern represents a one-way transport line; the particles can only jump from a long side of an ellipse to the apex of a neighbouring ellipse. Further, they follow the rim of ellipses with the same sense of rotation as the external in-plane magnetic field. An advantage of this feature is demonstrated in Fig. 13.6, which shows a junction added to a transport line. When the field is clock-wise rotated, the beads move around elements with the same sense of rotation and will therefore enter the loop of elements (Fig. 13.6(a)). Anti-clock-wise rotation will instead let the particles pass the junction without entering

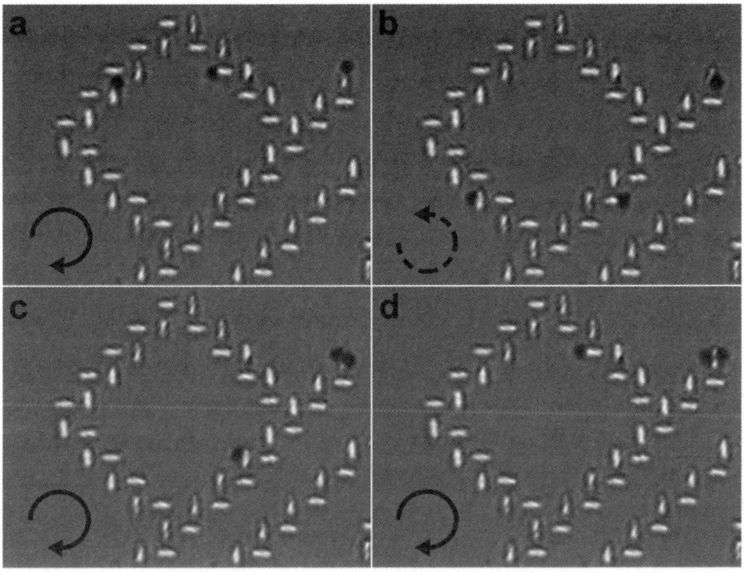

Fig. 13.6. Separation of magnetic beads (Dynabead M-270, diameter = 2.8 μm) by changing the sense of magnetic field rotation. The curled arrows in each figure show the sense of field rotation. (a) Two particles are trapped inside the loop. A third particle is located at the end position of the long transport line and is not further considered. (b) When the first particle is approaching the second junction, the field rotation is turned anti-clockwise. (c) The first particle avoided the loop and has continued moving straight to the end of the long transport line. The field is then switched back to clockwise rotation in order to trap the remaining particle in the loop. (d) The remaining particle is kept in the loop and the two particles are thus separated. Reprinted with permission from Ref. 15. Copyright (2005) Wiley-VCH Verlag.

the loop. Momentarily switching the sense of magnetic field rotation (Fig. 13.6(b–d)), makes it possible to separate beads one-by-one.

It is easy to imagine that life-science molecules can be transported and separated with the method described above. For this, we have used so-called Micromer-M beads (cf. Section 13.2.1.) with the bead surface covered by polyethylene glycol with amino function (PEG-NH_2). Proteins were conjugated to the amino groups using SPDP (cf. Section 13.2.2) and the beads were suspended in PBS buffer. In this case, the major obstacle to a successful bead transport was sticking forces between the bead proteins and the surface of the transport microchip.

Proteins can be classified as advanced polymers with more than 20 different amino acids as building blocks. These amino acids contribute both with charges and hydrophobic regions in the polymer and since amino acids are amphiphilic in nature the number of charges vary by pH. Electrostatic interactions, hydrophobic interactions and hydrogen bonds can thus be foreseen when proteins interact with surfaces, which makes the situation quite complex. In aqueous media the contribution of hydrogen bonding to the adsorption affinity is often marginal,[47] since these forces are constantly reorganising between different elements where hydrogen bonds can be formed. Moreover, our results show that the dominating sticking force, which may even inhibit bead transportation, is the hydrophobic force.[45]

In our experiment, a bead (radius $R = 2.5 \cdot 10^3$ nm) covered by IgG on PEG (total length of organic molecule $l \approx 50$ nm) has an area in contact with the chip surface of approximately $2\pi R\, l = 8 \cdot 10^5$ nm^2, corresponding to 1% of the total bead surface.[45] Alsteens *et al.*[48] report on chemical force microscopy where the interaction between a hydrophobic tip and a self assembled monolayer of CH_3- and OH-terminated alkanethioles is estimated. Using their result for 50% coverage of CH_3-alkanethioles and taking only the part of the tip in contact with the monolayer into account, the hydrophobic force becomes 0.02 nN/nm^2. Applying this result to our beads, the hydrophobic force between IgG covered beads and the chip surface will be of order 10^4 nN.

The possibility to manipulate magnetic beads by magnetic fields is used in several biotechnological applications, including single molecule force spectroscopy using magnetic tweezers,[49,50] methods where micromagnets are used to achieve a programmable assembly of magnetic beads[51] and microelectromagnets used for magnetic manipulation in lab-on-a-chip systems.[52] Magnetic forces on a bead fall off rapidly with displacement between field source and magnetic bead. Consequently, one can expect appreciable forces only when the bead is in reasonable proximity to the magnetic field source. Magnetic tweezers, where the field source is a single-pole electromagnet,[50] have been developed with pulling forces as large as 10 nN on a 4.5 μm bead at 10 μm distance from the pole piece. Forces on magnetic beads interacting with thin film micromagnets as well as with neighbouring magnetic beads were studied in Ref. 51, with the attempt to develop a method for programmable assembly of beads using magnetic microwell templates. Forces on a second bead in close proximity to a microwell with one bead already trapped, varied with the distance between the two beads; applying a large enough magnetic field, forces close to 1 nN on a 8 μm bead were predicted with a separation between the two beads equal to the bead diameter.

Our case is different from the applications described above in that the magnetic field from a magnetized element will not be uniform over the magnetic bead volume. This also implies that the force acting on a bead is non-uniform, in the sense that the part of the bead facing the magnetic element will experience a larger force than more remote parts of the bead. To estimate the force exerted by a saturated magnetic element on a superparamagnetic bead one needs to calculate the magnetic stray field $\boldsymbol{H}_s(\boldsymbol{r})$ outside the element. The force acting on a small volume element dV of the bead at $\boldsymbol{r}'$ is then given as $d\boldsymbol{F}(\boldsymbol{r}') = -\nabla \boldsymbol{E}$, where $\boldsymbol{E} = -\mu_0 \boldsymbol{M}(\boldsymbol{r}') \cdot (\boldsymbol{H}_s(\boldsymbol{r}') + \boldsymbol{H}_a) dV$ is the field energy, $\boldsymbol{M}(\boldsymbol{r}')$ the bead magnetisation and $\boldsymbol{H}_a$ the applied magnetic field assumed to be constant in magnitude over the bead volume. To obtain the total force acting on the bead, one integrates $d\boldsymbol{F}(\boldsymbol{r}')$ over the bead volume. Calculations have been performed for Permalloy magnetic elements of rectangular block geometry, with dimensions 2 μm × 6 μm × 0.1 μm, and magnetic beads with a diameter of 2.8 μm.[53] The beads considered in the calculations were polystyrene beads with iron-oxide nanoparticles evenly distributed inside the polystyrene matrix. With the Permalloy magnetisation saturated along the long axis of the element, the horisontal force component pointing towards the magnetic element varies in magnitude between a few tenth of nN to some nN as the distance between the centre of the bead and the short side of the element decreases from a few μm down to some tenth of a micrometer. At the same time, the vertical force component grows in magnitude, from a few tenth of nN to become as large as some tens of nN. Even though the calculations have been performed for an element geometry and bead size slightly different from what is used in the present study, the results of the calculations indicate the order of magnitude of the magnetic force exerted by a magnetic element on a superparamagnetic bead.

In conclusion, the hydrophobic force between a protein covered bead and the chip surface can be 100–1000 times larger in magnitude compared to the magnetic force due to a magnetic element.

The effect of hydrophobic forces can be reduced by modifying the surface with polymer brushes like polyethylene oxide (PEO) or derivatives of PEO.[54] Protein-surface interactions, where the surface is covered with PEO, have been studied in detail in Ref. 55. Both the polymer length and the surface density of the polymer are of crucial importance for good protein resistance. Direct force measurements between two adsorbed PEO surfaces in an aqueous solvent show that repulsion forces develop at certain separation distances due to steric repulsion phenomena.[56–58] The physics of terminally attached polymer surfaces has been studied theoretically by de Gennes.[59]

In this work, PEG derivatised with end groups of thiols has been used as a polymer brush. The gold-thiol binding is useful for chemical modification of a surface with organic molecules.[60] The transport microchip was therefore sputter coated with a 10 nm thick gold layer. Our experimental results show that the hydrophobic interaction between the protein conjugated magnetic beads and the surface of the transport microchip is to a large extent eliminated by chemical grafting of the bead transport microchip with PEG. Moreover, the bead transport was further improved by adding a small amount of surfactant (0.5% w/v of Tween-20) to the suspension. Using this arrangement, we have managed to transport four different proteins with various properties regarding isoelectric point and protein size, viz. Lyzozyme, α-Lactalbumin, Human Serum Albumin (HSA) and Immunoglobulin G (IgG). Protein levels down to one tenth of an atto mole were attached to the bead surface and successfully transported on the microchip. By adding a reducing agent like DTT to the suspension, the proteins are released instantaneously from the beads.

13.5. Conclusions and Outlook

Summarising, the VAM-NDA bioassay principle,[18,44] discussed in Section 13.3, is a lab-on-a-bead type magnetic biosensor scheme, relying on changes in Brownian relaxation of magnetic nanobeads[10] and the use of high performance molecular tools, viz. padlock probe target recognition[37,38] and RCA.[39,40] It may be applied for detection of several types of biomolecule targets; DNA (using padlock probes), RNA and proteins (using proximity probes[41,42]). Due to its non-optical nature, the VAM-NDA bioassay principle has the potential for future development of low-cost diagnostic test devices of over-the-counter and/or point-of-care type.

So far in the development work of the VAM-NDA method, we have used a commercial SQUID-magnetometer to record the dynamic magnetic properties of the samples. The used SQUID does not allow low enough nanobead concentrations that are necessary for improving the lower limit of detection of target molecules beyond the pM range. Changing the read-out format by developing a miniaturised platform based on a magnetoresistive sensor[8,61] or a micro-Hall probe,[7,62] is expected to considerably improve detection sensitivity. Moreover, by introducing a magnetic field sensor that covers a broader frequency interval than that of the SQUID, a larger range of bead sizes can be implemented thus allowing for improved multiplex detection strategies.

In Section 13.4, a method capable of controlled transport and release of proteins at well defined positions on the chip surface is described. Although demonstrated for proteins, the method is useful for any kind of molecule. To our knowledge, this is the first report of controlled transport of sub-attomole levels of molecules with micrometer precision. The transport system is flexible with respect to dimensions and can be up or down scaled by at least a factor of 10.

Potential applications of the system are in the field of separation technology and as analytical system since during transport on the chip, the protein covered beads are controlled one-by-one. A real time analytical system testing e.g. for allergic disease and monoclonal antibody-antigen interactions can be constructed where beads conjugated with antigens continuously are transported through a sample containing region of the transport microchip. With suitable miniaturised magnetic field sensors added to the chip, counting and numbering of the beads can be performed. This yields an effective encoding system, where beads carrying a particular antibody-antigen combination can be directed to a specific transport line and beads carrying other antibody-antigen combinations to other lines, thus forming the building block for an effective multiplex analysis system.

It may even be possible to combine the two here described methods. In a diagnostic system making use of miniaturised magnetic field sensors to detect the response of magnetic beads to AC and/or DC magnetic fields, it will be necessary to guide magnetic beads to come in close proximity to the field sensors. This can be accomplished by using the here described transport system for bioconjugated magnetic beads. An all magnetic multiplex analysis system can be constructed where different bead sizes are used for different target DNA molecules. With magnetic field sensors capable of single-bead-detection,[7,8] different bead sizes can be separated by the magnitude of the DC field induced bead magnetic moments and transported to different sites on the chip where the existence or non-existence of RCA-coils is determined from the Brownian relaxation behaviour of the beads. Quantitative analysis can be performed by either measuring the response of magnetic beads one-by-one as they pass by the magnetic field sensor or by collecting a comparably large number of beads close to the magnetic field sensor before measuring the AC magnetic field response of the ensemble of beads.

13.6. Abbreviations and Acronyms of Chapter 13

AC	Alternate Current
DC	Direct Current
DMSO	DiMethylSulphOxide
DNA	DeoxyriboNucleic Acid
DTT	DiThioThreitol
HFP	High-Frequency Peak
HSA	Human Serum Albumin
IgG	ImmunoGlobulin G
LFP	Low-Frequency Peak
MFM	Magnetic Force Microscopy
MRI	Magnetic Resonance Imaging

PEG	PolyEthylene Glycol
PEO	PolyEthylene Oxide
PBS	Phosphate Buffered Saline
RF	Radio-Frequency
RCA	Rolling Circle Amplification
RNA	RiboNucleic Acid
SPDP	*N*-succinimidyl 3-(2-pyridyldithio)propionate
SQUID	Superconducting QUantum Interference Device
VAM-NDA	Volume-Amplified Magnetic Nanobead Detection Assay
T_1-decay, T_2-decay	Relaxation processes in MRI
ω	Angular frequency of applied AC magnetic field
$\chi(\omega)$	Complex low-field magnetic susceptibility
χ_∞	High-frequency magnetic susceptibility
χ_0	Low-field static magnetic susceptibility
$m(\omega) = m'(\omega) - im''(\omega)$	Complex magnetisation
H_{AC}	Amplitude of applied AC magnetic field
τ	Characteristic relaxation time of a magnetic bead
τ_N	Néel relaxation time
E_b	Energy barrier (in connection to Néel relaxation)
K	Magnetic anisotropy constant
V_p	Magnetic nanoparticle volume (in connection to Néel relaxation)
$\beta = kT$	Thermal energy
τ_0	Microscopic relaxation time (in connection to Néel relaxation)

τ_B	Brownian relaxation time
V_B	Hydrodynamic volume of a magnetic bead
η	Dynamic viscosity of the carrier liquid of a ferrofluid
f_B	Brownian relaxation frequency
R	Magnetic bead radius
l	Length of organic molecule (in connection to PEG)
$\boldsymbol{H}_s(\boldsymbol{r})$	Magnetic stray field ouside Permalloy element
dV	Small volume element of a magnetic bead
$\boldsymbol{r}'$	Position inside a magnetic bead
$\mathrm{d}\boldsymbol{F}(\boldsymbol{r}')$	Force acting on a small volume element, dV, of a magnetic bead at $\boldsymbol{r}'$
$\boldsymbol{E}$	Field energy
$\boldsymbol{M}(\boldsymbol{r}')$	Bead magnetisation
$\boldsymbol{H}_a$	Applied magnetic field

Acknowledgments

We are grateful to M. Strømme, T. Zardan Gomez de la Torre, M. Nilsson, J. Göransson, L.-E. Johansson and K. Eriksson for collaboration and valuable discussions. Financial support from the Swedish Research Council (VR) and the Knut and Alice Wallenberg Foundation (KAW) is gratefully acknowledged.

References

1. S. Mornet, S. Vasseur, F. Grasset and E. Duguet, *J. Mater. Chem.* **14**, 2161-2175 (2004).
2. J. M. Perez, L. Josephson, T. O'Loughlin, D. Högermann and R. Weissleder, *Nature Biotechnology* **20**, 816-820 (2002).
3. J.-H. Lee, Y.-M. Huh, Y.-W. Jun, J.-W. Seo, J.-T. Jang, H.-T. Song, S. Kim, E.-J. Cho, H.-G. Yoon, J.-S. Suh and J. Cheon, *Nature Medicine* **13**, 95-99 (2007).

4. M. A. McAteer, N. R. Sibson, C. von zur Muhlen, J. E. Schneider, A. S. Lowe, N. Warrick, K. M. Channon, D. C. Anthony and R. P. Choudhury, *Nature Medicine* **13**, 1253-1258 (2007).
5. P. Moriz, S. K. Jones and B. N. Gray, *Int. J. Hyperthermia* **18**, 267-284 (2002).
6. T. Osaka, T. Matsunaga, T. Nakanishi, A. Arakaki, D. Niwa and H. Iida, *Anal. Bioanal. Chem.* **384**, 593-600 (2006).
7. G. Mihajlovic, P. Xiong, S. von Molár, K. Ohtani, H. Ohno, M. Field and G. J. Sullivan, *Appl. Phys. Lett.* **87**, 112502 (2005).
8. D. L. Graham, H. A. Ferreira and P. P. Freitas, *Trends in Biotechnology* **22**, 455-462 (2004).
9. Y. Amemiya, T. Tanaka, B. Yoza and T. Matsunaga, *J. Biotechnol.* **120**, 308-314 (2005).
10. J. Connolly and T. G. St Pierre, *J. Magn. Magn. Mater.* **225**, 156-160 (2001).
11. F. Vartdal *et al.*, *Tissue Antigens* **28**, 301-312 (1986).
12. H. H. Oberg, D. Wesch, J. Lenke and D. Kabelitz, *Scand. J. Immunol.* **64**, 353-360 (2006).
13. M. J. Lynch, C. G. Leon-Velarde, S. McEwen and J. A. Odumeru, *J. Microbiol. Methods* **58**, 285-288 (2004).
14. C. M. Bearden, B. K. Book, R. A. Sidner and M. D. Pescovitz, *J. Immunol Methods* **300**, 192-199 (2005).
15. K. Gunnarsson, P. E. Roy, S. Felton, J. Pihl, P. Svedlindh, S. Berner, H. Lidbaum and S. Oscarsson, *Adv. Mater.* **17**, 1730-1734 (2005).
16. T. Deng, G. M. Whitesides, M. Radhakrishnan, G. Zabow and M. Prentiss, *Appl. Phys. Lett.* **78**, 1775-1777 (2001).
17. B. R. Yellen and G. Friedman, *Adv. Mater.* **16**(2), 111 (2004).
18. M. Strömberg, J. Göransson, K. Gunnarsson, M. Nilsson, P. Svedlindh and M. Strømme, *Nano Lett.* **8**, 816-821 (2008).
19. G. Li, V. Joshi, R. L. White, S. X. Wang, J. T. Kemp, C. Webb, R. W. Davis and S. Sun, *J. Appl. Phys.* **93**, 7557 (2003).
20. T. Neuberger, B. Schöpf, H. Hofmann, M. Hofmann and B. V. Rechenberg, *J. Magn. Magn. Mater.* **293**, 483-496 (2005).
21. P. Majewski and B. Thierry, *Critical Rev. Solid State Mater. Sci.* **32**, 203-215 (2007).
22. B. Gillot, A. Rousset and G. Dupre, *J. Solid State Chem.* **25**, 263-271 (1978).
23. G. Hägg, *Allmän Och Oorganisk Kemi* (Almqvist och Wiksell, Uppsala 1969).
24. Micromod Partikeltechnologie GmbH, www.micromod.de .
25. Dynal Biotech, www.invitrogen.com .
26. Z. Ma, Y. Guan, X. Liu and H. Liu, *Polym. Adv. Technol.* **16**, 554-558 (2005).
27. K. Aurich, M. Schwalbe, J. H. Clement, W. Weitschies and N. Buske, *J. Magn. Magn. Mater.* **311**, 1-5 (2007).
28. J.-L. Gong, Y. Liang, Y. Huang, J.-W. Chen, J.-H. Jiang, G.-L. Shen and R.-Q. Yu, *Biosensors and Bioelectronics* **22**, 1501-1507 (2007).

29. C. W. Lee, K. T. Huang, P. K. Wei and Y. D. Yao, *J. Magn. Magn. Mater.* **304**, e412-e414 (2006).
30. G. K. Kouassi and J. Irudayaraj, *Anal. Chem.* **78**, 3234-3241 (2006).
31. J. Carlsson, H. Drevin and R. Axén, *Biochem. J.* **173**, 723-737 (1978).
32. M. Strömberg, K. Gunnarsson, H. Johansson, M. Nilsson P. Svedlindh and M. Strømme, *J. Phys. D: Appl. Phys.* **40**, 1320-1330 (2007).
33. P. Debye, *Polar Molecules* (The Chemical Catalogue Company, New York 1929).
34. L. Néel, *Ann. Geophys.* **5**, 99-136 (1949).
35. W. F. Brown Jr., *J. Appl. Phys.* **34**, 1319-1320 (1963).
36. W. F. Brown Jr., *Physical Review* **130**, 1677-1686 (1963).
37. M. Nilsson, H. Malmgren, M. Samiotaki, M. Kwiatkowski, B. P. Chowdhary and U. Landegren, *Sci.* **265**, 2085-2088 (1994).
38. U. Landegren, F. Dahl, M. Nilsson, S. Fredriksson, J. Banér, M. Gullberg, J. Jarvius, S. Gustafsdottir, O. Söderberg, O. Ericsson, J. Stenberg and E. Schallmeiner, *Comparative and Functional Genomics* **4**, 525-530 (2003).
39. A. Fire and S.-Q. Xu, Proceedings of the National Academy of Sciences of U.S.A. **92**, 4641-4645 (1995).
40. D. Liu, S. L. Daubendiek, M. A. Zillman, K. Ryan and E. T. Kool, *J. Am. Chem. Soc.* **118**, 1587-1594 (1996).
41. O. Söderberg, K. J. Leuchowius, M. Kamali-Moghaddam, M. Jarvius, S. Gustafsdottir, E. Schallmeiner, M. Gullberg, J. Jarvius and U. Landegren, *Genetic Engineering* **28**, 85-93 (2007).
42. O. Söderberg, M. Gullberg, M. Jarvius, K. Ridderstråle, K.-J. Leuchowius, J. Jarvius, K. Wester, P. Hydbring, F. Bahram, L.-G. Larsson and U. Landegren, *Nature Methods* **3**, 995-1000 (2006).
43. K. S. Cole and R. H. Cole, *J. Chem. Phys.* **9**, 341-351 (1941).
44. M. Strömberg, T. Zardán Gómez de la Torre, J. Göransson, K. Gunnarsson, M. Nilsson, M. Strømme and P. Svedlindh, *Biosensors and Bioelectronics* **24**, 696-703 (2008).
45. L.-E. Johansson, K. Gunnarsson, S. Bijelovic, K. Eriksson, A. Surpi, P. Svedlindh and S. Oscarsson, unpublished (2008).
46. http://math.nist.gov/oommf/software.html.
47. M. D. Joesten and L. J. Schaad, in *Hydrogen Bonding* (Marcel Dekker, New York 1974).
48. D. Alsteens, E. Dague, P. G. Rouxhet, A. R. Baulard and Y. F. Dufrêne, *Langmuir* **23**, 11977-11979 (2007).
49. K. C. Neuman and A. Nagy, *Nature Methods* **5**, 491-505 (2008).
50. A. R. Bausch, F. Ziemann, A. A. Boulbitch, K. Jacobson and E. Sackmann, *Biophys. J.* **75**, 2038-2049 (1998).
51. B. B. Yellen and G. Friedman, *Langmuir* **20**, 2553-2559 (2004).
52. K. Smistrup, P. T. Tang, O. Hansen and M. F. Hansen, *J. Magn. Magn. Mater.* **300**, 418-426 (2006).

53. Johan Pihl, 2004, *Field Manipulation and Design of Thin-film Structures for Programmable Motion of a Magnetic Bead*, UPTEC FO4 086, diploma work at Uppsala University.
54. W. Norde, *Colloids and Interfaces in Life Sciences* (Marcel Dekker, New York 2003).
55. S. I. Jeon, J. H. Lee, J. D. Andrade and P. G. De Gennes, *J. Coll. Interf. Sci.* **142**, 149-158 (1991).
56. J. Klein and P. Luckham, *Nature (London)* **300**, 429 (1982).
57. J. Klein and P. Luckham, *Macromol.* **17**, 1041 (1984).
58. P. Luckham and J. Klein, *Macromol.* **18**, 721 (1985).
59. P. G. de Gennes, *Macromol.* **13**, 1069 (1980); ibid. **14**, 1637 (1981); ibid. **15**, 492 (1982).
60. R. G. Nuzzo and D. L. Allara, *J. Am. Chem. Soc.* **105**, 4481 (1983).
61. P. P. Freitas, R. Ferreira, S. Cardoso and F. Cardoso, *J. Phys.: Condens. Matter* **19**, 165221 (2007).
62. A. Sandhu, Y. Kumagai, A. Lapicki, S. Sakamoto, M. Abe and H. Handa, *Biosensors and Bioelectronics* **22**, 2115-2120 (2007).

Chapter 14

DOMAIN WALLS FOR LOGIC AND DATA STORAGE APPLICATIONS

Colm C. Faulkner

Centre for Research on Adaptive Nanostructures and Nanodevices
Trinity College Dublin, Dublin 2, Ireland
E-mail: colm.faulkner@tcd.ie

14.1. Introduction

New properties emerge in magnetic nanostructures when dimensions become comparable to length scales such as the spin diffusion length, the magnetic exchange length, and the domain wall width. Additionally as dimensions are reduced, element geometry, and the proximity of interfaces strongly influences magnetic properties. Therefore new applications emerge.

Once it is technologically feasible to control the geometry of ferromagnetic structures on the nanoscale, it is possible to isolate and control the position of domain walls on these length-scales. Due to advances in fabrication technology, and improved spatially and temporally resolved magnetic microscopy techniques, magnetic domain walls (DWs) have been heavily investigated in recent years. The predicted breakdown of Moore's law in the semiconductor industry,[1] where the number of transistors on a fixed area of computer circuit board doubles every two years, means that emerging magnetic materials and technologies may play an important role in future logic processing devices, as well as improved data storage applications.

Devices which link both charge and magnetization were introduced in 1991 by IBM, in the form of hard disk drives using anisotropic magnetoresistance, where the resistance of a magnetic device changes according to the orientation of the magnetic layers. Many of these devices, utilizing the spin degree of freedom of the electron, and labelled "spintronic" devices may feature magnetic domain walls in the future.

Domain walls, which we can consider a well defined entity like a soliton, can be used to manipulate information using magnetic fields, electric current, or a combination of both. Therefore a treatment of domain wall mediated switching and dynamics in patterned magnetic nanostructures for future device applications is presented herein.

14.2. Theory

A domain wall is a spatially localized mobile boundary between two regions of opposing magnetization states. The propagation fields, velocity, and structure of magnetic domain walls were initially predominately studied in thin films, either experimentally,[2,3] or computationally, following from the classic study of magnetic domain walls travelling in wires and elongated iron whiskers.[4] Some recent advances in the experimental study of domain walls and domain configurations include the observation of domain wall motion with sub nanometre resolution (essentially at an atomic length scale) in garnets by Hall Probe,[5] magnetic force microscopy (MFM) tip stray field induced domain wall displacement,[6] and the direct experimental observation of vortex core type domain formations in 50 nm thick NiFe elements.[7] Spin polarized scanning tunnelling microscopy, has also been employed to map the in-plane and out of plane magnetization of Fe nano-islands in the vortex configuration to sub 10 nm resolution.[8]

Generalized modes of reversal for a ferromagnetic monodomain particle, such as the curling, buckling and nucleation modes were treated by Frei *et al.* in 1957.[9] Generally the wire structures discussed in this chapter are the soft ferromagnetic alloy Permalloy, $Ni_{80}Fe_{20}$, chosen for its low coercivity and magnetostriction, high Curie temperature, and widespread use in magnetic recording head sensors. Additionally Permalloy exhibits a strong anisotropic magnetoresistance effect (AMR), which is useful for point contact measurements. The length scales treated are typically 3–50 nm for wire thickness, 100–500 nm for wire width, and wire lengths of the order of microns, unless explicitly stated otherwise. Structures on these length scales are generally single domain at room temperature (geometry dependent), and reside above the super paramagnetic limit at 293 K. Planar wires are treated, as they are straightforward to fabricate by standard electron beam lithography, although much work has been performed on domain walls with a circular cross section.[10]

Typically an elongated Permalloy planar nanowire of these dimensions does not support a DW at remanence. They are bi-stable at remanence, having 2 allowable magnetization configurations, with spins aligned parallel to wire sides due to shape anisotropy. It is energetically more efficient for samples of these dimensions to reside in a ground state of 1 ferromagnetic domain. However a magnetic domain wall can be reproducibly situated in such samples, by methods including geometric modification,[11] current carrying strip lines[12,13] or an appropriate field sequence.[14]

In thin films the domain wall type depends on intrinsic material properties, but for a nanowire the geometry, as well as applied fields and currents have an important effect on the wall spin structure. For a given material, the DW type is governed by the wire geometry,[15] and the spin structure is the result of an energy minimization process. For soft magnetic materials such as Permalloy, the two basic domain wall types are transverse where spins rotate in the plane of the wire (either head-to-head or tail-to-tail), Fig. 14.1(a), and vortex where spins curl clockwise or counterclockwise around an out of plane vortex core (Fig. 14.1(b)). Generally a vortex DW occurs in thicker and wider nanowires. (Flux closure reduces surface charge and demagnetizing fields, leading to a lower energy ground state, than a transverse wall in a wire of identical dimensions).

14.3. Wire Switching

Magnetic nanowires are excellent candidates for device applications, because a very high level of control is exercised over magnetization reversal mechanisms, switching fields, and increasingly the reversal rates. First consider high aspect ratio, single domain ferromagnetic nanowires. Magnetization reversal is generally triggered by a nucleation event, typically at an imperfection or wire end, followed by DW propagation, leading to rapid domain expansion and the reversal of the entire nanowire. Data plotted in Fig. 14.2 is for 6 element arrays of elongated planar Permalloy nanowires from 60–500 nm in width. For the wire dimensions studied, internal dipolar fields control the coercivity, H_c, i.e. it depends on the demagnetizing field of the sample. H_c drops rapidly with increasing wire width, and excellent control of the switching properties of the wires is evident from the data plot. However it is noted, that as the wire width is reduced to dimensions of the order of 100 nm, desirable for device applications, the coercive field becomes very high – of the order of 100's of Oersted. From Stoner-Wohlfarth analysis, an absolute upper bound of H_c for an elongated particle with uniaxial anisotropy is given by an analytic expression of the form:[17]

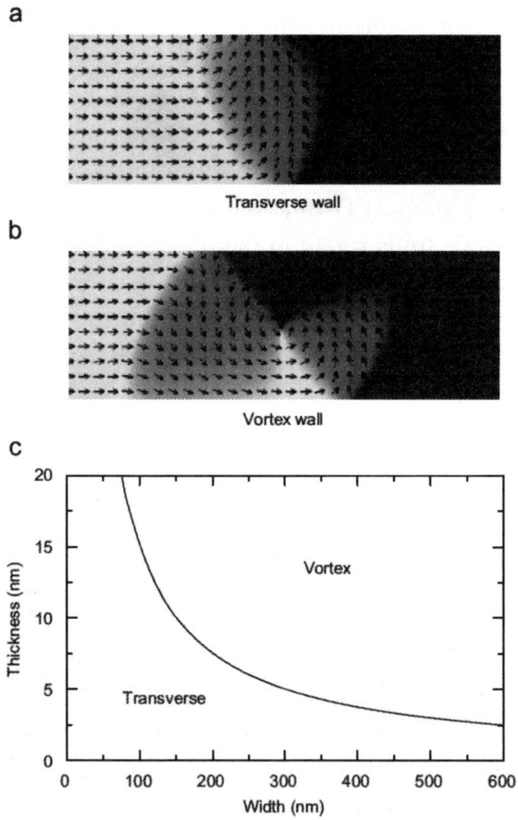

Fig. 14.1. Micromagnetically determined spin structured of a transverse head-to-head (a) and vortex domain wall (b) in a thin Permalloy strip. The transverse domain wall resides in wires of reduced thickness and width, relative to the vortex wall (c).[16]

$$H_c = \frac{8M_sT}{W} \cdot S \tag{1}$$

where M_s is the saturation magnetization, T the thickness, W the wire width, and S is a finite length shape anisotropy factor. This is the magnetostatic energy barrier associated with reversing magnetization of the entire wire in-plane, parallel to the short axis – reversal by coherent rotation. The expression is appropriate in the thin film limit, where sample magnetization is constrained in-plane, and micromagnetic configurations such as vortices with out of plane spins are not energetically favoured.

$$S = \left[\frac{(\rho - \frac{1}{\rho})}{\sqrt{1+\rho^2}} \right]. \tag{2}$$

Here, $\rho = L/W$ is the wire in-plane aspect ratio. The discrepancy in switching data between the experimentally measured results and the analytically calculated results shown is attributed to thermal effects and sample side wall imperfections inherent in the fabrication process. In short we note that excellent control over ferromagnetic nanowires is possible, but switching fields are very high as critical dimensions are reduced, requiring more power for magnetic switching to occur. However, wire geometries can now be engineered such that one or more DWs reside in the system in the absence of field. Then magnetization reversal can be effected reproducibly at low DW propagation field values, as opposed to the intrinsic nucleation field of a wire.[11,18]

14.4. Domain Wall Propagation

When an external magnetic field is applied to a static domain wall in a nanowire, the wall feels a change in potential, and local pinning can be overcome. In theory a DW in a perfect planar nanowire, with no roughness or defects, will propagate at a negligible field, and recent experimental reports demonstrate that the DW can propagate over reasonable length scales with negligible power dissipation,[19] and propagation fields of circa 10 Oe, an order of magnitude or lower better

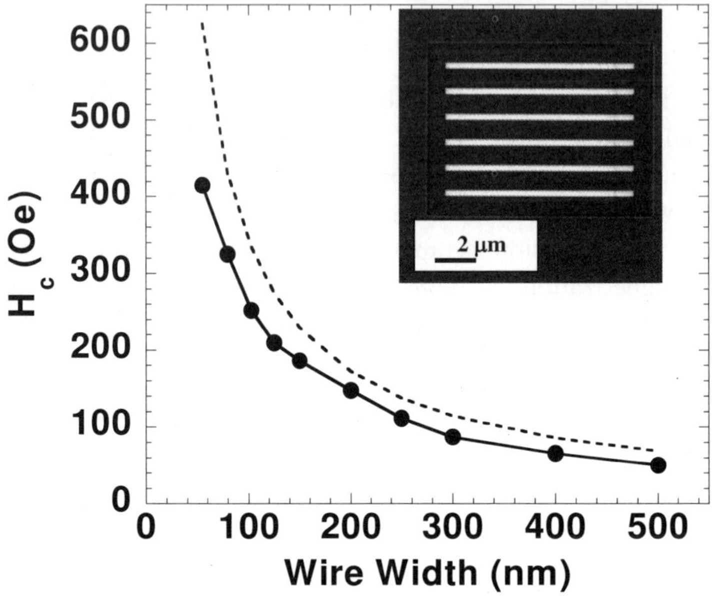

Fig. 14.2. Experimentally measured switching field, Hc, [—•—], for $Ni_{81}Fe_{19}$ planar nanowires of the following dimensions: width = (60–500 nm), length = (9 µm), film thickness = (5 nm). The dashed line (-----) represents an analytic solution of Hc. Inset is a secondary electron image from a W = 300 nm 6-element planar wire array.

than nucleation fields for these structures, [20] see also Fig. 14.2. We note also that power gain is a critical consideration for device applications. It is therefore desirable to isolate a domain wall in a magnetic nanostructure, and investigate domain wall propagation, as a distinct type of magnetization reversal process.

14.5. Domain Wall Injection

In the last decade advances in nanolithography have enabled the fabrication and investigation of magnetic nanostructures with increasingly complex geometries. It is now possible to engineer the dimensions of magnetic structures in 3 dimensions on the nanoscale, and therefore engineer the switching mechanism with great precision. Using the giant magnetoresistance effect in a single NiFe/Cu/NiFe nanowire, Ono and co-workers detected the propagation of a single domain wall in a high aspect ratio nanowire, at temperatures from 77–160 K.[18] The trapping of a single head-to-head domain wall about a notch in a submicrometer NiFe(200 Å)/Cu(100 Å)/NiFe(50 Å) wire was described.

In order to separate DW nucleation and propagation fields, and to control the position of a DW in a nanostructure with deep sub-micron accuracy, structures are fabricated with large low aspect ratio pads attached to thin elongated nanowires. A high aspect ratio structure is ferromagnetically "soft", having a low switching field. A DW can be localized at the junction between the 'soft' and 'hard' regions of the patterned structure, and then selectively propagated into the thin nanowire. Nucleation pads of relatively large lateral dimensions (~μm) relative to a wire afford a technique to artificially introduce domain walls into nano-wires for experiment, at applied fields below the wire critical nucleation field. Several groups have utilized domain wall injection pads, as a domain wall source to perform transport, propagation, and pinning studies on magnetic domain walls. One of the first examples of this kind of experimental geometry is the work of Fournel *et al.* in samples of (CoNi/Pt) defined by electron beam lithography.[21] The technique of controlling magnetization reversal of a continuous wire structure, by field mediated domain wall reversal was extended to 150 nm wide NiFe wires investigated by GMR,[22] with measurements by MFM,[23] MOKE,[24] and the ballistic Hall effect[25] subsequently performed. A variety of pad geometries (circular, square, diamond) were investigated in the GMR measurement of trilayer structure switching.[26]

By tuning the injection pad geometry (rectangular, circular, diamond shaped, triangular), pad aspect ratio, and the ratio of pad to output wire thickness, the DW injection field can be very well controlled, at fields markedly lower than the injection field of a similar padless wire structure.[11] For example in the data presented here (Fig. 14.3), the elongated DW nucleation pad, aspect ratio 5:1, switched at $H_C = 44$ Oe, and time averaged magneto-optical Kerr effect hysteresis loops, measured locally with a 5 μm laser spot diameter, give evidence of

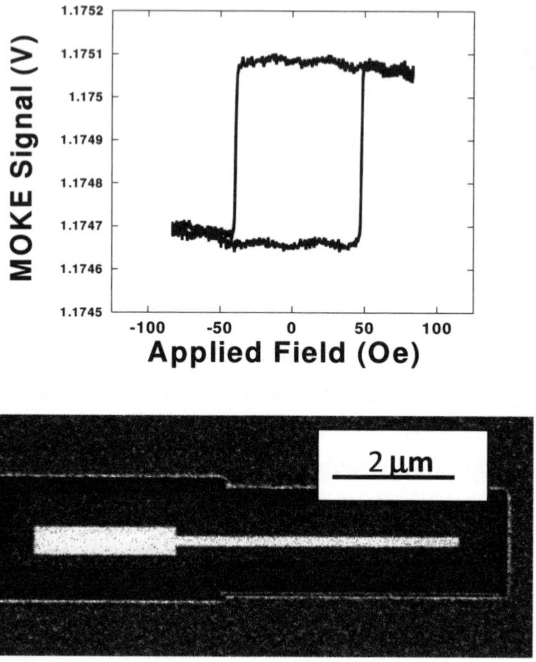

Fig. 14.3. Locally measured time averaged M-H loop on a 5 nm thick, 200 nm wide Permalloy nanowire, with a 600 nm, 3 μm wide domain wall injection pad at the left end, and corresponding micrograph of the sample.

reproducible domain wall switching at all points along the wire at 44 Oe. This controlled injection and propagation of domain walls at low field values relative to the nucleation field of structures without DW injection pads has even been demonstrated in wires kinked at 30°.[11]

These switching values are markedly lower than the nucleation field of a pad free control structure — 150 Oe. Domain wall motion occurs in the direction of applied field. Magnetization seeks to align itself with the applied field, and this re-alignment is affected by the propagation of the domain wall boundary. By modifying the geometry of wire ends with a domain wall injection pad, not only can the reversal field and mechanism for magnetization reversal be engineered, the initial reversal event can be localized to one wire end. Thus the direction of magnetization reversal by field controlled domain wall propagation can be controlled.

14.6. Rounded Corner Structures

Intentionally introducing an asymmetry into magnetic nanostructures affords more control over the magnetization reversal process, and is therefore an avenue to future applications. Ferromagnetic nanostructures with rounded corner features are an attractive experimental system for fundamental investigations,[14,27–29] and possible applications such as magnetic random recording memory — MRAM.[30] Once a DW is present in a nanowire, the DW propagation field is relatively low, being only limited by the intrinsic wire coercivity, and not, for example, a DW pad depinning field from a pinning centre, or a domain wall nucleation field.

In rounded corner structures (Fig. 14.4), domain wall nucleation and propagation can be easily separated, by applying a series of external fields in different directions. If a DW can be isolated in a rounded wire, many of its properties, such as velocity,[27] mass,[31] or response to spin polarized current[32,33] can be probed. The L-shape,[34] or corner geometry is a geometry in which a domain wall is easily stabilized and isolated.

A rotating magnetic field is used to propagate DWs along wires that change direction and turn corners.[35] The phase of the transverse field, Hy, and the longitudinal field, Hx, must be separated by $\pi/2$, generating a field vector that rotates in an anti-clockwise direction. A DW may propagate around a corner structure if the field and corner are of the same chirality or "handedness". Care should be taken to match the x and y field components to the corner geometry, so a domain wall is not pinned at a corner.

Figure 14.4 depicts a micrograph of a continuous planar Permalloy wire structure on a Si substrate, fabricated by subtractive focused ion beam milling. Upon application of a suitable magnetic field in the 'y' plane, a domain wall is nucleated in the domain wall injection pad, 'A', and propagates to corner 'B'. The domain wall is confined within the corner, until an 'x' field propagates a domain wall along the structure from 'B' to 'E'. With no component of external field in the x-direction, the wall remains at the corner. A rotating magnetic field of correct chirality can depin the domain wall from corner ‘B’ at a very low depinning field. Figure 14.5 is a typical hysteresis loop measured at point ‘C’ showing domain wall switching at ~10 Oe, time averaged over 1,300 field cycles. This very reproducible switching, and evidence of very low DW reversal fields has also been evidenced micromagnetically and proposed for data storage applications.[37] Experiments can be performed in the long output wire, at lower fields than those possible when using solely a DW injection pad and output wire (Hinj ~40 Oe).

14.7. Domain Wall Localisation/Trapping/Point Contacts

A domain wall in a nanowire can be considered as a quasi-particle. The study of DW propagation is closely related to the study of varying energy landscapes and pinning sites, such as grain boundaries, defects, edge roughness, sample thickness inhomogeneity, and dislocations. This topic in nanoscale magnetic systems is of interest for a number of reasons.

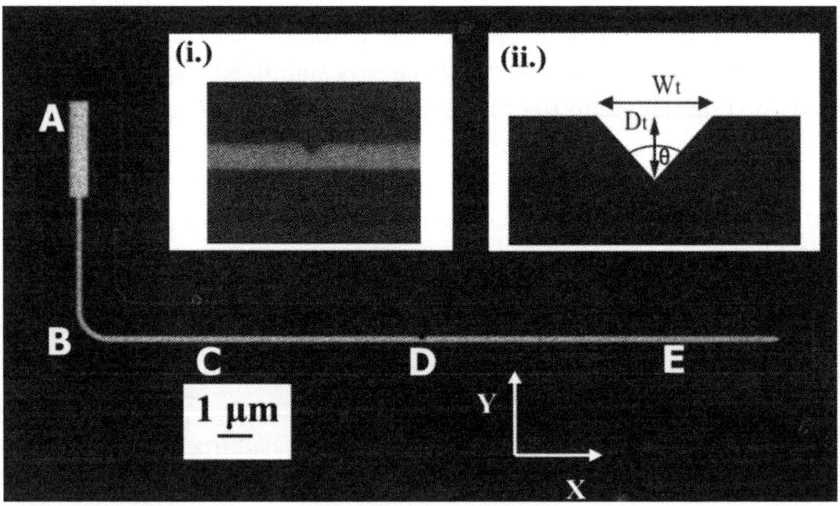

Fig. 14.4. Backscattered secondary image of a rounded corner geometry NiFe planar wire structure [light grey] on a Si substrate [dark gray]. ‘A’ indicates a DW injection pad, ‘B’ a smoothly rounded corner, ‘C’ and ‘E’ are hysteresis loop measurement positions, and ‘D’ a lithographically defined defect. Inset i-ii show the trapping centre in more detail.[36]

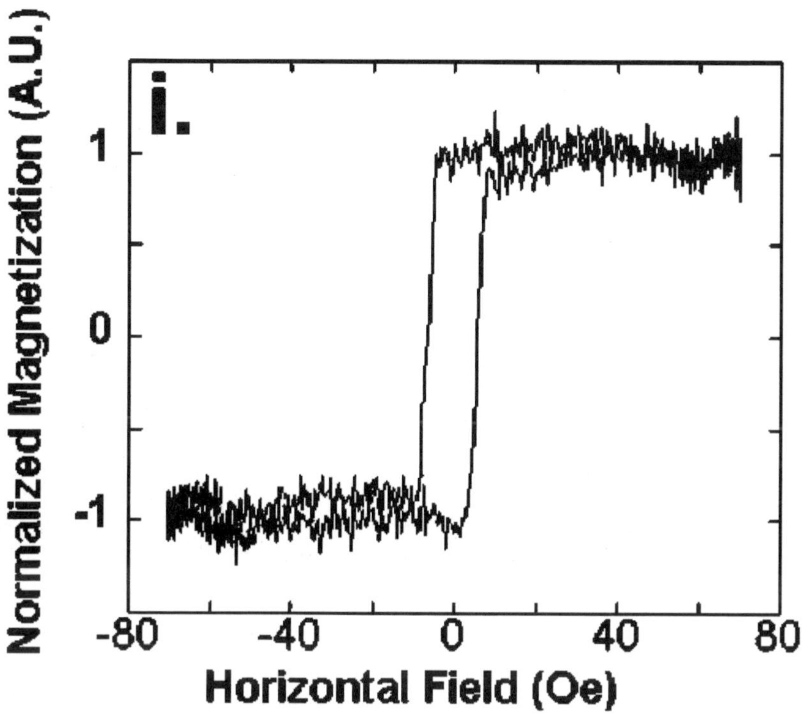

Fig. 14.5. Hysteresis loop evidencing reproducible domain wall propagation from a corner structure at extremely low field value.[36]

Firstly, fabrication limits for electron beam lithography and ion beam milling techniques[38] are continuing to improve, making possible the study of magnetization reversal processes in very complex wire geometries.[39,40] As the dimensions of nanomagnets continue to decrease, the role of defects, interfaces, and surface effects in these quasi 1-D systems becomes more important,[41] and new physical phenomena may be detected. Secondly, in the drive towards higher density magnetic random access memory systems, workers have found that magnets of reduced lateral dimensions have higher switching fields, costing more power and dissipating more energy per reversal event. One method of circumventing this problem is to use a 'domain wall trap' system,[37] now extended to the notched ring or 'doughnut' system,[42] where a domain wall always resides in the system. Thus magnetization reversal is achieved over low field values, necessary for depinning and propagating a domain wall, as opposed to the high field values needed to activate nucleation of magnetic domains and associated magnetization reversal. In addition traps afford a means of controlling the position of domain walls at low field values for the purpose of spin current experiments,[43] room temperature magnetoresistance measurements, dynamics experiments, or research into magnetic tunnelling processes.[44]

Different geometrical variations generate a potential landscape that exerts forces on the domain wall. By tuning the geometry, the energy landscape can be controlled. Domain walls can be localized in a wire by bends or sharp L-shapes,[34] traps, protrusions, injection pads, wire junctions and intersections.[35,45–47] For example a perpendicular wire on a long Permalloy nanowire, in a T-shaped junction, can act as a DW gate, either blocking a DW or letting it propagate down the horizontal wire, based solely on the magnetic orientation of the vertical wire.[47] Another method of creating a potential barrier is by subtractively milling a thin line from the top of the wire by focused ion beam.[48] Figure 14.6 shows bi-stable DW mediated switching at a trapping site, with sharp 2-step switching transitions desirable for applications. Figure 14.7 shows the increase in switching field with trap depth for M-H loops measured post DW interaction with a trap, for samples similar to Fig. 14.4.

14.8. Domain Wall Protrusion

The concept of a domain wall notch or trap has been extended to protrusions which extend outwards from the wire side instead of cutting a section out of the wire sidewall.[20,29,52] The energy potential of these structures was recently investigated and compared to notches.[20] Depending on the system characteristics the pinning centre can be either attractive, or repulsive. The term "anti-notch" has also been coined,[29] and it is noted that a wall may take positions in front of, partially in front of, or inside the trap.

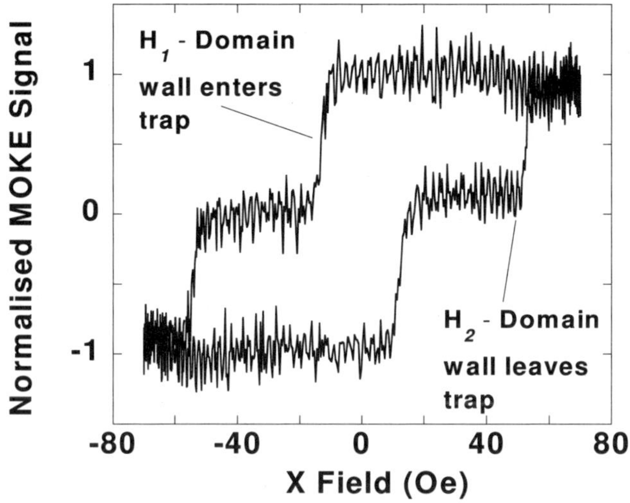

Fig. 14.6. Hysteresis loop taken at position D, with the MOKE laser spot positioned over a single nano-trap of depth 125 nm and width 240 nm. Loop was averaged for 60 s – for 1600 field cycles.[19]

14.9. Domain Wall Chirality

A 180° domain wall, also known as a transverse wall can point up or down at rest. Applying transverse fields changes the axial field needed to inject a domain wall from a pad of low coercivity material,[49] and Bryan also observed by single shot MOKE measurements drastically different switching fields for a wire output under identical field conditions. Evidence was presented for different domain wall states being present at junctions between domain wall injection pad and the output wire, leading to different switching field for the output wire.

Hayashi *et al.* successfully imaged 4 different wall type at same notch by magnetic force microscopy, measuring markedly different de-pinning field for each.[50] Subsequently it was proposed that by breaking the 1-D symmetry of an elongated nanowire with a notch, or other modification, domain wall chirality can be determined, and leveraged for potential applications in a memory device.[51] A notch poses a different pinning potential to a transverse domain wall depending on the wall chirality.[20] I.e. at a fixed field, walls of a given chirality may propagate through a notch, and wires of the opposite chirality may be pinned. This chirality filtering effect applies to vortex domain wall types[52] and the head-to-head or tail-to-tail type found in thinner films.[20] Furthermore a domain wall trap can act as either a potential well or barrier depending on the chirality of the domain wall.

14.10. Domain Wall Dynamics

The success of many DW devices depends on fast and reproducible DW mediated magnetic switching, because the processing speed of these devices depends on the domain wall velocity. The velocity of a DW driven by an in-plane magnetic field in a nanowire is limited by the Walker field, H_w. At low fields, below H_w, the DW velocity, v, increases linearly with applied field, $v = (\gamma_0\Delta/\alpha)H$, where γ_0 is the

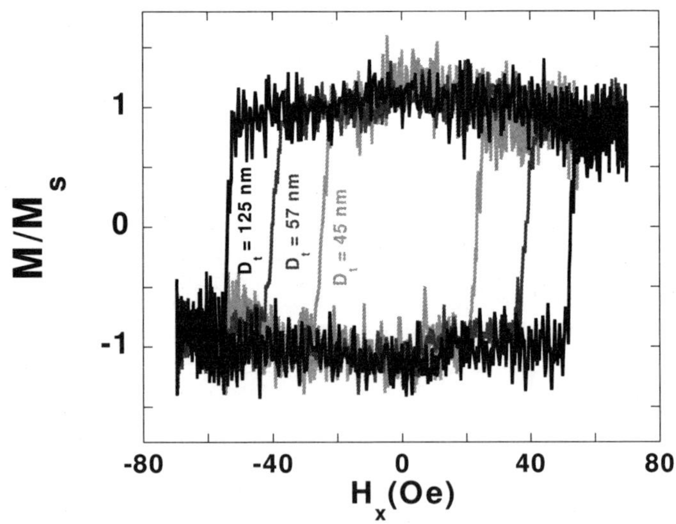

Fig. 14.7. Time averaged locally measured M-H loops at measurement position E, for trap depths from 45 nm–125 nm.

gyromagnetic ratio multiplied by the permeability of a vacuum μ_0, Δ indicates the wall width, and α the Gilbert damping parameter. This is the viscous regime. Above this critical field threshold the domain wall propagation mode changes, and velocity does not increase linearly with field. Below H_w the domain wall structure remains stable as it propagates, above H_w this is not the case. The DW moves in an oscillatory fashion, periodically stopping and starting in a turbulent propagation mode, leading to an abrupt decrease in velocity. The domain wall cants away from the direction of propagation at increasing fields, and above H_w, the canting angle ϕ precesses continually. At the same time the chirality of the domain wall also periodically changes. The wall then alternates between an in-plane Néel-like structure, and an out of plane Bloch like vortex structure, and propagates at reduced mobility in an oscillatory fashion.

In tri-layer structures,[18] the GMR effect was utilised to elucidate the switching mechanism mediated by an individual domain wall in a NiFe wire at low temperature (100 K), yielding an experimentally determined damping parameter, α = 0.63. Using a high-resolution longitudinal magneto-optical Kerr effect (MOKE) system, Atkinson *et al.* successfully isolated an individual domain wall in different positions of a 2-corner planar NiFe – wire.[27] Domain wall velocities were locally probed in the structure, by localized single shot Kerr effect laser measurements.[53] Current pulses were applied by micro-strip line, and wall velocities up to 1500 ms^{-1} were measured,[27] giving a damping constant, α = 0.053, comparable to extended thin film studies in Permalloy.[2]

By time and spatially resolved MOKE,[54–56] magnetic transmission X-ray microscopy,[57] or resistance measurements,[58] Walker breakdown has been successfully investigated in the lab, and an enhancement of domain wall velocity in the presence of both current and external field has been recorded.[55] The oscillations of domain wall position and chirality can be further elucidated micromagnetically (Fig. 14.8). It is also suggested that field pulse rise time affects the domain wall velocity.[56] Much focus has been on in-plane magnetized systems, but investigations into DW dynamics in perpendicular anisotropy systems have also been performed.[59]

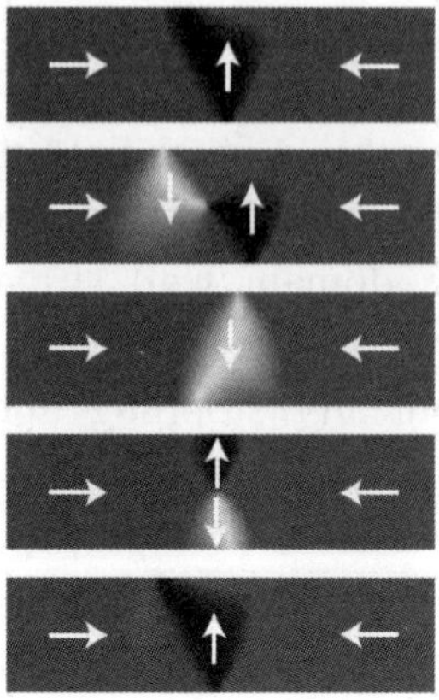

Fig. 14.8. Micromagnetic simulation of the temporal evolution of domain wall spin structure and chirality above the Walker field. A 200 nm wide, 10 nm thick Permalloy wire was simulated, with $\alpha = 0.001$, and moving boundary conditions to negate any end effects.[58]

14.11. DW Velocity Enhancements

14.11.1. *Transverse field*

The Walker field represents a fundamental limit to the DW propagation field, and limits the operating frequency of any proposed DW devices. A number of treatments/techniques have been proposed to increase this threshold, enhancing the DW velocity. Reducing the thickness of the nanowire can increase the field at which Walker breakdown occurs.[60] The use of a field perpendicular to the direction of domain wall motion[61] has also been shown to increase or decrease the domain wall velocity, and the Walker breakdown field. In addition the direction of transverse field affects velocity and H_w.

14.11.2. *Roughness*

Nakatani *et al.* proposed that edge roughness, on a ~9 nm length scale, in elongated NiFe wires could enhance domain wall velocity, stabilizing the domain wall at velocities above the theoretical Walker limit, by acting as an additional channel for energy dissipation from a system, and off-putting the onset of domain wall oscillation, and the formation of vortex type wall configurations until higher Walker fields.[60] This seemingly counterintuitive claim was well supported by micromagnetic simulations. For example, Miltat and co-workers demonstrated that rough edged wires increase H_W, showed that the defect sizes affects H_w, and that the DW velocity above the Walker breakdown field is greater for wires with edge fluctuations of 6–7 nm length scales.[60]

14.11.3. *Current assisted*

When spin polarized currents are injected into a nanowire, the domain wall velocity can be enhanced or reduced, depending on the direction of current flow,[55,62] by up to 110 ms^{-1}.

14.11.4. *Ion irradiation*

For perpendicular magnetic anisotropy systems, such as Co/Pt multilayers, an enhancement of domain wall velocity with Ga focused ion beam fluence was demonstrated experimentally.[63]

14.11.5. *Out of plane field*

It has been demonstrated micromagnetically that a strongly out of plane magnetized underlayer, such as the perpendicular magnetic anisotropy (PMA) material FePt, to a soft magnetic strip leads to fast domain wall motion for a wide range of driving fields, and can suppress the onset of the Walker effect.[64,65] The PMA underlayer helps prevent the formation of vortices at wire edges, meaning oscillatory domain wall motion occurs at higher fields.

14.12. Spin Torque

Berger proposed that electric current should apply a torque to a magnetic domain wall in 1978.[66] As a spin-polarized electron passes through a magnetic domain wall, its spin direction follows the magnetization of its local environment, and it therefore rotates from the spin ''up'' state to spin ''down''. Since each quantum of angular momentum, ħ, must be conserved, angular momentum is transferred to the spins of the domain wall. This transfer of momentum creates a torque on the wall magnetization and can apply a force to the wall similar to pressure applied to a macroscopic object. In this manner, the position of a DW in a nanowire may be manipulated by electric current.[28,33,67,68] Current controlled displacement of DWs, or current assisted switching of other magnetic nanostructures, enables localized control of magnetic structures without effecting neighboring element. No magnetic strip lines are needed during device fabrication, and fast and reproducible switching can result. For example Vernier and co-workers demonstrated that a domain wall can be propagated along a 120 nm wide Permalloy wire, and even around a corner, solely using current. Generally, above a critical current, the DW moves in the direction of the flow of electrons. The actual domain wall spin structure can also be modified by this application of current pulses.[69] Progress must be made in reproducibly controlling the domain wall propagation at lower critical current densities [of the order of $\sim 10^8$ A/cm^2 for Permalloy], more suitable for device applications.

Initial investigations on this topic were performed quasi-statically[28,33] however for technologically relevant applications, walls must be moved on short timescales. Amplification of domain wall motion with a set of current pulses, using a resonance effect has recently been demonstrated,[70,71] making possible the motion of domain walls in a given potential at lower currents, and at reduced power. The key is to tune the current pulse to the precessional period of the domain wall. Tuning the length and separation of current pulses enhanced this effect. In some of these investigations oscillatory domain wall motion has been observed, in which the wall moves against the flow of electrons.[70]

Current causes Joule heating, which can create or annihilate domain walls, or change domain wall spin structure, e.g. chirality, or domain wall type.[69] Allenspach and co-workers recently demonstrated that domain wall chirality can be better controlled by spin current, if NiFe nanowire is fabricated on a Fe underlayer.[72] Much work will be directed on current controlled magnetic switching in the coming years, and critical current densities will be reduced. New materials and geometries will be investigated. For example, current induced domain wall motion has already been demonstrated in promising future semi-conducting spintronics material systems like alloys of GaAs and Mn (albeit at low temperatures).[73] A proposed technology that may yield even higher data storage densities in the future is a storage paradigm using magnetic vortices.[74]

14.13. Domain Wall Mediated Data Storage

Magnetic storage of binary information, requires an energy barrier between two opposing directions of magnetization, which is large enough to resist thermal effects.[25] For applications such as sensors or data storage[37] sharp domain wall depinning transitions, simple switching mechanisms, and statistically robust magnetic switching are needed. Ring elements are an attractive system in which to study domain wall motion, due to the device symmetry, and the ease of positioning a domain wall in these structures by using an external field. The ring geometry has been proposed for MRAM devices,[30,75,76] analogous to magnetic memory flux closure devices in the 1950's. Changing the ring aspect ratio,[76] or adding a notch,[77] gives further control over the magnetization reversal.

14.14. DW Racetrack Memory

Parkin and co-workers have proposed a domain wall racetrack memory device, where long tracks of 2 or 3-d wires on silicon, each holding up to 100 domain walls act as high performance, reliable, low cost, solid state memory storage devices.[68] A memory storage paradigm with a high number of vertical storage strips gives a much greater data storage density than traditional 2-d geometries. Domain walls would be spaced by either notches, wire width modulations, or wire property modifications.

Domain walls in this paradigm are moved by short pulses of spin polarized current, and the read in/out is performed by a nanoscale spintronic device, such as a magnetic tunnel junction. Magnetic fields cannot be used to transport domain walls in this memory device/shift register. If a magnetic field were applied, eventually domain walls would be forced into each other and annihilate. Therefore spin polarized current is used to displace the domain walls. The direction of motion of the domain walls is independent of the magnetic charge of the wall, so an entire sequence of walls can be shifted around the loop, when a critical current is exceeded. Parkin and workers have already reported on displacing pairs of domain walls by spin polarized current alone.[68] A high concentration of domain walls in each wire would reduce the cost of the system. Controlling position of domain walls by short pulses of current is desirable for efficient operation of this device type, as has been previously discussed.

Challenges for this architecture include reducing the critical current density, increasing the density of domain walls in a given wire, and fabrication challenges related to a 3-d geometry. The current densities reported in the literature can result in domain wall spin structure change, domain wall chirality change, and Joule heating up to the Curie temperature of Permalloy – 850 K.

14.15. Domain Wall Logic

Recently a quantum cellular automata (QCA) Boolean logic paradigm was proposed by Cowburn and Welland,[78] based on propagating information in chains of 110 nm diameter ferromagnetic dots. These strings of single domain dots have been successfully combined into logic gates at room temperature.[79] The QCA paradigm evolved into a DW logic scheme whereby information is transmitted and manipulated in continuous planar ferromagnetic nanowires at room temperature, by controlled manipulation of DWs.[35,80] This DW logic scheme is analogous to the bubble logic scheme of the 1970's.[81] Through very precise control of wire geometries, domain wall mediated switching can be repeatedly controlled. Wires treated here are typically 5 nm thick and of the order of 200 nm wide, so do not support vortex walls, or 360° domain walls, which can make switching mechanisms less reproducible.

In the context of a possible logic schema, the reproducible transfer of information, in the form of magnetization vector orientation in a wire, over tens of microns mediated by field controlled domain wall propagation has been demonstrated. Controlled information transfer by domain wall propagation was demonstrated in elongated wires, and around corners in the presence of an appropriate external field sequence and direction. Single nanowire switching by domain wall can be controlled not only by magnetic field, but by element geometry by intentionally introducing artificial trapping centres in wires.

Some advantages are non volatile data retention, radiation hardness, reduced energy consumption, a relatively simple planar geometry, compatability with existing spintronics technologies such as MRAM and the prospect of high integration densities. Disk drives require mechanical movement, domain wall logic does not. Opposing magnetization directions in a wire correspond to the logical "0" and "1" states. Logic operations are performed using wire junctions linked by continuous planar wires. The 4 principal operations that have been experimentally demonstrated to date include logical NOT, logical AND, signal fan out (1 DW is split into 2), and signal crossover (where DW's cross a wire at right angles to the direction of DW propagation.) The preferred magnetization direction of the wire structures presented herein is parallel to the wire edge.

The two magnetic orientations supported in elongated submicron wires represent the Boolean 1 and 0. Note that if there is a 180° curve in a wire, and the magnetization is continuous throughout, then the absolute direction of magnetization is different before and after the bend. Therefore Allwood and workers[35] assigned logical "1" to wire magnetization in the same direction as DW propagation, and logical "0" to magnetization direction opposing DW motion. The extent of DW propagation is limited by 90° and 180° corners in the nanowire networks. In this manner the rotating field permits clear representation of different logic states, and a direction of information propagation in the logic system is well defined.

14.16. NOT

Signal inversion, and therefore NOT gate functionality was realized at room temperature, in a 3-terminal continuous wire junction, by field controlled domain wall motion (Fig. 14.9).[35] This operation was subsequently further investigated and optimized.[82,83] Locally probing the magnetization states in continuous wire loops with 3-terminal junctions, in a 27 Hz rotating field, workers were able to reproducibly invert the

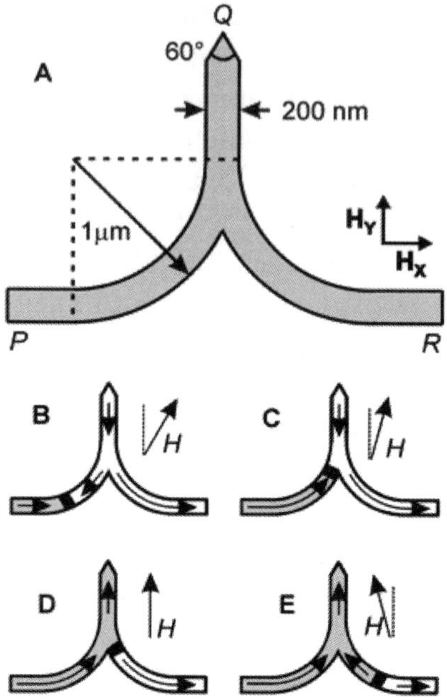

Fig. 14.9. Schematic of a typical NOT gate geometry. 'A' depicts a typical device geometry and magnetic field directions. 'B'-'E' indicate how the magnetization vector in the devices respond to an external rotating magnetic field.[35]

logical state of the device output, over up to 100,000 repetitions. The rotating field, which is in the plane of the samples, functions as the power supply and clock. Initial NOT gates were connected in a closed loop, such that the output is connected back round to the input.

The operation of the devices was subsequently verified and imaged by magnetic force microscopy.[83] Note that excellent geometrical control is exerted over these patterned thin film structures by a customized FIB fabrication process,[38] and that small changes to the junction geometry [on ~25 nm length scales] gives reproducible control over the device operating field values, and the device performance tolerances.

It is possible to link a series of gates together to perform more complex logic functions. A series of NOT gates were concatenated in a loop to form a shift register, where complex bit patterns can be stored or reproducibly transmitted about the circuit. Another useful aspect of this device geometry is that the gate is symmetrical, so can be operated in one of 2 directions. The concept of a domain wall NOT gate has subsequently been investigated by other workers, and recently NOT gate functionality was demonstrated by current driven domain wall motion in a set of nanocontacts fabricated from Invar (a Nickel Steel alloy).[84]

14.17. AND/OR

Additional logic gates are required to compliment the NOT gate for computational operations to be performed. A gate with 2 or more inputs is required. For a logical AND/Or-like function, 2 magnetic nanowires unite at a junction, with one output wire in a three terminal configuration.[45,85] Domain walls interact at junctions. The switching field of the AND gate output wire depends on whether there is 0, 1, or 2 DWs contained in the input wires. The switching field at the output wire is markedly higher when there is one domain wall propagating into the junction, that when 2 domain walls propagate into the junction. The operation of any potential domain wall device strongly depends on the narrow contact geometry of the wire junction. Varying wire widths, and wire angles at terminal junctions affords control over the switching fields and field operating ranges of any multi-terminal domain wall devices.[86] Furthermore 3-terminal devices of this nature have recently been demonstrated in multi-layered NiFe/Cu films suitable for GMR-type characterization and current driven output wire switching has been demonstrated.[85]

14.18. Fanout/Cloning

It is important to split signals in logic architectures, such as a multi-terminal domain wall circuit. This process where domain walls are duplicated at 3-terminal junctions has been dubbed 'cloning'.[87] Consider a planar nanowire DW conduit that splits into two arms in a tuning fork-type geometry, as shown schematically in Fig. 14.10. If a DW propagates into the junction, and the junction nucleation field is high, then the DW expands across the junction, bubble-like, and injects a DW into both of the device outputs.[80,87] Effectively 1 domain wall is cloned into 2. By integrating these tuning fork junctions with other gates, it was demonstrated that this cloning operation was reproducibly effected over

Fig. 14.10. Schematic of the domain wall cloning process.

13,000 field cycles, and that two of these cloning junctions can be connected in series. The device performance or figure of merit is well controlled by subtle modifications to the Y shaped device geometry. For example if the wires are wide the junction nucleation field is very low, meaning high field cannot be used. If the junction wires are too narrow, a DW cannot be successfully propagated through them, due to intrinsic sidewall pinning effects, and the onset of superparamagnetism.

This concept has been extended by concatenating cloning gates in series, as shown in Fig. 14.11. The device shown was tested in an anti-clock-wise rotating field applied in the plane of the samples. A domain wall is injected from the NOT gate output wire (loop top of Fig. 14.11) left into cloning junction, C1. From here one domain wall propagates through 2 corners and domain wall interconnect to C2 where it is divided. The output wires from C2 are connected to the input wires of junctions C3 and C4. Therefore 2 walls propagate to gates C3 and C4, where they are split again. Local MOKE measurements taken at positions 1-5 verified this domain wall cloning process. One domain wall in a single nanowire is cloned into 4, giving 4 times digital amplification.

14.19. Crossover

The signal cross-over element is a 4-terminal device where in the appropriate field sequence, a domain wall can cross an orthogonal wire. I.e. 2 nanowires cross over each other in the same plane. See Fig. 14.12, where the main logic gates, and their analogous counterparts in CMOS technology are displayed.

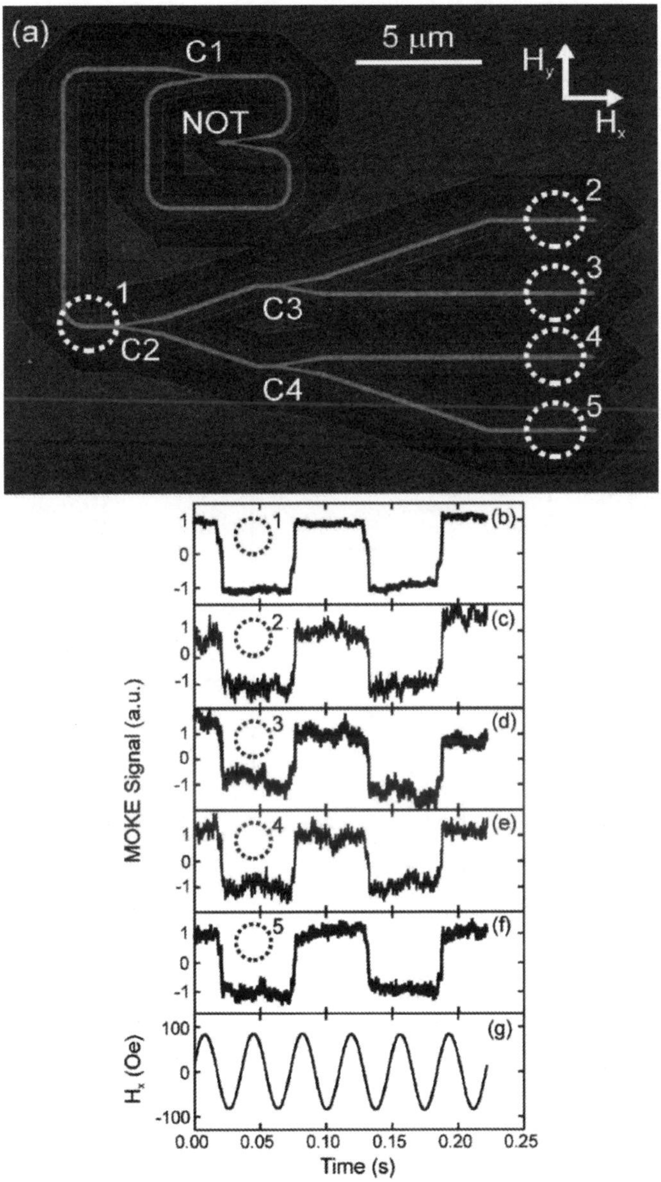

Fig. 14.11. 'A' is an electron image of 4 domain wall cloning junctions linked together, labelled C1-C4. Plots 'b'-'f' show magnetization switching data for MOKE measurement positions labelled 1-5, and the concomitant magnetic field trace is shown in 'g'.[87]

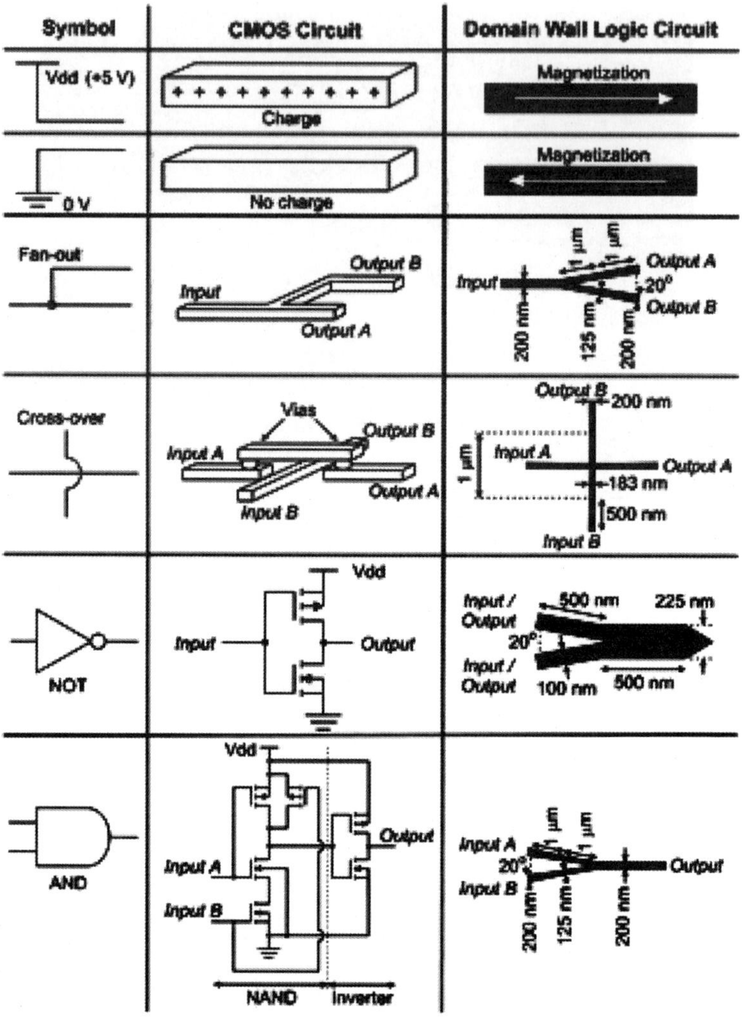

Fig. 14.12. Domain wall logic elements.[80]

14.20. Data Input

Any practical circuit must be able to receive data from external sources, and data must also be read out from the system. For the DW logic architecture, writing of information may be realized by current carrying wires, or possibly heating,[88] but layering of devices leads to further fabrication complexity. However a simple planar field writable method was recently described and experimentally verified,[87] using low coercivity areas of magnetic material, similar to a domain wall injection pad. DWs can be reproducibly and selectively nucleated into a structure without compromising the performance of other logic gates.

14.21. DW Diode

In the DW logic paradigm, the direction of information flow is defined by the handedness of the rotating magnetic field and the orientation of the corners. It is necessary to have a diode device for selectively filtering information in one direction only, so direction of logic signals can be controlled. (For example to prevent the unwanted meeting and annihilation of 2 domain walls). A domain wall diode was demonstrated experimentally in triangle like protrusions at the edge of a Permalloy nanowire.[82,90] This diode or ratchet effect was demonstrated for a number of these protrusions concatenated in series,[13,91] and has also been documented in a simplified asymmetric DW trap geometry.[92] In these structures, in a geometry dependent field range (for fixed temperature), below the wire nucleation field, a domain wall will propagate through the feature in one direction only.

Since the feature is asymmetric, Fig. 14.13, the DW is presented with a different energy landscape, depending on the direction of propagation. Therefore for a fixed field a DW may propagate through a patterned triangular feature, for one field parallel to the wire long axis, but not the other. By varying the device geometry, the field bias switching effect can be tuned.[90,92]

Current driven DW motion has been documented in these devices.[91] By magnetic force microscopy measurements (MFM), current driven DW magnetization reversal was evidenced in one direction, but nucleation of more domains walls, and a higher critical current density was measured when current was applied in the opposing direction. Since it is known that due to stray fields from the tip, MFM is an invasive measurement technique, these devices were also characterized by the giant magnetoresistance effect.[13]

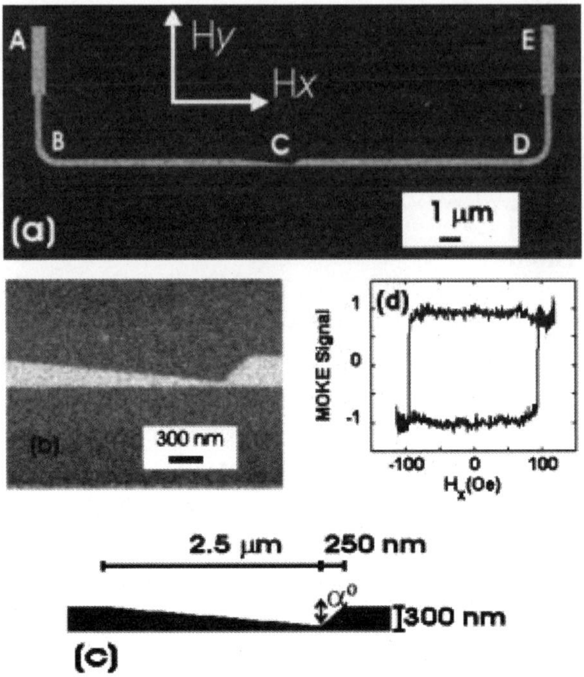

Fig. 14.13. 'A' shows a geometry where DWs were propagated in different directions, through a central trap. 'B' shows a magnified trap electron scan. 'C' elucidates a diode geometry in more detail.[92]

14.22. Outlook

Domain wall mediated switching of planar nanowires for device applications has been discussed. Domain wall traps have been shown to control domain wall spin structure, position, propagation or depinning field, and direction of propagation in planar nanowires. Further investigations into domain wall devices of increasing complexity can be expected, leading to incorporation in commercially viable devices and future scientific breakthroughs.

References

1. G. E. Moore, Electronics **38** (1965).
2. S. Konishi, S. Yamada, T. Kusuda, IEEE Trans. Magn. **7**, 722 (1971).

3. C. E. Patton, Z. Frait, C. H. Wilts, J. Appl. Phys. **46**, 5002 (1975).
4. K. J. Sixtus, L. Tonks, Phys. Rev. **37**, 930 (1931).
5. K. S. Novoselov, A. K. Geim, S. V. Dubonos, E. W. Hill, I. V. Grigorieva, Nat. **426**, 812 (2003).
6. X. B. Zhu, P. Grütter, V. Metlushko, B. Ilic, Phys. Rev. B **66**, 024423 (2002).
7. T. Shinjo, T. Okuno, R. Hassdorf, K. Shigeto, T. Ono, Sci. **289**, 930 (2000).
8. A. Wachowiak, J. Wiebe, M. Bode, O. Pietzsch, M. Morgenstern, R. Wiesendanger, Sci. **298**, 577 (2002).
9. E. H. Frei, S. Shtrikman, D. Treves, Phys. Rev. **106**, 446 (1957).
10. W. Wernsdorfer, B. Doudin, D. Mailly, K. Hasselbach, A. Benoit, J. Meier, J.-Ph. Ansermet, B. Barbara, Phys. Rev. Lett. **77**, 1873 (1996).
11. R. P. Cowburn, D. A. Allwood, G. Xiong, M. D. Cooke, J. Appl. Phys. **91**, 6949 (2002).
12. A. Himeno, T. Ono, S. Nasu, K. Shigeto, K. Mibu, T. Shinjo, J. Appl. Phys. **93**, 8430 (2003).
13. M. Himeno, S. Kasai, T. Ono, J. Appl. Phys. **99**, 08G304 (2006).
14. E. Saitoh, H. Miyajima, T. Yamaoka, G. Tatara, Nat. **432**, 203 (2004).
15. R. D. McMichael, M. J. Donahue, IEEE Trans. Magn. **33**, 4167 (1997).
16. G. S. D. Beach, M. Tsoi, J. L. Erskine, J. Magn. Magn. Mat. **320**, 1272 (2008).
17. S. W. Yuan, H. N. Bertram, J. F. Smyth, S. Schultz, IEEE Trans. Magn. **28**, 3171 (1992).
18. T. Ono, H. Miyajima, K. Shigeto, K. Mibu, N. Hosoito, T. Shinjo, Sci. **284**, 468 (1999).
19. C. C. Faulkner, M. D. Cooke, D. A. Allwood, D. Petit, D. Atkinson, R. P. Cowburn, J. Appl. Phys. **95**, 6717 (2004).
20. D. Petit, A. V. Jausovec, D. Read, R. P. Cowburn, J. Appl. Phys. **103**, 114307 (2008).
21. F. Fournal, Y. Chen, F. Carcenac, N. Essaidi, H. Launois, V. Kottler, C. Chappert, IEEE Trans. Magn. **34**, 1027 (1998).
22. K. Shigeto, T. Shinjo, T. Ono, Appl. Phys. Lett. **72**, 1116 (1998).
23. Y. Yokoyama, Y. Suzuki, S. Yusaa, K. Ando, K. Shigeto, T. Shinjo, P. Gogol, J. Miltat, A. Thiaville, T. Ono, T. Kawagoe, J. Appl. Phys. **87**, 5618 (2000).
24. W. Y. Lee, C. C. Yao, A. Hirohata, Y. B. Xu, H. T. Leung, S. M. Gardiner, S. McPhail, B. C. Choi, D. G. Hasko, J. A. C. Bland, J. Appl. Phys. **87**, 3032 (2000).
25. P. Lendecke, René Eiselt, Guido Meier, Ulrich Merkt, J. Appl. Phys. **103**, 073909 (2008).
26. K. Shigeto, T. Okuno, T. Shinjo, Y. Suzuki, T. Ono, J. Appl. Phys. **88**, 6636 (2000).
27. D. Atkinson, D. A. Allwood, G. Xiong, M. D. Cooke, C. C. Faulkner, R. P. Cowburn, Nat. Mater. **2**, 85 (2003).
28. N. Vernier, D. A. Allwood, D. Atkinson, M. D. Cooke, R. P. Cowburn, Europhys. Lett. **65**, 526 (2004).

29. C. W. Sandweg, N. Wiese, D. McGrouther, S. J. Hermsdoerfer, H. Schultheiss, B. Leven, S. McVitie, B. Hillebrands, J. N. Chapman, J. Appl. Phys. **103**, 093906 (2008).
30. M. Kläui, C. A. F. Vaz, J. A. C. Bland, L. J. Heyderman, Appl. Phys. Lett. **86**, 032504 (2005).
31. E. Saitoh, H. Miyajima, T. Yamaoka, G. Tatara, Nat. **432**, 203 (2004).
32. N. Vernier, D. A. Allwood, D. Atkinson, M. D. Cooke, R. P. Cowburn, Europhys. Lett. **65**, 526 (2004).
33. M. Kläui, C. A. F. Vaz, J. A. C. Bland, W. Wernsdorfer, G. Faini, E. Cambril, L. J. Heyderman, Appl. Phys. Lett. **83**, 105 (2003).
34. D. A. Allwood, N. Vernier, G. Xiong, M. D. Cooke, D. Atkinson, C. C. Faulkner, R. P. Cowburn, Appl. Phys. Lett. **81**, 4005 (2002).
35. D. A. Allwood, G. Xiong, M. D. Cooke, C. C. Faulkner, D. Atkinson, N. Vernier, R. P. Cowburn, Sci. **296**, 2003 (2002).
36. C. C. Faulkner, PhD Thesis, University of Durham, 2005.
37. D. McMichael, J. Eicke, M. J. Donahue, D. G. Porter, J. Appl. Phys. **87**, 7058 (2000).
38. G. Xiong, D. A. Allwood, M. D. Cooke, R. P. Cowburn, Appl. Phys. Lett. **80**, 1195 (2002).
39. M. Herrmann, S. McVitie, J. N. Chapman, J. Appl. Phys. **87**, 2994 (2000).
40. F. Cayssol, D. Ravelosona, C. Chappert, J. Ferré, J. P. Jamet, Phys. Rev. Lett. **92**, 107202 (2004).
41. J. Gadbois, J.-G. Zhu, IEEE Trans. Magn. **31**, 3802 (1995).
42. M. Kläui, C. A. F. Vaz, J. A. C. Bland, W. Wernsdorfer, G. Faini, E. Cambril, J. Appl. Phys. **93**, 7885 (2003).
43. J. Grollier, P. Boulenc, V. Cros, A. Hamzić, A. Vaurès, A. Fert, G. Faini, Appl. Phys. Lett. **83**, 509 (2003).
44. L. Gunther, B. Barbara, Phys. Rev. B **49**, 3926 (1994).
45. C. C. Faulkner, D. A. Allwood, M. D. Cooke, D. Atkinson, R. P. Cowburn, IEEE Trans. Magn. **39**, 2860 (2003).
46. F. Cayassol, D. Ravelosona, J. Wunderlich, C. Chappert, J. P. Jamet, J. Ferré, J. Magn. Magn. Mat. **240**, 30 (2002).
47. D. Petit, A. V. Jausovec, H. T. Zeng, E. Lewis, L. O'Brien, D. Read, R. P. Cowburn, Appl. Phys. Lett. **93**, 163108 (2008).
48. A. W. Holleitner, H. Knotz, R. C. Myers, A. C. Gossard, D. D. Awschalom, Appl. Phys. Lett. **85**, 5622 (2004).
49. M. T. Bryan, D. Atkinson, D. A. Allwood, Appl. Phys. Lett. **88**, 032505 (2006).
50. M. Hayashi, L. Thomas, C. Rettner, R. Moriya, X. Jiang, S. S. P. Parkin, Phys. Rev. Lett. **97**, 207205 (2006).
51. L. K. Bogart, D. S. Eastwood, D. Atkinson, J. Appl. Phys. **104**, 033904 (2008).
52. K. J. O'Shea, S. McVitie, J. N. Chapman, J. M. R. Weaver, Appl. Phys. Lett. **93**, 202505 (2008).

53. D. A. Allwood, G. Xiong, M. D. Cooke, R. P. Cowburn, J. Phys. D. **36**, 2175 (2003).
54. G. S. D. Beach, C. Nistor, C. Knutson, M. Tsoi, J. L. Erskine, Nat. Mater. **4**, 741 (2005).
55. G. S. D. Beach, C. Knutson, C. Nistor, M. Tsoi, J. L. Erskine, Phys. Rev. Lett. **97**, 057203 (2006).
56. K. Weerts, P. Neutens, L. Lagae, G. Borghs, J. Appl. Phys. **103**, 094307 (2008).
57. G. Meier, M. Bolte, R. Eiselt, B. Krüger, D.-H. Kim, P. Fischer, Phys. Rev. Lett. **98**, 187292 (2007).
58. M. Hayashi, L. Thomas, C. Rettner, R. Moriya, S. S. P. Parkin, Nat. Phys. **3**, 21 (2007).
59. C. Burrowes, D. Ravelosona, C. Chappert, S. Mangin, E. E. Fullerton, J. A. Katine, B. D. Terris, Appl. Phys. Lett. **93**, 172513 (2008).
60. Y. Nakatani, A. Thiaville, J. Miltat, Nat. Mater. **2**, 521 (2003).
61. M. T. Bryan, T. Schrefl, Del. Atkinson, D. A. Allwood, J. Appl. Phys. **103**, 073906 (2008).
62. M. Hayashi, L. Thomas, Y. B. Bazaliy, C. Rettner, R. Moriya, X. Jiang, S. S. P. Parkin, Phys. Rev. Lett. **96**, 197207 (2006).
63. J. P. Jamet, J. Ferré, P. Meyer, J. Gierak, C. Vieu, F. Rousseaux, C. Chappert, V. Mathet, IEEE Trans. Magn. **37**, 2120 (2001).
64. J.-Y. Lee, K.-S. Lee, S.-K. Kim, Appl. Phys. Lett. **91**, 122513 (2007).
65. A. Kunz, S. C. Reiff, Appl. Phys. Lett. **93**, 082503 (2008).
66. L. Berger, J. Appl. Phys. **49**, 2156 (1978).
67. A. Vanhaverbeke, A. Bischof, R. Allenspach, Phys. Rev. Lett. **101**, 107202 (2008).
68. S. S. P. Parkin, M. Hayashi, L. Thomas, Sci. **320**, 190 (2008).
69. M. Kläui, P.-O. Jubert, R. Allenspach, A. Bischof, J. A. C. Bland, G. Faini, U. Rüdiger, C. A. F. Vaz, L. Vila, C. Vouille, Phys. Rev. Lett. **95**, 026601 (2005).
70. L. Thomas, M. Hayashi, X. Jiang, R. Moriya, C. Rettner, S. Parkin, Nat. **443**, 197 (2006).
71. L. Thomas, M. Hayashi, X. Jiang, R. Moriya, C. Rettner, S. Parkin, Sci. **315**, 1553 (2007).
72. A. Vanhaverbeke, A. Bischof, R. Allenspach, Phys. Rev. Lett. **101**, 107202 (2008).
73. M. Yamanouchi, D. Chiba, F. Matsukura, H. Ohno, Nat. **428**, 539 (2004).
74. S. Kasai, Y. Nakatani, K. Kobayashi, H. Kohno, T. Ono, Phys. Rev. Lett. **97**, 107204 (2006).
75. J. G. Zhu, Y. F. Zheng, G. A. Prinz, J. Appl. Phys. **87**, 6668 (2000).
76. F. J. Castano, C. A. Ross, A. Eilez, J. Phys. D **36**, 2031 (2003).
77. M. Kläui, J. Phys.: Cond. Mat. **20**, 313001 (2008).
78. R. P. Cowburn, M. E. Welland, Sci. **287**, 1466 (2000).
79. A. Imre, G. Csaba, L. Ji, A. Orlov, G. H. Bernstein, W. Porod, Sci. **331**, 205 (2006).
80. D. A. Allwood, G. Xiong, C. C. Faulkner, D. Atkinson, D. Petit, R. P. Cowburn, Sci. **309**, 1688 (2005).
81. A. Hubert, R. Schäfer, Magnetic Domains (Springer, Berlin, 1998).

82. D. A. Allwood, G. Xiong, M. D. Cooke, C. C. Faulkner, D. Atkinson, R. P. Cowburn, J. Appl. Phys. **95**, 8264 (2004).
83. X. Zhu, D. A. Allwood, G. Xiong, R. P. Cowburn, P. Grütter, Appl. Phys. Lett. **87**, 062503 (2005).
84. P. Xu, K. Xia, C. Z. Gu, L. Tang, H. F. Yang, J. J. Li, Nat. Nanotech. **3**, 97 (2008).
85. C. Nam, Y. Jang, K. S. Lee, B. K. Cho, Nanotechnol. **19**, 015703 (2008).
86. D. A. Allwood, G. Xiong, M. D. Cooke, C. C. Faulkner, D. Atkinson, R. P. Cowburn, J. Appl. Phys. **95**, 8264 (2004).
87. D. A. Allwood, G. Xiong, R. P. Cowburn, J. Appl. Phys. **101**, 024308 (2007).
88. D. A. Allwood, G. Xiong, R. P. Cowburn, J. Appl. Phys. **100**, 123908 (2006).
89. D. Atkinson, R. P. Cowburn, Appl. Phys. Lett. **85**, 1386 (2004).
90. M. T. Bryan, T. Schrefl, D. A. Allwood, Appl. Phys. Lett. **91**, 142502 (2007).
91. M. Himeno, S. Kasai, T. Ono, Appl. Phys. Lett. **87**, 243108 (2005).
92. C. C. Faulkner, D. A. Allwood, R. P. Cowburn, J. Appl. Phys. **103**, 073914 (2008).

AUTHOR INDEX

SUBJECT INDEX